SURVIVAL
OF THE FITTEST
Understanding Health and
Peak Physical Performance

By the same author

SHADOWS ON THE WASTELAND:
Crossing Antarctica with
Ranulph Fiennes

SURVIVAL

OF THE

FITTEST

Understanding Health and
Peak Physical Performance

Mike Stroud

JONATHAN CAPE
LONDON

First published 1998

1 3 5 7 9 10 8 6 4 2

© Mike Stroud 1998

Mike Stroud has asserted his right
under the Copyright, Designs and Patents Act 1988
to be identified as the author of this work

First published in Great Britain in 1998 by Jonathan Cape,
Random House, 20 Vauxhall Bridge Road, London SW1V 2SA

Random House Australia (Pty) Limited
20 Alfred Street, Milsons Point, Sydney,
New South Wales 2061, Australia

Random House New Zealand Limited
18 Poland Road, Glenfield,
Auckland 10, New Zealand

Random House South Africa (Pty) Limited
PO Box 337, Bergvlei, 2012 South Africa

Random House UK Limited Reg. No. 954009

A CIP catalogue record for this book
is available from the British Library

ISBN 0-224-04485-0

Papers used by Random House UK Limited are natural,
recyclable products made from wood grown in sustainable forests.
The manufacturing processes conform to the environmental
regulations of the country of origin.

Typeset in Bembo by MATS, Southend-on-Sea, Essex
Printed and bound in Great Britain by
Mackays of Chatham PLC, Chatham, Kent

Contents

Illustrations

To my Father
who introduced me to the great outdoors.

Preface

THIS BOOK is about the advantages and disadvantages of our evolutionary inheritance. On the positive side, we all have a resilience and physical strength that far exceed people's expectation. On the negative, there is a propensity to develop serious health problems.

There are many ways in which to push the human body. Some of the chapters relate to performance in the challenges of sport – the activities that men and women have evolved to take the place of hunting and gathering – while others concern our abilities to cope with extreme conditions, examining them through tales of those who have sought challenges in the wilder parts of our planet, or who have met life-threatening situations through accident or misadventure. In addition to descriptions of human strength, some chapters cover human weakness – the heart attacks, strokes, diabetes, and other illnesses that arise from our modern lifestyles and diets that are far from those for which evolution designed us.

The ideas expressed in these pages were developed through an education in anthropology and medicine, and a career as both a hospital doctor and researcher into human performance and survival for the Ministry of Defence, but they are illustrated by more unusual personal experience. I have been fortunate enough to take part in many physically demanding events and expeditions in some of the most savage and beautiful regions of the world. These gave me the chance to explore my own and others' physical and psychological limits, and have led me to the belief that by taxing some of our hidden evolutionary strengths we can avoid many of our inherited weaknesses, and so remain active and healthy well into old age.

Mike Stroud
January 1998

INTRODUCTION

★

The Dawn of Fitness

IT WAS a bitter morning. The sun rose with some promise of radiance, orange rays glinting on the hoar frost, but a few minutes later the group entered the marshlands where wreaths of mist emanated from dark unfrozen water. The herald of warmth became a mere dull disc hanging in the misty silence, and the deep cold persisted. As the morning passed the sunlight became no stronger, although its pale presence made the task of tracking easier. The hunters moved forward at a greater pace, running mixed with walking, depending on terrain. The running was welcome – at least it kept the chill at bay.

The men had been following the beast for the best part of three days, moving as fast as they could over the bare, snow-covered plains or pushing through huge areas of naked birch woods. Now they were chilled, tired and hungry – desperately in need of food to replace the fuel that their muscles had burned away. They had seen little that they could eat since starting out. A few nuts, one or two fungi, that was the sum of it – not enough to sustain them if the hunt were to last for many days more. Nevertheless, they still felt reasonably strong, drawing on their bodies' fat stores that had been laid down months before when the hot summer sun had shone and the land had been a place of warmth and plenty. All that now seemed far away – distant dreams of comfort to offset pain as they waded across yet another icy river.

In their dreams, the whole of summer was a time of pleasure but, in reality, it had been far from easy. Well before the sun had reached its annual zenith, good water had become hard to find and their long trips in search of food had been very difficult. Two of the tribe had succumbed to the heat even before the day of shortest shadows, and more had died later after the return of longer nights. Yet, as with all painful memories, the truth of that time was now relegated to the subconscious. Their thoughts dwelt now on warmth, comfort and recollections of a bounty of food.

Grubs had been plentiful that summer and nuts and berries had coated the trees right up to the first frosts. As well as getting stores into their huts for winter, everyone had been able to eat well right through the leaf-fall, putting on extra fat which now gave them a layer of insulation against the seeping chill and provided a reserve of fuel to keep them moving. Without it, the hunters would have been forced to give up days before.

The mammoth had been wounded from the very beginning, caught by their spears as it wandered into a carefully prepared trap. But although maybe four or five of the sharp flint blades had found their mark, they had served only to weaken the animal. Now, nearly four days later, it still showed tremendous endurance and there were few signs of it slowing down. That was commonly the way. Every winter, when the persisting snows finally forced the men to try to hunt for food, they would go out in their groups and often return empty-handed. They almost never killed anything cleanly, and when they did try to bring down a mammoth, usually they merely injured it and then spent many days harrying the beast before there was any chance of success. Often, they would lose their quarry in the marshes or become so exhausted from the chase that they had to give up. A mammoth was a huge prize but one that was never gained without great cost.

On this occasion, the hunters had not seen the animal since the previous evening and all they could do was to follow the trail that wound through the soft snow. The men were flagging, but they could tell from both the sharpness of the tracks and the recently bent reeds that the beast must also be tiring. They were definitely gaining on it. With this realisation came the onset of fear as each began to think of the final confrontation. Past experience did not make it a prospect to relish, but the burst of adrenaline that fear induced within them renewed their strength, and the line of men moved across the land in a steady lope. Suddenly, there were excited calls from the left. Two of the group were shouting, pointing with their spears and waving their clubs. They had obviously seen the animal not far away. The prize was close – food, fur and bone were there for the taking, and their families would be thrilled if they returned well laden. It would take many days to get back with their loads but at least all their efforts would have been worthwhile.

A few hundred yards away an area of high reeds was active in the cold and windless afternoon. The men on the hill were pointing there, clearly able to see their quarry, and those down below paused to assess the situation. The oldest member of the group had been to this place before and quickly described it to his companions. He knew that in front of them the marshy ground became a thin peninsula, a finger of ever softer mud that extended out into a deep dark lake. It was a strange mere, a pool with a vile taste that he did not understand and one which for some reason remained unfrozen through most winters. The mammoth would be trapped by open water on three sides. It was an ideal place for victory.

The men from the hill rejoined them and they moved off in a line, edging warily along the narrowing mud banks, their stone-tipped spears ready for the kill. They had no misconceptions about the effectiveness of their weapons and they needed space to manoeuvre if the animal turned on them. As they approached the narrowest point of the isthmus, they were particularly vulnerable and so were keen to cross to where they could see that the banks widened out into an area filled with reeds. At the same time, they needed to move slowly in order not to panic the beast. On the wider head of the peninsula they would have their best chance. It was now crushingly silent and each man wondered what the animal would do next. Their hearts beat palpably in their chests and their bodies became as taut as springs.

Without warning, and before they reached the wider reed beds, the mammoth re-appeared. Trapped, wounded and enraged, it charged, determined to rid itself of the persistent pests that had stung and chased it for so long. Although they were prepared, the huge animal charged so fast that there was little they could do to avoid it. Confined by the mud banks, they were trapped as surely as they had meant to trap their prey, and they knew that now no number of spears would stop the mammoth in its tracks. The realisation that they would be crushed brought immediate action: they all turned and fled. It was nearly two hundred spans back to the wider lake shore where rocks and trees would offer some protection – much too far. While only halfway to safety, their legs began to fail them and the pounding of tree-thick limbs drew close behind.

Most of the hunters made it, throwing themselves from the mammoth's path just as they thought they could run no more. Two did not. One, a young man of only sixteen summers, was felled as he tried vainly to keep up. He had never been one of the faster runners and some had said that he should never join a hunt. Now they were proved right as, his head crushed like fruit, he met his end, but at least it was quick and merciful. The other victim, the older man who had known the place before, was not so fortunate. At fifty or sixty summers, he was slower now than when he had been in his prime and perhaps he too should have stayed at the camp, although that would never have been practical. It was many years since the tribe had last travelled in the region and he was the only one of the elders who remembered it well and could still move fast enough to go out with the hunt. At least, nearly fast enough, for although he had not been killed outright, he was far worse off. His chest had been crushed by one of the mammoth's feet and he lay on the ground, panting for breath and gasping with pain, while the great beast disappeared in the gathering twilight.

The uninjured men emerged from the rocks and examined their two companions. The younger, quite dead, could be left but the older man was more of a problem. There was nothing they could do to help him, and if he did not die from his injuries he would die from the cold. Nevertheless, they could not abandon him to the wolves and felt an obligation to try and get him back to his family. A stretcher of branches and reeds was hastily constructed so that they could drag him slowly home. To make their journey would take all of seven sundowns, perhaps more, and unless they were lucky there would be little or nothing to eat on the way. It would consume all their strength and determination, but each man knew that they had no choice. Each had survived much worse before.

ONE

★

Facing the Challenge

FOR MANY PEOPLE, the marathon represents the ultimate in human endurance, but perceptions are changing. Not long ago, the ability to complete the distance was thought to be a capacity confined to superhumans alone, but now it is accepted that ordinary people can complete such races. It is a fact witnessed by the tens of thousands who run through the streets of our major cities every year. Indeed, many now see the marathon as at the lower end of the endurance spectrum. Where, for decades, there were only a few scattered races providing greater tests – the Tour de France or the Hawaiian 'Iron Man' – numerous ultra-distance events are now staged all over the world. Simultaneously, there has been an increased interest in activities such as mountaineering and prolonged expeditions to far flung regions. From city streets to mountain tops, more and more people are prepared to pit their fitness and mental stamina against prolonged physical demands. As they do so, it becomes clearer that both men and women have extraordinary endurance capacities.

I first met Mary Gadams in 1994 while running in a multi-marathon across more than 120 miles of Saharan sand. After that race, she read my book, *Shadows on the Wasteland*, which described my crossing of Antarctica with Sir Ranulph Fiennes. That was a journey made on foot, without re-supply of food or fuel, and with no help from other men, animals or machines. At the time, it was the longest, self-sustained walk in history. Taking this as a proof of my doggedness, Mary became convinced that I should be one of her team as she hatched plans to enter the first ever Eco-Challenge. She telephoned from the United States one quiet Sunday afternoon – domestic routine and the indulgence of a late roast lunch shattered by her invitation to wake up, train, and push myself hard once more.

The Eco-Challenge was a race conceived to take competitors be-yond their limits. It would cross more than three hundred miles of demanding back-country in the south-west of the United States using a variety of modes. Much would be running, hiking or mountain biking, for which I could offer both navigational skills and a reasonable level of endurance fitness, but the race was also to include horse-riding, Canadian-style canoeing and white-water rafting, at all of which I was a complete novice. Even the rock climbing, at which I counted myself quite adept, would include the use of artificial techniques to climb huge rock walls. Ascending clamps, pitons, and tape steps, all were unfamiliar to me. I was intrigued but daunted.

Each team had to comprise five persons, with at least one male and one female, and all participants had to complete every event. Here was another cause for concern for, despite my proven slogging skills, I did not see myself as in the same athletic league as her other choice of team mates. I had seen Mary herself running in the Sahara and knew that she was better at great distance than I was. Bill, who had already agreed to join her, had come second in the U.S. National Triathlon Championships, which made him pretty exceptional, and Whit, the other male, held the record for the fastest ascents of the highest peaks in every U.S. state. Most extraordinary of all, was the second woman in the team. Helen Klein was one of the greatest distance athletes the world has ever known.

When Mary persuaded Helen to join her, Helen was already 72 years old and a great-grandmother. Nevertheless, she was as fit as, if not fitter than, most people fifty years younger. She had already shown a phenomenal prowess at extreme distance running – a sport that she took up at the ripe age of 55 to compete with her distance running husband who was fourteen years her junior. I guess that that age discrepancy in itself denotes an unusual biological phenomenon. Helen was always young for her years, and as she became older, the difference between the calendar and reality went on increasing.

After taking up jogging, Helen steadily built up to running marathons. Soon she could beat all comers of her age, and so she progressively increased her distances to discover that she could also perform well at much longer events. Here she could provide competition to anybody, whatever their age or sex. In a way, this was not surprising, for it has been known for many years that beyond the non-stop 100-mile mark, women are as good as men. However, it is unusual to find top-class ultra-distance competitors over the age of 50 and unheard of for an athlete to continue successfully into their seventies. Helen broke all rules. By the time she joined Mary's team for the Eco-Challenge, 17 years of running had led her to completing more than 75 marathons and 150 ultra-marathons.

Even more worrying for me was that Helen showed little sign of slowing down with her advancing years. She could still run 100 miles in a little more than twenty hours – the equivalent of running close to four consecutive marathons at a pace that was respectable for a young woman. As a veteran, she was truly in a league of her own. Her personal motto was 'I'd rather wear out than rust out', and although I was only 40, I would clearly have trouble keeping up with her. Yet, despite my expressing such reservations, Mary was persuasive. I agreed to go for it and, following a now very late lunch, went out for a lengthy jog. The experience cost me violent indigestion and several days of stiffness. There was a long way to go but life felt good.

★

The horses were brought down from the hills on the evening before the race began. You could tell from their wild neighing and the sharp reports as they kicked at the fencing that it had been a long time since they were last cooped up. Speaking to the wranglers, we learned that during the whole winter and the spring the horses had run wild on the high pastures. It did not bode well for a docile mount, and when I walked round the corrals to see for myself that these were the 'fully trained trekking horses' described by the race organisers, I was far from reassured. A Palomino in particular caught my eye. Bucking, kicking and even biting at its companions, it looked as if it had never been broken-in, let alone been used for hacks. It also possessed the strangest eyes I had ever seen on a horse – blue, almond-shaped and unearthly pale against the browns of its nose. They looked as if they came from a wolf rather than any equine ancestor, and I prayed that our team would not be making closer acquaintance.

My prayers were not answered. As dawn broke the next morning it was a strange scene that filled the slopes of the desert-brown hills. The fifty teams of five persons each were now mixed with the one hundred and fifty horses that were even more excited than on the previous evening as they were saddled and assigned to riders. The first stage of our race was to be about twenty-five miles with three members of each team riding while the other two ran alongside. Our route stretched away from the high ground on which we stood across mainly flat scrub-covered country dotted with 'buttes' of red, towering sandstone rock that looked liked movie-set castles. Those familiar with horse-riding well knew that the stage would be hard on man and beast while the uninitiated, like myself, still thought it would be easy if rather frightening.

Although in the few months before joining the team I had considerably improved my equestrian skills, this was no great feat in view of my starting point. My only previous experience with horses had been a brief attempt to mount one many years before, and that had ended with my falling straight over the far side as the animal shrugged me off. It had seen a novice coming and decided that it had no interest in even walking with such uncoordinated baggage. Now, after attending evening riding classes, I believed that I could get on without such an immediate exit – until I saw that my mount was to be the horse with the ghostly eyes. At that moment I backed down and suggested a change of plan. It seemed safer for one of the more experienced equestrians in the team to take charge of this wilful beast, and since neither of the other two horses looked much better, I was to start out running. In doing so I would join Helen, who was also an inexperienced rider.

We got away at 6 a.m. – not an orderly start but a wild stampede. In retrospect, this was inevitable. The horses were unused to being saddled and mounted and many of the riders were of a poor standard. Why nobody was seriously hurt I do not know. As the countdown was completed, one hundred and fifty animals took off across the desert in a frenzied galloping group, most of the riders totally out of control and hanging on for dear life. With the rough going it was only moments before there was a scattering of bad falls. Some of the contestants dropped at the feet of horses coming on behind; others became entangled in the improvised straps and ropes that were used to attach their kit and were dragged along as bouncing sacks. It was surprising that nobody was seriously hurt, and after about ten minutes, the horses became calmer and spread out as things settled down. The teams which had not sustained a fall regrouped and went on, the rules insisting that all five members stayed close together at all times through the ten days that lay ahead. Behind us, small groups were dotted over the starting plain, some nursing their bruised team mates, others waiting as horses and their supposed riders circled each other warily in a taut-reined waltz of mutual suspicion.

Once the horses had settled, both my own and Helen's courage returned and after about ten miles we swopped with Bill and Whit and took to the saddle. The horses were still restless, and I could not get my mount to trot for any distance without him getting progressively faster. Before long I would find myself at a canter or full gallop, which terrified me. Then I realised that I had all the space in the world and it did not seem to matter. Careering across the open desert was a very different thing to cantering along some confined British bridle path and, with American-style saddles, it was more comfortable to go fast anyway. Rising to the trot was virtually impossible with an armchair for a seat and so slow speeds, other than a walk, gave a very uncomfortable ride.

Unfortunately, the horse refused to stand still at any time. This was a problem because the team had to stay close together, but every time I tried to wait for the others to catch up, my horse would start pawing the ground, snorting and threatening to rear right up and rid itself of the pest that kept holding it back via the bit. Within a few seconds of stopping, I would be forced to kick the animal on to stop it throwing me. In the end, the only way I could keep in touch with my companions was to gallop away from them for a couple of minutes, and then back towards them again. I had to do this all the way to the first checkpoint at the 14-mile mark.

At fourteen miles we were halfway through the first riding/running stage and there was a veterinary inspection. The vet was unhappy about the way that my horse was overheating, probably due to the considerable extra distance we had travelled together. To be safe, he pulled it out of the race, and so we set off again with just two animals. This was actually no disadvantage. Many of the teams who did not lose horses that were weakening were to regret it later. At the pre-race briefing the night before, it had been pointed out that any team was only as fast as its slowest member, but few had recognised that this was likely to be one of the animals. Although men and women are clearly much slower than horses over a few miles, when it comes to tens of miles, things can change. Humans make incomparable endurance machines.

After the checkpoint, Whit, Mary and I took to running and the terrain became more tiring. We were gaining height up a gently winding valley, entering a range of barren, orange-cliffed hills. It was a glorious place in which to be. Extraordinary natural rock walls and pinnacles were made mysterious by evidence of past Indian settlements. On one cliff beside the trail aboriginal art figures were carved in the ochre stone. You could feel the eyes of history gaze down as we passed beneath.

It was a little way past these carvings that we encountered our first real difficulty. Helen wanted to stop her horse and get off for a pee. She asked me to hold the animal while she retired briefly behind a rock. The horse, that ghost-eyed Palomino once more, was docile enough until she came back to reclaim it. I was holding it by the bridle as she started to climb back into the saddle, but as she did so, all hell broke loose. The horse seemed to think that its day's work was over, for as soon as Helen got a foot in the stirrup it reared, neighing, whinnying, kicking and bucking wildly. Helen was thrown some distance but landed unhurt. I, on the other hand, felt my head meet one of the beast's flailing hooves and immediately blood trickled down the side of my face. I wanted to let go and try to staunch the flow, but the horse set off across the valley floor, and I knew that it would go for ever if I did not hang on. Losing your horse meant immediate disqualification, so I kept hold as it dragged me forward, thrashing its head from side to side to try to rid itself of my unwelcome weight.

The whole episode lasted for only twenty seconds or so, though it felt much longer. Holding a galloping horse while being dragged across stones and through scrub turned out to be terrifying but, once committed, letting go seemed an even more frightening option. At least while I held the bridle I could keep my head clear of the pounding hoofs and rapidly passing rocks, and even the Palomino eventually realised that dragging a man by the mouth and halter was not much fun. It finally came to a halt, maliciously lining up with a large and vicious thorny bush with which I now made intimate acquaintance.

Whit ran up and took the halter while I struggled to free myself from the spines. He looked startled when he saw the vivid scarlet of my previously white head-band, with the blood running down from beneath it to my jaw. The blood was already beginning to congeal and must have looked as if it covered a deep wound down half my face. Whit seemed reassured when I spoke and wiped some of the blood away. Although the cut was just a scalp wound an inch or two in length, it was deep enough to need some stitches.

They would have to wait. The rules allowed medical help at the camps between different stages but not any assistance out on the course. We had to cover the rest of the distance quickly so that I could get my cut cleaned and fixed up. With a headache and bruises, and feeling as if I had been through a couple of rounds with Mike Tyson, I went back to the trail – running, loping, walking and finally riding once more to make the first full staging point in the early afternoon. There I handed over my horse to our two-person support team and received seven stitches in my scalp, inserted with only perfunctory local anaesthetic. It was uncomfortable but had the advantage of speed. Night was just a few hours away and we had to move on. We ate and drank well and then took on what we thought would be adequate supplies to see us through the second stage. It was to be a long run-come-hike through deep canyons in the range of mountains we could see ahead before once again crossing open, scrubby desert.

We set off as soon as we could so as to put as much distance as possible behind us before nightfall. Part of the canyon route was a 'dark zone', considered to be too dangerous to negotiate without light. We wanted to reach the start of it before resting so that we could enter the difficulties first thing the following morning and then complete the canyon section by the evening. In that way we would be able to cross the hot desert in cool darkness and, if we went without sleep, make it to the end of Stage 2 by early the day after. With the entire distance looking like less than one hundred miles on the map, we reckoned that under 48 hours was a realistic time. As it turned out we were hopelessly wrong, although all the other teams seemed to make the same mistake.

Soon after leaving the checkpoint, we entered the canyon system. We were still a little way from the difficulties that would be barred in darkness, but even here the canyons were winding, narrow, dark and, worse, were largely filled with bitterly cold water. The water was deep enough to force us to swim for quite long distances, pushing our backpacks with our food and emergency equipment ahead of us, shivering violently as we did so. Furthermore, where the canyon bottom was not flooded, the way was jammed with rock falls covered in slimy mud. It was not, therefore, until after midnight that we arrived at the control point at the start of the difficult section and lay down to rest on a narrow ledge above the river, and it was only two or three hours before we were up again to eat and pack before daylight.

It took the next day to struggle through the dark zone gorges. They were so narrow that it felt as if we were in caves rather than clefts. We swam much of the way, and by the time we had completed the difficulties we were utterly exhausted, and there were still many more miles of easier canyons to go. In the end, a second night was nearly over before we emerged mud-coated and somewhat hypothermic at the desert's edge. We were only halfway to the second stage-point, and later in the day we would run out of food if we did not cut back. We decided to spin it out, and with no breakfast set off once more. The mountains were soon left behind and the bad-lands surrounded us. Above, the sun rose into a disturbingly cloudless sky.

The desert was another nightmare. There were a host of river springs and water holes shown on the map but it turned out that they were either mythical or poisonous – marked with those skulls on posts seen in old Western films. As we set out from the next control point a race official warned us that only Well Springs was safe to drink from. In any case, most of the others were dry. With these springs more than forty miles away and the temperature rising to 30°C in the shade – with no shade anyway – we could not carry enough water to run the distance. We had to walk rather than trot, and it was going to take a very long time.

The terrain was also more difficult than we had anticipated – little easier than the canyons left behind us. Our maps suggested that the desert section would be large open plains, interspersed with empty smooth-floored mountain passes and some truly giant but dry canyons. The contour lines on the maps, however, did not reveal that the entire route was criss-crossed by narrow, steep-sided dry gullies which constantly sent us in search of a safe way down into them and then back up often vertical loose far sides. It made progress very slow indeed. Having risen at 6 a.m. from just one hour's sleep by the side of the last river, we had not reached the springs by 2 a.m. the following morning, and were still less than halfway to the next checkpoint and more food supplies.

Late that second night we ran out of water as well as nourishment. We were parched, and with the springs now less than ten miles away, we all wanted to press on. Helen, however, was in no state to persevere. Although her endurance was living up to her legend, she was used to both eating and drinking on the run and resting between long days of effort, and the lack of food and rest were beginning to tell on her in a way she had never experienced. Seeing her sway down the trail in the darkness, we had no choice but to halt. But we had no grounds for complaint. We were still in the top twenty per cent of the field, and when we met other exhausted teams, the sight of our great-grandmother made their jaws drop. We loved her for it.

★

There is a common belief that prior to this century everybody died young because the state of medical knowledge was so poor. It is a misconception. If our ancestors reached adulthood, there was a fair chance that they would reach the middle-age of our standards and some would even survive into old age. After all, three score years and ten is a term from the Bible. That is not to say there has been no increase in the *average* human lifespan. Clearly there has, although most of the increase followed changes in public wealth and health rather than any improvement in medical practice. The introduction of clean water and universal sewerage in the Victorian era did more for average longevity than anything doctors have ever produced. The increase in average age of death also reflects a decline in the mortality of infants, children and women in childbirth rather than of older individuals. Throughout history, and even pre-history, the life expectancy of a forty-year-old has not changed as much as most people imagine. At least, it did not do so until the last twenty years or so.

In these last two decades, the situation has finally altered. Recent medical advance has now influenced the lifespans of the elderly enough to make a difference. In nations with high health spending, the three score years and ten have moved up to and beyond a round four score. But why has this change come so late? The modern medical era should have borne useful fruit well before the last twenty years, particularly with the discovery of antibiotics and their successful treatment of most infections. It seems that every useful medical discovery was offset by detrimental changes to health occurring with alterations in lifestyle. Although we learned to cure diseases that had once been fatal, new and deadly conditions moved in to replace them. Instead of dying from lungs infected with pneumonia or ravaged by TB, we succumbed to bronchitis and lung cancer related to our smoking; instead of cholera and typhoid in our gut, the incidence of bowel cancer rose; and instead of rheumatic fever damaging our hearts in childhood and then killing us later through the scarring of our heart valves, we perished from heart attacks as our arteries became clogged. Overall, millions of us continued to meet an early grave – not through the natural hazards of life, but through smoking, the wrong diet and a lack of exercise.

★

We lay on the desert track with the sky above veiled with stars, although their beauty was hard to appreciate when legs and bodies were so very tired, uncomfortable and sore. We carried sleeping bags for the cold nights but the ground was hard and stony and our tongues cleaved to arid throats, coated in sandy dust. As we tried to rest in the darkness, we could comfort ourselves with the knowledge that the springs were now quite close. We would stop for just two hours for Helen to recover slightly and then move on well before sunrise. In the cool of the early morning, we would be able to lope along, and a couple of hours at most would see dawn at the springs.

Well Springs – what visions that name engendered. Streams tumbling down grassy hillsides, water bubbling over stones and pools in which we could bathe our painful feet. Beside them we would sit and share the very last of our food – a one person pack of breakfast cereal. Then, fully refreshed, we would push on over the last part of the desert to the next staging point. With luck we would be there by that evening, and then we could eat properly.

Yet again, hopes were not matched by reality. Although we arrived at the site of the springs quite quickly, we could find no water. At first we thought that we must have been mistaken about our position and spent some time taking bearings from dimly-seen hills. These confirmed our grid reference. We stood and listened carefully but could hear no running water, just the crushing silence of isolation. Quartering the area carefully, we came across another team that had arrived the evening before. They too had failed to find the water and were waiting for morning light. Ominously, they had already unpacked their emergency radio to call for help if they could not find water. One of them was in a bad way from dehydration, and they knew they could go no further.

The sun rose quickly, and once there was more light we began the search again. After half an hour or so, we heard a shout from some distance. Down a small side valley someone called 'It's here!' We all ran towards the cry – to see, to drink, and to relax. What disappointment. Instead of even one rippling brook, one burbling waterfall, or one small pool, there was just a puddle – a puddle a couple of yards across and only a few inches deep. To make matters worse, it had been trampled by cattle and the water was filled with their droppings. It was a ghastly addition to a liquid that was already seething with small larvae and other swimming creatures. A hard decision had to be made. Should we give up here or extend our trust in our sterilising tablets a long way beyond their specification?

It crossed our minds that we should stop – but that was not the answer for which we had come so far. Not only had we suffered the days of racing already achieved, but we had been through all those months of training and preparation. We could not simply give up because of squeamishness. In silence we filled our cups, dissolved our tablets and tentatively sipped the resulting vile, brown, iodine-flavoured filth. Then, still silent, we prepared our bottles for the rest of the day and made ready to go on. Only Helen actually put the stuff on her small share of muesli and ate it while chatting amiably to the other team who sat in abject depression. Only our 72-year-old faced the situation with unflagging humour and then brightly called for the off. As we left, the others called HQ to throw in their towel.

Our second day across the desert went no better. We walked through the heat, drinking reluctantly from our still limited water supply and heading for the next checkpoint where there would be bottles of mineral water and cans of coke to refresh us. But again, the maps and terrain matched poorly and so a nightmarish zig-zag took the remaining twenty-five miles up nearer to fifty. We had to rest again, and lay that night by the trail in a dusty canyon. It was lunchtime the next day before we eventually reached the second staging point. Instead of 48 hours from the first, it had taken four days and nights, and by the time we got there we all had bad stomachs and diarrhoea from the contaminated water. Even so, we were still in the race, while several others had dropped out. Many of the teams had been unable to face the cowshit water.

We remained at the second stage-point for a couple of hours while Helen was patched up enough to go on. Although young in heart and limb from years of exercise, she was not so young in skin. Hers had seen a long life of Californian sunshine and so her skin was fragile and paper thin. Now it needed to be bandaged to cover the cuts, blisters and cacti scratches carried with her.

The next stage, on mountain bikes, covered about thirty miles of unmade tracks to reach the edge of another canyon system. The ride went well for the first couple of hours, but then disaster struck. Heading down a steep winding hill on a loose gravelly surface, Helen lost control – a result of her gamely performing another sport with which she was hardly familiar. She hit the ground just ahead of me and I braked to a stop. She was quite still and did not stir for several seconds. I was off my bike and at her side before her eyes even flickered.

As a doctor, I knew what I should do. This was a woman who had hit her head hard enough to lose consciousness, albeit briefly. Such a patient should be admitted to hospital for observation, for a blow to the head can cause serious problems later. If there is slow bleeding within the skull, the pressure can build up without signs until quite suddenly the victim can lapse back into unconsciousness. There may then be only minutes before the pressure rises enough literally to squeeze the hind part of the brain out through the hole in the base of the skull. If this happens, it is fatal. Helen had been out for a few seconds, but afterwards was entirely lucid. She said she felt fine and had no headache. While she was raring to go, I faced a dilemma. I knew what I should do but, as a contestant, I did not want to do it. In the end I left it up to Helen herself. She had been a nurse in younger days and knew exactly why people went to hospital in such circumstances. It did not surprise me that she opted to go on, asking me to keep an eye on her through the night to ensure that she did not lapse into a coma. I agreed, and as soon as I had performed a little first aid we were once again on our way.

Above one eye and down one leg, Helen displayed some impressive lacerations, pouring blood. They should have been stitched but that was not an option. I cleaned the wounds as best I could and applied paper adhesive closures carefully. They would leave her with scars but she would bear them with pride. The whole episode served to emphasise her extraordinary mental as well as physical strength.

★

In addition to the misconception that in the past people died young, another false premise directs our thinking to the likely consequences of ageing. In western societies, it is accepted that as decades pass we inevitably become less active. It is assumed to be a natural part of senescence but, from an evolutionary point of view, this cannot be true.

The events I described in the prologue – the multi-day chase of the mammoth – were pure fiction. They were guesswork, portraying one aspect of life from perhaps ten thousand years ago. The image fits the popular pictures of Stone-age life in northern Europe, but anthropologists believe that the idea of early men as a spear-wielding mammoth hunter must be misleading. Women probably hunted also and, away from winter, the hunt would have been a minor part of life anyway. Antelope are very fast and wary and, as Jared Diamond has pointed out in his book *The Third Chimpanzee,* poking a buffalo, rhinoceros or mammoth with a stone-tipped spear would be tantamount to suicide.

Most of the early hunter-gatherer populations must have depended more on gathering than hunting. Grubs, worms and insects would have been added to a diet of fruits, nuts, roots and berries, and so they were not vegetarians. Rodents and birds would also have been caught on occasions, but success with larger quarry must have been rare. The return of the hunting party bearing rats and rabbits is a somewhat less macho image than one bearing elk and elephant.

Nevertheless, gathering foods would have entailed hard work. Men, women and children must have travelled widely and almost constantly, searching for plants, tracking with the hunts that did occur, and crossing whole regions with the change of seasons. Activity, extremes of climate and an almost constant threat of starvation must have been an integral part of life. To survive, everyone must have been extremely physically fit, whatever their age.

The need to stay fit would have begun to change with the advent of civilisation and the development of farming. Early in the process, most people would have remained physically active, working on the land. Later, however, more and more individuals would have become more sedentary as they took to trading and administration, and not until very much later, with the arrival of industrialisation and transport, would levels of activity have markedly diminished. Only in recent years, in the age of silicon, has inertia become the norm. A century ago, most people would have been active through most of their childhood, teenage years and adulthood, but in the last few decades the situation has been transformed.

Today activity levels for most individuals diminish with age in line with their changing commitments. While young children, our games keep us active for much of every day – although even this is threatened by the lure of the 'Megadrive'. Then, as we go through high school, spontaneous games decline and some days see little in the way of vigorous movement. Early adult jobs or university education see activity become more sporadic still, confined to an hour or two a week at a favourite sport. Then it stops altogether. In the developed world, most men and women over the age of 30 abandon physical exertion altogether and turn instead to the inescapable attractions of the motor car, television and the Internet. At around the age of forty, some do briefly resume exercise in an attempt to maintain some semblance of a youthful body, but few stick at it for long. Soon they rejoin the sad but inexorable drift towards sloth – the normal four-stage progression of accepted adult ageing described by Sue Lim in a BBC radio programme as: 'Lager, Aga, Saga and Gaga'.

When I make this point to friends or colleagues, I am accused of exaggeration and, yes, I admit, many adults do undertake something. They tend a garden, perform some D.I.Y., push trolleys round the supermarket, and may even walk the dog. Yet none of these things will stress the heart, lungs and muscles in ways for which they were designed. Even if spared immobility through bad joints or other illness, middle to old age sees those who do not have physically demanding jobs lose their physical fitness to such a degree that they even fail to meet the minimal demands of life in the late twentieth century. They become too infirm to get by without help. Of course, they may say 'So what? We have no need to push ourselves at all.' Few realise that they are turning a blind eye to the advantages that exercise can bring. Meanwhile, stories of capacities such as Helen's become more and more extraordinary.

★

The end of the cycling delivered us to the lip of Snake Canyon – another dark zone. In many ways we were grateful for that and settled down for our first night's sleep of more than a couple of hours.

Unfortunately, it soon began to rain and the bivvy covers of our sleeping bags – being thin to minimise the weight of our backpacks – were totally inadequate for a desert downpour. Where twenty-four hours before we would have killed for this water, we now lay soaked, miserable and cursing in the cold. By sunrise we had slept little and now faced another formidable obstacle.

Snake Canyon was quite magnificent. More than two thousand feet deep and winding away to join the distant Colorado river, it formed a huge chasm at our feet. Five separate ropes, more than six hundred feet long, disappeared into the chasm from the canyon lip, dropping into a huge void down a wall that could not be seen even if you lay down and peered over the dizzying drop. We were above a huge, overhanging half-dome, and after donning climbing harnesses and helmets, we attached ourselves to the ropes and gingerly stepped back, down and out into space.

None of us had ever done anything on this scale before. The ropes had been specially made to rig this, the highest free abseil the world had ever seen. Helen, in particular, had had only limited training – abseiling inside a California gym – but in yet another demonstration that age need not limit your possibilities if your attitude does not, she stepped over the brink with a smile and a wave for the cameras. Then she raced down her rope faster than the rest of us.

The rappel delivered us on to a huge ledge part of the way down the canyon wall, and from there we traversed left to turn a steep corner. I went first and emerged on a small rock platform from which a rope spanned an enormous side canyon. It was about seventy yards long and pulled tight to an anchor point on a ridge beyond the deep cleft in front of us. That ridge was just a narrow blade of rock, and I could see a team ahead making their way along it, traversing the broken sandstone teeth. From their nervous and slow progress, it was obvious that another canyon lay on the far side, perhaps matching the thousand-foot drop at our feet. One thing was certain: this was not a course for the faint-hearted.

Helen came up behind me, exhaustion in every movement. The last few days had taken their toll, but as she reached the edge and stopped, she smiled as usual. There was no way that this was going to phase our granny. Casually she began to ready herself for the crossing, clipping into the belay point and examining the ropes that spanned the abyss. Then she sat down for a moment, obviously thinking about her situation. Up until now, Helen had only read about Tyrolean traverses, but here she was about to cross one of the most exposed spans I had ever seen. On the ridge ahead there were more of these rope bridges spanning the gaps between the huge rock teeth. There was no doubt that we were in the middle of some of the most dramatic rope lengths the climbing world had ever seen. While I checked her harness and safety kit, she quietly ate a powerbar and summoned more of that mental strength.

When I signalled that all was ready, she got to her feet, drew a deep breath and, on the count of three, jumped, and I pushed. She flew out over the void with a whoop of combined fear and glee, speeding down the rope with her pulley running free. Then, after perhaps a hundred feet, she began to slow as she reached and passed the mid-point of the span, eventually coming to a halt dangling right over the centre of the canyon. Now she would have to work to get herself up the other side – pulling hand over hand up-hill.

However hard you tension these ropes, your weight stretches and bows them, and moving beyond the middle is demanding. Helen began the job but the further she went, the steeper the rope became. I could see her fighting her fatigue. Although her legs were strong from years of running, she had never taken on challenges needing this sort of upper body fitness. It was telling, and her face showed the pain. If she did let go she would be back in the middle of the canyon and weaker than before. We all shouted encouragement.

'Go, Helen, go!'

At last her hand touched the far side and painfully, slowly, she managed to clip a karabiner to the anchor point. There she rested briefly before pulling herself in. She struggled to her feet and sat down with chest heaving. Fighting to regain both breath and emotions, she raised her hand and waved me on.

One by one, we all followed, and once established on the rock blade, we began to move along it – walking, crawling, and climbing. We remained attached to ropes at all times, but although this granted safety, they did not remove the overwhelming sense of height. I had climbed for much of my adult life but had never been in such an exposed position, and for people like Helen and Mary, who had never known climbing, it must have been totally overwhelming. The broken blade was also not the end of the rock course. From the last gigantic tooth that thrust out into a bend of Snake Canyon, we descended by a series of abseils surrounded by nothing but air. Altogether there were eleven more descents to reach the river floor below and so it was late in the day when we finally came to the canyon bottom and once again were swimming, wading and scrambling towards the next goal. We would rendezvous that night with our support team and then take to the water for much of the rest of the race. There were four more days in which to complete the course.

★

For our hunter-gatherer ancestors, inactivity at any age must have run counter to evolutionary development. Older members of groups may have become somewhat slower as they aged, but they could not have foregone an active lifestyle. If they had done so, they would simply have been left behind. Perhaps this is what did happen. When older individuals, past useful breeding age, needed a share of food and held things up, their families and fellows simply abandoned them. Yet such a cynical view runs counter to logic. The rejection of the elderly would not have assisted group effectiveness in a world where nothing was written down. All wisdom resided within the elders, for it was they who knew where water might be found in the once-every-thirty-year drought, or where the herds had gone when they failed to follow their usual migration routes. They would also have made invaluable baby-sitters, not for trips to the Stone-age entertainment complex but on a daily basis while the younger adults sought for food. Group survival must have depended to some extent on grandparents and on selection of genes conveying longevity.

There is a mistaken belief that it is impossible for nature to favour genes conveying an increased lifespan. It follows a simplistic interpretation of Darwinism, that natural selection works only on individuals. If that were the case, any mutation that gave benefit beyond breeding age could not become established through selective advantage, since only genes granting an individual greater personal breeding success could become more common. A gene improving lifespan beyond active breeding years would therefore fail. In reality, however, natural selection acts on groups as well as individuals. Throughout early history our ancestors would have lived in small bands based around extended families who shared many genes. A gene that made someone live longer might not lead to that man or woman having more progeny, but might still make his or her group more likely to survive. Since other group members were close relatives, they would be likely to be carrying the same gene and hence, in the end, natural selection could favour the longevity gene. Nevertheless, it could only do so if there were no costs. The survival of the elderly had to come without automatic physical weakness.

The selection of physical fitness must have occurred at all ages. Those who were more capable of going the distance to find the foods, and of moving fast when hunting or escaping the animals that turned on them, would have had a better chance of surviving at all times of life. Similarly, those who could remain strong when too hot, cope when too cold, or keep moving when short of food and water, would have been the ones to live. More important, they would have bred more successfully. Genes conveying resilience of all types must have been passed from generation to generation over hundreds of thousands, even millions of years. The result was inevitable. Through the passage of eons, most of the peoples of our ancestry were cut down by the blind sword of selection to leave a species that was supremely fit and adapted for survival in the wild environments of the Earth.

★

The white-water rafting was to take place on a section of the Colorado river known as Cataract Canyon, where serious rapids with Grade 5 wave trains, drops, stops, and rocks abound. Our inflatable rafts were very small compared to those used by commercial companies taking tourists down the canyon with guides. We too were going to have professional help for this risky part of our journey. The clock would be stopped for the time we spent on the white-water section of the river and then started again once we were below the major rapids, where the guides would leave us to paddle on down river to our exit point.

Even our team's expert guide, a young woman called Molly, expressed some concern about taking us down the rapids in a smaller boat than she had ever used before, especially with an amateur crew which included a 72-year-old great-grandmother. Still, gamely she pushed her reservations aside and set off with us. During the long journey down to the first of the rapids, she maintained her delightful sense of humour and taught us some paddling drills. We needed to recognise the commands she would shout when the real test came the following day. Instructions to paddle right . . . left . . . forwards . . . backwards . . . had to be obeyed without question, immediately and with every effort.

The night was spent on an island in the river, with plenty of time for sleep, although again we were disturbed by rain. In the middle of the night, we had to get up to create a shelter from our upturned raft supported by the paddles, but this worked well and, despite a downpour, we slept until 5 a.m. As we arose the rain stopped, and the morning became clear. The appearance of the sun lightened the tensions we felt, and spirits were high as we cast off.

Around us were several other teams who had spent the night nearby. A few were now so far ahead that they had run the rapids before sundown the previous evening. With little paddling, we were all swept downstream by the increasingly powerful flow of the huge river. Emerald green banks swept by, contrasting harshly with the vertiginous, red-rock walls behind, but slowly they became smaller and then disappeared altogether. The canyon had narrowed as the walls closed in, plunging from the sky above directly into the speeding waters. The boats moved faster and the flow became a flood, with white foam flying. A distant roar echoed from the walls ahead, becoming steadily more insistent. Suspense rather than the volume of sound stifled our chatter.

Soon we were moving through the fifteen-mile rapids at great speed. Up in the bow, Helen was hanging on with both hands, jamming her shoes where the rubber walls met the floor to keep herself stable. Behind her, we sat two on each side with our feet hooked under rope lines that were rigged across the floor to keep us in-board as repeatedly our boat was engulfed in water. Molly sat as high as possible on the back of the raft, to enable her to take a safe route ahead. Every second or two she issued a taut command, steering deftly between the rocks and drops with a combination of our propulsive efforts and short thrusts from her own larger broad-bladed paddle.

Steering is a loose term for what can be achieved in a fourteen-foot raft in a maelstrom. With the river pouring hundreds of thousands of gallons a second through the narrow gorge – seemingly much of that into our boat – it was essential for us to avoid the worst of the waterfalls, rocks, or tree trunks jammed across the river. Manoeuvres were limited to turning the boat and paddling with maximum effort towards one bank or the other, as we hurtled sideways towards the next danger. All the time, we were trying to gain a part of the stream that would not lead to destruction. Just before a drop or a wave train was entered, the boat had to be turned bow forwards to ride the violence without over-turning. As we did so, we had to continue our frantic paddling to ensure that we were not caught in vicious undertows or giant whirlpools. Throughout, our tiny craft was more under than on the water. Every torrent-filled drop made me wonder if we could make it, but each time we popped up successfully. With every fall. Helen gave a whoop of approval – a delightful sound soaring above the roar of water but foreshortened as she disappeared beneath another icy green wave. It was a wet roller-coaster gone mad – a ride that made the best of Disneyland pale in comparison.

Around us, many of the other boats were less successful. More than half of them capsized or lost crew members overboard, and soon the river was full of people travelling rapidly with no transport while out of reach swept their empty boats turned-turtle. Between the rapids were occasional eddies in which to stop and rest, and here overturned boats could soon be sorted out, although not necessarily with every crew member back on board. Some of the boatless travellers failed to make the eddies, despite their best efforts at swimming, and on they swept down long sections of huge rapids in freestyle. It must have been an extraordinarily exciting but dangerous experience. Fortunately, no one was hurt.

At last we came out at the end of the cataracts. It was early afternoon and our arms were wiped out from their efforts. With two days to complete, those arms had much more yet to do. The remaining segments of the race would all depend on upper body power. Just to exit the canyon, we would need to climb thousands of feet of sheer cliffs, and beyond that there would be a long canoe trip. Worse, before we even reached the roped climb out, there were 40 miles of slow and sluggish river to navigate. Though a little peaceful tranquillity was appealing, a fourteen-foot inflatable would not be my choice of craft for paddling long distances.

Molly left us to the task, and with the race clock re-started we paddled through the remainder of the afternoon and on into the night without stopping once until we finally reached the exit point at 3 a.m. Here the darkness forced a stop – welcome for shoulders, arms and blistered hands, but all too short. At 6 a.m. a rosy glow caught the upper walls of the canyon and signalled that it was safe to start the climb.

Standing at the foot of the cliff was both frightening and awe-inspiring. It rose in a series of buttresses to disappear from sight far above, the thin climbing ropes connected to the rocks by tiny steel pegs. Even from below, it was obvious that while some of the ropes were long enough to span whole sections of rock, with ledges making a connection to the next rope, others crossed unbroken vertical or overhanging sections that were far too long for single rope lengths. There the rope-to-rope transfers would have to be made while hanging free. It would be utterly exhausting for the climbers among us, let alone for Helen and Mary. There were no doubts in my mind. We had come far, but this challenge would be beyond us, Whatever her determination, Helen could not scale this wall.

Through every moment of that long climb, I thought Helen would fail. It took me nearly two and a half hours to make the ascent, and by the time I reached the top my strength was decimated. Far below, Helen's distinctive helmet was just a dot, even through binoculars. Whit was climbing with her and carrying most of her load while patiently guiding and giving advice. He could not help her physically, and every foot of her ascent was an enormous strain on her desperately thin arms. Yet her will would not be broken, and inch by painful inch, she gained height as the sun crossed the sky and blazed down upon her slowly moving figure – a human spider crawling up a gigantic web.

Bill and Mary were with me, looking down as Helen drew closer. It was agonising to watch her changing from one rope to the next, with all her weight hanging from her thin arms. Clutching two jumars – the friction devices that slide up a rope but not back down – she climbed slowly. Arm up . . . heave . . . leg up . . . push, and repeat. When she came to a peg that connected the ropes to the rock, she had to clip in to a tape loop attached to it, by using a karabiner joined to her harness. It entailed pulling herself completely upright with just one arm while reaching for the loop with the other arm in which she held the karabiner. It was vital to clip into the peg loop with a short line so that the connection to your climbing harness would then take all your weight. Only then could she release the jumars which held her so that they could be re-attached to the rope above the peg. Time and again, Helen would fail at this exhausting manoeuvre. Either the strength would go from her pulling arm or her fingers would prove too weak to open the safety gate on the karabiner. She would reach up, try to clip in, and then fall back several feet with a violent jolt. Then, after a short rest, she would climb up and try once more, only for the process to repeat itself. It was pitiful to witness, and the strain showed in every line of her face. At the same time, to see this small old woman dangling from ropes on the gigantic rock wall was inspiring. Everybody added their will to her own as she pushed onwards, but with the sun's heat adding dehydration to her weakness, she became ever slower as three and four hours went by.

In the end, I was to be proved incorrect. My conviction that this was a task beyond even Helen's capability did not take account of the power of will. Five hours passed and Helen was still going. The sun reached its zenith – six hours on a baking wall with no water before she came within a few hundred feet of the top. We could actually hear her cursing at herself in most ungrandmotherly language each time she failed and fell back, but she would not give up. Then, at shortly before three o'clock, she finally pulled herself over the lip and lay quietly face down at the edge of the canyon. For a few minutes we let her be to release her tears. Then we picked her up and hugged her before continuing on our way.

★

What physical differences separate us, the modern humans, from our early roaming ancestors? Many would imagine there is a large gulf between us, but I contend that it is closer to a crack. Hunter-gathering lasted until about ten thousand years ago and in many parts of the world went on until much more recently. With no fundamental change in lifestyles, selection would have continued to generate men and women who were perfectly suited for physical activity and hardship. Furthermore, underlying these capabilities, people would also have been perfectly adjusted at the biochemical level, with harmony between their lifestyle, diet and health.

Can we say the same for modern societies? Ten thousand years is not long in terms of the evolutionary clock and so there has been virtually no time for significant genetic change. Yet civilisation has led to huge alterations in our environment which have put us out of step with our evolutionary background. There is now a discrepancy between our biological design and what we do and eat. It is a discrepancy with considerable implications.

On the positive side, it can explain phenomena like Helen. Admittedly she may be at the extreme end of a biological distribution of fitness but nevertheless she illustrates our potential. I believe that most of us have an astonishing natural strength which includes the ability to walk and run considerable distances every day, and we also have the capacity to survive in conditions that are scarcely imaginable.

On the negative side, the discrepancy between our lifestyle and our evolutionary design is at the root of many of our problems. The advent of civilisation brought with it a radical change from our ancestors' diet of fruits, roots, vegetables and grubs. With improved tools and more success with hunting, the proportion of meat in our diet must have expanded a little over the last 100,000 years, but it would not have been until the successful domestication of animals around 10,000 years ago that the consumption of meat and dairy products increased rapidly. The speed of change would have been too fast for our biological systems to adapt, and so it is fair to assume that the changes must have upset a number of our metabolic controls. Furthermore, with the farming of animals and planting of crops, we lost many of the uncertainties in food procurement.

A reliable food supply has many implications for health. Obviously it is beneficial in avoiding death or illness due to starvation, but there are some disadvantages when the supply becomes too good. Our bodies are genetically tuned for constant activity and set up to face a predominantly vegetarian and varied intake. They expect to meet seasonal shortages and frequent famines which no longer appear. As a consequence, our health and our hunger control mechanisms are deeply confused. Ten thousand years ago there would have been no fat people around, and problems which now dominate our hospitals – heart disease, strokes, diabetes and even some types of cancer – would have been rare.

★

We left the top of the canyon in orange evening light, heading for our last support team rendezvous. It was a simple ten-mile run, downhill on tracks and narrow roads. This was a delight. Although our arms and backs were tired after the paddling and the climb, our legs were fresh, for it had been nearly three days since we had walked. We could happily trot and relax, admiring the glorious scenery that surrounded us. For the first time I noticed the enormous numbers of blooms in the desert terrain – multi-coloured bells, as delicate as any you might find in a flowered English meadow, mixed with spectacular funnels of orange, red and pink bursting from harsh-spined cacti. The paths weaved their way between them, giving the appearance that they had all been laid out deliberately – the desert equivalent of a Japanese garden.

It was sundown when we reached the last stage-point which lay on the banks of Lake Powell, an enormous artificial expanse of water, formed from the damming of this part of the upper Colorado canyon system. As darkness fell we sat overlooking the placid waters and ate a good meal. Beside us, loaded and ready to go, were our two Canadian-style canoes. Mary and Whit would paddle one of them (Helen installed between to get some rest) while Bill and I would take the other with all our kit, food and drink. By the time we were ready to go, it was pitch dark, and with the moonrise still some hours away, navigation would not be easy.

The map showed that we were in a broad flooded side canyon which we had to follow for seven miles before joining the main lake. Once there, it would be fairly easy to maintain the right line, for the course was dead straight for another twenty miles or so and all we had to do was to follow a compass bearing through the night. Beyond that the canyon wound again but, once there, we would have the moon and then dawn to light the way. The difficulty would be finding the main lake in the first place. The side canyon had its own branches which ran in different directions. It would be easy to go up any of these by mistake, thinking it was a bend in our course. There was no current to guide us and the branches were much the same width as our winding route.

No sooner had we started than another problem became evident. The darkness was so intense that hills were a mere extra blackness against starlight. It was impossible to judge their size or distance and we had little idea how fast we were moving. Sometimes we would fix on a distant hill and make it our target for the next hour's paddling, only to find it was a big rock a hundred yards away. At other times we took evasive action to round a rock that turned out to be a distant hill. We tried to follow the map, noting every change in bearing so that we could mark it as the turning of another corner, but soon we became deeply confused and only learned of an error after we had taken a heading that was impossible for our course. Then we would turn round to try again. Once or twice I am sure we covered the same section of water several times in different directions, and in the end it took more than seven hours' paddling to arrive on the main lake at about 3 a.m. We had slept for less than two hours in the previous forty-five, and for the rest of the time we had worked with arms and backs. We still had more than thirty miles of calm water to paddle and a stiff wind began to blow against us.

The moon rose as we reached the main lake. It helped to confirm our position, and the rest of the night was simply long and fatiguing. Our bodies now protested bitterly, despite the rhythmic paddling putting one into a kind of trance. Every so often the trance overcame the pain and eyes would close as minds drifted away. The dream would then be broken by waking with a jerk and realising where we were. It all reached a climax at around 5.30 a.m. when my boat suddenly lurched to one side and almost overturned. At the same time, there was a splash behind me. Bill had gone overboard. He had fallen asleep and dropped off the canoe.

We pulled him from the water and stopped on the bank for some food. The sky was beginning to lighten and one by one the stars went out. Slowly the tops of the high rock walls changed from black through red to gold. It was the beginning of our last day of effort, and we knew now that we would complete the course. There were fewer than twenty miles of paddling before it would all be over. Refreshed in spirit if not in strength, we returned to our boats and went on.

Four hours later, with no further stops, we came to the line. We were not the first by any means. Indeed some teams had arrived more than twenty-four hours previously and another sixteen teams had made it in the interval. Still, at eighteenth out of fifty teams we, with our great-grandmother, were elated. The welcome we received said it all. As we disembarked from our canoes loud cheers greeted Helen. For her, it was an extraordinary moment of triumph.

TWO

★

Out of Africa

IN ORDER to understand fully the origins of our resilience and the problems that arise from conflicts between biology and lifestyle, we must examine our evolution more closely. For hundreds of years it has been clear to many thinkers that man is one of the animals. Even by the second century A.D., the Greek physician Galen had dissected various creatures and recorded that human beings are most closely related to the monkeys. It was obvious to him that our bones are similar in shape to theirs, as are our organs and muscles. Since then, the reason for this relationship has become clear. The emergence of Darwin's theory of evolution explained how different families of animals came into being and now, a century and a half later, the evidence is conclusive. We humans are full members of the sub-group of primates known as the apes.

The apes include gorillas, chimpanzees, orang-utans and gibbons, and at different times ideas have varied as to which of these is our nearest relative. Today a majority of the public believes that this dubious honour must go to the gorilla, although not long ago some experts would have advocated the orang-utan. New techniques, however, have shown that neither candidate fits the bill. Studies comparing our DNA with those of our primate relatives have provided an unequivocal answer – the chimpanzees are our closest cousins.

Some six or seven million years ago, in Africa, there was a population of small apes that were the common ancestors of both man and the chimp. The population must have then divided into two groups which started to live apart and tended not to interbreed. The cause of this division is not entirely clear but the idea most popular among anthropologists is that the two groups started to occupy different environmental niches. At around that time the African climate became both hotter and drier, and it seems that this led to a thinning of the forests and an increase in the areas of savannah. With a reduction in the number of trees, some of the apes started to live on the ground and then moved out on to newly open grasslands. Others remained more arboreal, living within the forests. The result was that the two groups were exposed to different environmental demands and this led to their slow evolutionary divergence. The tree dwelling group slowly evolved towards the two species of modern chimps. The ground dwellers embarked on the path that would eventually lead to man.

One of the changes that life on the savannah appears to have prompted was a move to becoming upright, and various suggestions have been made to explain this. An early interpretation was that the change occurred in order to free our hands for tool use but the fossil evidence does not support this. The change was already well established before the first evidence of rudimentary tool-making and expansion of brain size four million years ago. Another suggestion is that the move to the savannah meant that the groups had less access to shade. With the sun high in the African sky, heat stress would have been a considerable problem and the adoption of an upright stance would have reduced exposure by limiting the surface area under direct rays from the sun. The same radiant heat argument has also been used to explain why we became bare-skinned except for the tops of our heads, but if the lack of shade was the driver towards naked uprightness, it seems odd that other species, living in equally hot and sunny circumstances, have not found the same necessity for change. Another theory suggests the erect posture was adopted in order to reach fruits in the smaller trees and bushes that were growing in the new hot, dry weather and were too frail to be climbed. These ideas have led to the arguments being reversed. Instead of the use of tools bringing us to our feet, it seems more likely that it was the coming to our feet that promoted the use of tools, which in turn drove the rapid development of brain size and intellect – features that were to become the greatest assets of humanity.

It might be thought that an increase in intelligence would confer considerable survival benefits for any animal but, without simple tools, intelligence may be less of a survival advantage than one imagines. Even if you are a lion with the capacity to understand astrophysics, your advantage over a more stupid lion is fairly limited. Your knowledge would be of little help in mating or fighting, and although your generally increased intellect and canniness would give you occasional benefits on the hunt, it would not necessarily make you more successful than rivals who were faster or fitter. It therefore seems likely that selection for intelligence exerted only a modest effect on the development of most species and that many modern animals are little more intelligent than their ancestors. At the same time, increased intelligence does promote considerable survival advantages once it can utilise technology. Once our ancestors were upright, the freeing of their hands and the consequent use of tools made intellect a much more powerful asset, and the chances of survival for brighter members of a group increased considerably. The evolution of intelligence and technology must have accelerated together.

Although the divergence in intelligence and posture between the ancestral chimps and humans is striking, it would not have happened quickly. Mutations are infrequent and the majority of them fail to become established in populations. They arise from mistakes in the copying of the DNA code as it is passed from one animal to its offspring but many occur in parts of the DNA that are redundant. These result in no change of either structure or function in the animal that inherits them. Transmission to future generations is entirely a matter of chance, and with all the other animals of the group passing on the 'normal' form of the DNA, most mutated sequences in non-critical regions of DNA fail to become established in the population.

Of course some mutations do occur in a region of DNA that *is* active, and there is a change in a gene that actually encodes one of the animal's proteins. All proteins are made up of long chains of linked amino acids and the mutation causes the substitution of the wrong amino acid somewhere in the chain. This tends to alter the protein's shape, normally critical for its function. Most proteins either contribute to an animal's structure or operate as an enzyme to control one of its metabolic systems. In either case, their exact shape is usually crucial for their role. In fact the amino acid sequence is close to being inviolable, and because of this, most mutations in critical DNA regions are immediately fatal. They also go nowhere.

Only in rare instances does a mutation lead to an amino acid substitution that actually improves a protein's function. If this occurs, natural selection can come into play to increase the chances of that mutation becoming widespread in the population. Nevertheless, it must be remembered that the new mutation will become universal only if the descendants of the animal that first had it are the sole survivors. This happens either through exceptional random chance or when the advantage conferred by the mutation is really very considerable. Many advantageous mutations fall by the wayside and the overall pace of evolutionary change is slow.

At first sight, it might be imagined that there were large genetic differences between ourselves and the chimpanzees, perhaps suggesting that the acquisition of new mutations was somewhat easier than I have intimated. This is not the case. In reality, the differences between our genetic codes are tiny and direct comparisons have demonstrated that 98.4 per cent of chimpanzee and human DNA are identical. This is less of a difference than that seen between many animals designated as being of the same genus, and some anthropologists believe that it is mere pride that stops us redefining our position in phylogeny. In *The Rise and Fall of the Third Chimpanzee*, Jared Diamond points out that we are actually another species of chimp rather than a member of a separate grouping. Certainly we are more related to the chimps than they are to gorillas.

Despite many areas of debate, the steps that occurred during the period that separates us from the chimps are becoming clearer. By three million years ago, our upright ancestors had divided into at least two different lines, with several species in each line. There were heavily built versions of early humanoids of which the best known is *Australopithecus Robustus*. They survived for a very long period with little obvious development of either anatomy or technology but eventually their luck ran out, and they were eliminated by rivals more than a million years ago. The lighter built versions fared better. The best known of these is *Australopithecus Africanus,* but there were many others, differing only slightly. All were bi-pedal. They lived in various locations in Africa, and by two and a half million years ago one group had evolved large enough brains to allow the scientists of today to place them in the human genus. They are known as *Homo Habilis*.

When *Homo Habilis* appeared in the fossil record all early humans were confined to the African continent. They made tools that were probably good enough to give them a real advantage, and with that their development accelerated. By around 1.7 million years ago, they had evolved further into *Homo Erectus*, which had an even larger brain and presumably greater intelligence. Perhaps it was this that granted them the flexibility and imagination to move, and around one million years ago *Homo Erectus* began to spread and became established in other regions of the globe. It would be easy to imagine that these populations were the forerunners of the diverse races of today but, although there is considerable debate about the point, with claims that Australian Aboriginals are descended directly from *Homo Erectus,* it is probable that the ancestors of all current humans did not leave the African cradle until very much later. The confusion arises because, oddly, the precise history of our ancestors in the last million years is more difficult to interpret than that of earlier times. The later groups wandered much more widely than their predecessors, creating considerable confusion in the fossil record. However, a group that left Africa in a migration around 100,000 years ago was very much closer to all modern races.

Following this second African exodus, three lineages of *Homo* seem to have inhabited different parts of the world. One, in particular, fits the popular conception of early cave men. These were the Neanderthals – men and women who must have looked quite different to ourselves, with muscular bodies, flattened foreheads, heavy bony eye-brow ridges and faces that were prominent. Around 100,000 years ago, they were widespread and their remains have been found in caves over much of Eurasia. It is likely that most Neanderthalers actually lived in temporary shelters and that their discovery in caves is merely a quirk of preservation. Nevertheless, it is doubtful that they did much in the way of complicated construction. Although their brains were large, indeed larger than our own, they appear to have been very inferior intellectually. They made only primitive tools and had little or no art or culture. A long way from modern man, they finally disappeared around 40,000 years ago. Quite why remains a matter for speculation, but it seems most likely that they were in direct competition with other species. It is certainly possible that their extinction was due to our own ancestors embarking on the road to genocide.

Another of the three human species that lived 100,000 years ago was limited to parts of Asia. Examples have been found in the caves of Java and elsewhere in the Far East, but relatively few remains are known and so the development of their anatomy, tools and culture is not clear. They too were descendants of the first *Homo Erectus* migration but, like the Neanderthalers, they died out and left no descendants.

It is the third type of early human that holds the key to our modern existence. These are the Cro-Magnons, first recorded in African sites of around 130,000 years ago. They were very widespread on that continent, with the best fossil records found in the caves of South Africa. Around 100,000 years ago, it was some of the Cro-Magnons that undertook the second dispersal from Africa with the result that their remains began to appear over much of Eurasia, and later groups moved to the Americas and Australasia. The Cro-Magnon remains present a picture of advanced tool-makers who must have looked very like us. They hunted successfully, albeit for quite small prey, and must have gathered the remainder of their diet. Early in their history they did not display much in the way of advanced culture, but around 40,000 years ago this changed quite suddenly.

There is considerable speculation as to the trigger for the acceleration in Cro-Magnon development. The explosion seems to have been cultural rather than physical, for there are no simultaneous changes in their fossil records. One possibility is the discovery of language. Although it has been shown that groups of wild chimps use a limited vocabulary of calls to co-ordinate actions such as hunting, their capacity for communication is very restricted. In part, this can be ascribed to their inferior intelligence, although studies in the last few decades have shown that domesticated chimpanzees can use signing or other techniques to represent items or actions separate in time or space. They therefore possess an intellect adequate for some language, but if sounds are to be used to represent concepts, a wide range of vocalisation is required. When compared with humans, chimpanzees are restricted in this respect, due to the different anatomy of the mouth, pharynx and larynx. It is probably these differences that have limited the development of the chimps' communication.

Unfortunately, anatomical differences dictating vocal abilities leave no fossil imprint, and so we cannot define the time at which early humans developed speech. Nevertheless, we can surmise that once this did occur it would have precipitated a marked step forward – stimulating the further selection of intellect in the same sort of way that freeing of the hands and the use of tools had done. With no change in underlying brain complexity or innate intelligence, language would have launched a new era of social interaction and group success. Was it therefore speech that launched Cro-Magnon on the path to modern man?

How different are we from the Cro-Magnons or, more important, from their hunter-gathering descendants of immediate pre-civilisation? The answer is important, for if we are genetically very similar, it would support the supposition that we are designed for a lifestyle at odds with our own. Evidence can be gained from the DNA. The fact that we share 98.4 per cent of our genes with the chimpanzees means that only 1.6 per cent of our genes must account for all the obvious differences between us – intelligence, erect posture and speech. The remainder of our physiology and metabolism must therefore be very alike, and indeed this is the case. Biochemically, we could not be much closer cousins. Since the fossil record shows that the joint ancestors of chimps and man lived between five and seven million years ago, it must have taken around three million years for the DNA of the two groups to diverge by just one per cent through the acquisition of new mutations. There is no reason to assume that this rate of DNA change is not typical for all primate or hominid groups through the ages, and so the 40,000 years that separate modern humans from the mid-Cro-Magnon years represents a genetic difference of little more than 0.1 per cent − a difference within the range of variation that exists between modern human races.

This type of speculation, indeed my earlier description of slow change through the blind selection of mutations, can be viewed as overly simplistic. In his recent book, *Lifelines,* Steven Rose argues that the reality is far more complex and that evolution acts on complete individuals within their environmental framework rather than by a mechanistic, constant 'selfish gene' process as advocated by Richard Dawkins. Rose may be right and, if he is, then changes in our genes could have occurred somewhat faster in the last 40,000 years, when rapid development of intelligence may have pressured the selection process. Still, I doubt if it could have made much difference to the overall rate of biological adaptation and, as I have mentioned previously, it was only ten to fifteen thousand years ago that modern man moved away from Cro-Magnon life and began the domestication of cattle and the setting up of more permanent homesteads. The genetic differences between modern humans and those early neolithic farmers must therefore be tiny, probably less than 0.05 per cent, and so if it were possible to extract late hunter-gatherers from the past, I believe that they too could be trained as fighter pilots, doctors and lawyers. Of course, the converse is also true. We, the pilots, doctors and lawyers of today are almost identical to those men and women who dwelt in caves and shelters. We are born superbly designed and equipped for action and survival in a harsh world − but unfortunately not for the one in which we live.

THREE

★

A Gene for the Fastest

WHEN Linford Christie or Donovan Bailey go down on the blocks for the sprint to glory or dismay, they command bodies that are designed for moving rapidly. Since birth their genes have dictated that their leg muscles would be packed with a specialised, fast-contracting type of fibre, but that endowment has not been enough. Like all world-class athletes, they have put in endless training to go so far, building upon their birthright to improve their muscles' size, strength and short-term fuel supplies. Their training incidentally also honed their hearts and lungs to the point where they can now pump round twice as much oxygen as can an average man's. Yet, even boosted by a pre-race priming of adrenaline, heightened by the expectancy of a hushed stadium, it is not this exceptional oxygen delivery that will make the difference. The athletes will work too hard and the encounter will be too brief to use their aerobic fuel systems. They are the ultimate speed machines, with legs set up to produce a short burst of power that is dependant solely upon the chemical energy stored and ready within them. As far as those legs are concerned, the whole race could be run were their hearts to stop with the shot from the starting pistol. Their brains, of course, would be less flexible.

The gun is fired and a single pressure wave of sound is received by the competitors and passed, almost by reflex, to their motor cortex. There it triggers the conscious drivers of movement that hit the neural switches of the 'go' command which then stamp on the accelerators. Muscle engines explode into life as more than one hundred thousand fibres in each limb hurl the competitors forward. The sheer power of each of their huge muscle groups is then blended into a perfection of movement that can only be appreciated in the later slow-motion video. They burst from the blocks with the balletic grace of a Nureyev but the acceleration of a Ferrari.

Once away and into their punching run, most of the control is switched to the auto-pilots of the lower brain-centres programmed with patterns of muscle firings burned in by experience. The learning goes right back to the first tottering steps of their childhood, but unlike most of mankind, who simply added the everyday experiences of walking and running to such infant movements, their training has led to perfection. They have spent days, weeks, years to reach this level, and to the extraordinary power of their muscles has been added deliberate assessments of running style. With effort written in near frozen expressions of will, the athletes are now capable of the ultimate sprinting achievement – winning the Olympic hundred-metre Gold.

★

Although the high levels of activity undertaken by our ancestors led to the selection of men and women who were all physically fit, it did not confer the same type of athletic ability on every individual. Athletes of today cover the spectrum between those designed for short-term speed or power events, such as sprinting or weight lifting, and those designed for endurance events, such as the mile or the marathon. Underlying these specialist capabilities are variations in physiology which have only recently been understood.

All animals are designed for movement. As evolution led to larger and more complex organisms, they could no longer remain reliant on the limited food sources that lay beside them or washed slowly by in the primitive seas. They needed to travel from place to place to find more to eat, and for that they needed a means of locomotion. Over millions of years, all the higher genera developed a system based on muscles – specialised contractile organs that could propel them around the waters, the land or the air. In the primates, those muscles and the skeletons through which they operated, became the greater part of the body. Today nearly three quarters of our weight is dedicated to movement.

A useful analogy is to compare our locomotion systems with those that drive modern vehicles. In cars or trucks the energy for movement is created by burning fuel in the cylinders and then utilising the expanding explosive power to move a system of levers, then cogs, to turn the wheels. At the same time new fuel and air are delivered by fuel pump and carburettors and the waste products of burning – carbon dioxide, water and a few more noxious chemicals – are cleared away through an exhaust system. Excess heat is lost through the cooling system.

In man, as in all other large animals, the single combustion engine is replaced by numerous separate muscles which act like servo-motors seen in more complex man-made machinery. As with internal combustion engines, each muscle runs by combining fuel with air, although they can also run for short periods off chemical storage of energy. Instead of the inefficient explosion, the burning is much better controlled and the muscles produce a contraction power stroke rather than an explosive outward one. The power released is then harnessed by a series of levers which grant us motion directly rather than it being transformed into rotation. Evolution has never devised animals on wheels, probably because all parts of an organism need to have a blood supply and, unlike electricity, blood cannot be passed through brushes to or from a rotating axle. It is indeed difficult to conceive of how a living, wheeled locomotion system might have evolved. Furthermore, wheels have no advantage on uneven surfaces, and only in very recent times has our intellect developed sufficiently to allow us to smooth the earth's surface at will. Now, joined to inorganic wheels, the man or woman on a bicycle is probably the most efficient motion machine the world has ever seen.

Altogether we have around two hundred and fifteen pairs of muscles arranged symmetrically down right and left hand sides of our bodies. Most of these move bones at joints around which there are usually several different muscles with varied actions. The movement of each thumb, for example, is controlled by nine different muscles whose functions are complementary, acting to assist one another. Complementary does not mean that they necessarily move the joints in the same direction; in fact, every joint must have muscles set up to work in opposition since, as a muscle can only pull and never push, one could never reverse a movement without such an arrangement. Opposing muscles also allow smooth movement by tensioning and controlling both sides of an action, and they grant joints increased stability.

When, in 1660, the Dutchman, Anton van Lewenhooek, looked down one of the world's earliest microscopes, he saw that muscles are made up of tens of thousands of elongated fibres. Each of these is a single cell, with some of them among the largest cells in nature. Even in man, only a medium-sized animal, one fibre can be more than 30 centimetres long, running the entire length of large muscles such as those in the thigh. Yet, despite their length, these fibre cells are only a few thousandths of a millimetre thick. They therefore have the shape of an extraordinarily thin cylinder. Although each muscle fibre is actually capable of independent contraction, in reality they are covered by fine fibrous tissue that joins them into bundles. These bundles are then further linked to form the muscles themselves, with the fibrous coverings blending to form their tendons. As a result, thousands of fibres can pull on the skeletal levers simultaneously and can generate extremely large forces. These have led to extraordinary attainments. Individuals have cleared high jump bars 60 centimetres above their heads and pole vaulted well over 6 metres; they have run 100 metres in 9.8 seconds and 100 miles in eleven and a half hours; they can score 501 on a dart board with just 9 darts and achieve maximum breaks of 147 on the snooker table; they have even generated a force equivalent to 442 kilogrammes with a single bite. We may not be the fastest or strongest animals in the world but we are surely the most adaptable.

★

Inside each muscle fibre there are the structures common to all living cells – the systems that fuel, repair and re-build them as necessary – along with the specialised contraction mechanisms. The contraction systems lie within fibrils, even tinier cylinders that run the length of each fibre in a manner similar to that in which the fibres run the length of muscles. Fibrils can also contract independently if artificially isolated, but it took the invention of the electron microscope to understand how this occurs.

When your brain wishes to move a muscle, it transmits electrical impulses down the nerves of the spine and out along controlling nerve branches where they pass along the surface membranes of the muscle fibres. These membranes are folded in such a way that the electrical signal is carried close to all of the fibrils within. Each fibril is made up of chains of thousands of tiny contractile units which consist of two end plates with two different types of protein filaments connecting them. Actin filaments are very thin and extend from attachment points on each of the end plates towards the central part of the unit where they lie adjacent to one another, and in parallel. In the central region the thin actin filaments are interspersed with thicker filaments of myosin in a perfectly regular pattern so that each end of any myosin filament is surrounded by six actin ones extending in from opposite end plates.

When the muscle is relaxed, there is no contact made between the actin and myosin although hundreds of tiny extensions come off the side of the thicker myosin and almost touch the actin ones. It is these molecular-sized arms that are the key to movement. On the arrival of the nerve impulse, calcium is released around them from specialised microscopic holding tanks. It has two dramatic effects. First, it makes the tips of the tiny myosin arms sticky, so that they attach to the nearest actin filament, and then it makes the arms deform and bend, as if hinged, at the point where they extend from the main myosin filament. Because the arms at each end of the myosin bend towards the middle, the stick and swing action pulls the surrounding actin filaments fractionally in towards the centre. With them come their attached end plates and so the electrical impulse has resulted in a shortening of each unit.

Of course the shortening of an individual contractile unit in response to one nerve signal is too small and weak to be of any use on its own, but it occurs in thousands of units linked together in each fibril, in thousands of fibrils that make up each fibre, and in thousands of fibres that make up each muscle. The single electrical trigger can therefore result in a powerful muscle twitch. Yet muscle twitches alone, however strong, are also useless when it comes to controlled movement. It is necessary for muscles to shorten steadily, and this is achieved by repeating the triggering process rapidly. No sooner have the myosin arms stuck, bent and pulled, than they release again as the calcium is quickly pumped back into the storage tanks. Then a second electrical signal arrives before there has been any time for things to relax, and a new stick and pull cycle is initiated. With the brain capable of triggering 100,000 shortenings per second, the result can be dramatic. The muscle exerts a steady shortening pull, and by regulating the frequency of firings and the number of fibres triggered, the brain can exercise precise command over considerable forces.

The mechanisms that allow muscles to contract need a supply of energy to do their work, and this is achieved in the same way as the body fuels all processes of life – from the breakdown of chemical bonds within compounds. All chemical bonds contain energy but we can only use the types found in foods such as carbohydrates, fats and proteins. The energy in these comes originally from photosynthesis in plants – the process powered by radiation from the sun in which leaf chlorophyll bonds carbon dioxide with water to make sugars.

Between the plants using sunlight to trap energy in sugars and our utilising that energy for action there are a series of pathways. The first set of pathways lie outside our bodies and can be either straightforward or complex. In some cases we simply eat the plants and so ingest the sugars directly into our system. In others, however, the plants themselves convert sugars into different substances, or we eat animals that have consumed the plants and made other conversions. Overall, we ingest sunshine energy converted into a whole variety of chemical bonds in proteins, fats or more complex carbohydrates – molecules which then enter a second set of chemical pathways inside our bodies which are also of varied complexity.

At their most simple, the sugars or fats are absorbed and transported directly to muscle cells where the chemical bonds are broken and the energy is released and used for movement. More often, however, the energy from our food goes through a host of other intermediate reactions inside us, frequently spending time in either fat stores or body structures. Nevertheless, it eventually meets the same outcome. The sunshine that was incorporated into sugar bonds within the leaves of plants eventually runs our muscles as it does our other metabolism. As this happens, carbon dioxide and water are released once more. Indirectly, we are all solar powered.

The chemical energy within our food needs to be changed into a form that can be used at the molecular level in cells. Evolution has chosen a molecule known as adenosine triphosphate (ATP) for this purpose and it exists very widely in the body, fuelling the majority of cellular processes. ATP can provide energy conveniently because it is easily split by an enzyme to release a small packet of energy, and the splitting enzymes can be located wherever that energy is required. In muscle they form the very tip of the myosin arms and one molecule of ATP is used by each arm for each stick and pull cycle.

Muscle fibres come in two different types depending partly upon the nature of the ATP splitting enzyme. One type, fast fibres, can split ATP extremely quickly and so allow contractions of great speed, whereas the other type have a slower working system. There are, however, a number of other differences between fast and slow fibres.

When the gun fires at the start of a 100-metre sprint, the athletes' pre-made muscle stores of ATP can provide only enough energy to last for one or two seconds. In order to counter this, the muscles have a number of ways of replenishing the ATP supply. Fast fibres contain relatively greater stores of another simple substance, creatine phosphate, that will also split easily to release energy which can be used to regenerate ATP instantly. Yet even fast fibre stores of creatine phosphate are limited, only providing for ten to fifteen seconds of maximum effort, and although this is enough to decide the outcome of any 100-metre race, the levels start to decline from the moment the race commences, and if they were the only means of producing more ATP quickly, the runners would slow down well before the line. This does not happen. The creatine phosphate system is actually just a short-term back-up to give time for more complex energy production systems to come on line. These can regenerate much larger amounts of ATP for much longer periods and are of two types – anaerobic and aerobic.

Anaerobic metabolism generates energy to maintain muscle ATP without using oxygen – a huge advantage since there is no need to rev up the heart and lungs, before the process can go into action. The system can take over energy provision very quickly and so predominates in fast fibres. At the same time, using anaerobic systems has a downside. The energy is generated by the conversion of glucose into lactic acid, and the chemical bonds in lactic acid contain nearly as much energy as the glucose bonds in the first place. The process therefore releases very little energy from each glucose molecule and so glucose in the muscles is used up quickly. Although the glucose can then be replaced from a store in the form of glycogen, this takes time.

Glycogen consists of lots of glucose units linked in chains. When anaerobic work first starts, an enzyme splits the units off efficiently, so that the sprinters' muscles can go on with full anaerobic energy generation. But if high levels of work are sustained for more than a few seconds, the lactic acid formed from the process begins to accumulate. This poisons several enzyme systems, including the one splitting glucose from the glycogen. Within a 100-metre race, such lactic acid effects hardly matter, but go on for more than 200 metres, and even top class sprinters begin to falter. It is lactic acid build-up that explains why world records for 100 metres are well under ten seconds, for 200 metres are less than twenty, but for 400 metres are around 43 seconds and for 800 metres close to 100 seconds.

Slow fibres use predominantly aerobic pathways to generate energy. In these, glucose, fat or sometimes protein are 'burned' with oxygen to give carbon dioxide and water which are very simple molecules containing little chemical energy. Aerobic systems can therefore release far more energy than anaerobic ones, and since lactic acid is not produced, they can go on fuelling muscles for extended periods. Effective as they are for prolonged work, the rate of energy production is limited by delivery of oxygen by the heart and lungs and subsequent removal of carbon dioxide. They sustain only moderate work levels compared to the anaerobic systems.

The key to being a good sprint or power athlete is to have muscles with plenty of fast fibres in them, whereas the key to being good at endurance events is to have lots of slow fibres. The preponderance of one type of fibre or the other in your muscles is dictated by inheritance. It is your genes that make you go fast or far.

★

Although most anthropologists believe that our ancestors became erect to improve their reach, the requirement for speed has also been proposed as the primary reason beside those other theories of heat stress and tool use. This seems unlikely. Most four-legged animals run much faster than we do, and even chimpanzees and gorillas can set a fair pace with their knuckle walking or running. Nevertheless, when our ape ancestors first moved on to the savannah, they would have been more specialised brachiators than the large apes of today, with fore-limbs that were very useful for climbing but probably poor for aiding movement on the ground. It is possible that getting up on two legs could have conferred some speed advantage, and so the hypothesis must remain in frame.

Whether or not it influenced our posture, a turn of speed would have always had survival advantage. The faster our ancestors could run, the more successful they would have been at catching quarry or escaping predators. Yet, because of the specialised division between fast and slow fibres, and the consequence that absolute speed can only be gained at the cost of reduced endurance, the balance of advantage must have varied with local circumstances.

When most of us run hard, even for a few seconds, we rapidly feel our legs turning to jelly as the lactic acid builds up. Indeed, many of us experience this when doing no more than climbing a flight of stairs. Obviously we are not material to win an Olympic sprint medal, and although this is partly because we are untrained, it is also likely that we were never born for sprinting. Good sprinters may have running muscles consisting of as much as 90 per cent fast fibres whereas the average person has closer to a 50 per cent fast and 50 per cent slow mix. This gives most people a balance between speed and distance cap-abilities which must have been optimal through much of our evolu-tionary past. After all, it is no good being able to run like crazy for the few seconds needed to kill a quarry if you do not have the endurance to track it down in the first place.

If natural selection favoured a fast/slow mix for most people, why have some individuals ended up with muscles containing so many fast fibres? While the speed of DNA change means that the 100,000-year-old African Cro-Magnon population must have been almost identical to all modern humans, this does not mean all evolutionary development ceased during this period. After all, there are obvious racial differences today. However, since 100,000 years is but a short time in evolutionary terms, the only features likely to have changed much are those for which simple mutations of just one or two genes can create a quite definite survival advantage. An obvious example is colour.

Skin pigmentation is fairly simply inherited and strongly selected for by climate. No doubt all man's early African ancestors would have been dark, with melanin in the skin providing vital protection from excessive sun-induced damage. After men moved to cooler, less sunny regions, a high skin melanin would have had distinct disadvantages. Vitamin D production, needed for bone metabolism, relies upon processes activated in the skin by sunlight and, if the sunshine is weak and the skin dark, rates of synthesis can be seriously compromised. More northerly dwelling individuals would have had a considerable advantage if they were paler skinned, and so selection would have favoured the pale skin genes. Even so, rickets caused by vitamin D deficiency was common in north European races right up and into this century, and would have been commoner still if it had not been offset to some extent by the invention of farming around 10,000 years ago. Not all vitamin D is manufactured by body chemistry and some is obtained from meat and dairy products. Domestication of goats, sheep and cattle probably spared the northern dwellers many problems, although today they have arisen once more. With recent technology allowing more rapid migrations, vitamin D deficiency is now seen frequently in dark-skinned vegans who have moved to low sunshine climates with no time at all to adapt – especially among migrants to Britain from the Indian sub-continent. An opposite problem has also arisen for peoples with low skin melanin levels migrating at speed to areas of sunshine: skin cancer is increasingly found in white Australians and South Africans, and even among over-enthusiastic northern holiday makers.

Like skin colour, the predominant fibre type in muscles is a simply defined genetic feature for which local environments might pose strong selection pressures. It is particularly notable that a high proportion of Afro-Caribbeans are among successful world-class sprinters, and it is known that the average Afro-Caribbean has a considerably higher fast fibre count than other races. So it seems likely that their ancestors lived in situations where sprinting ability was a particular advantage compared to the need to run great distances.

The environment of West Africa, where Afro-Caribbeans originated, would have been a situation with plenty of food for gathering and plenty of animals that could be better hunted, or perhaps avoided, with a burst of speed. In these circumstances the advantages would lie in being able to run fast rather than far. In contrast, mountain environments, such as those of Ethiopia, have produced many of the world's greatest long distance athletes. In such regions there would have been far fewer animals to hunt and much less food to gather. The early men and women would have done better to move more slowly but have the capacity to rove far and wide in search of food.

44

Although differences in ancestral environment could underlie the current dominance of Afro-Caribbean runners in world sprinting, it is also possible that the selection pressures responsible were of more recent origin. Many of the current top class sprinters would have had ancestors transported by slave traders to the Caribbean or the United States around two hundred years ago. Those men and women may have been unwittingly selected for speed as opposed to endurance by ships' captains who deliberately picked the strongest-looking individuals, for muscularity is associated with high fast fibre counts. Furthermore, once committed to those inhuman transport ships, additional selection for strength might have taken place when the weaker succumbed to the extreme hardships, and even after reaching the plantations the same sort of factors may have operated.

Support for the significance of this new process of 'unnatural' selection is derived from the fact that it is the American and Caribbean populations of Negroes that do so well at speed events rather than the present populations of West Africa. At the same time, their different social situations may also have influenced the situation, and indeed the whole phenomenon of Afro-Caribbean sprinting success is arguably a social issue. With the prejudice that still exists in western societies against black-skinned people, success in sport is one of the few ways of rising from the bottom of the heap, and many experts believe that such a spur could easily account for all their sprinting success. To my mind, however, this simply does not fit the facts. There is plenty of prejudice against many other races in modern society, yet other minorities are not so successful in the world of sprinting. Furthermore, if Afro-Caribbean success was entirely socially stimulated, why have they been less successful at races over longer distances?

Of course, by discussing racial differences from a genetic point of view I tread on sensitive ground and open myself to accusations of Malthusian thinking. My argument, however, is not prejudicial against any race and I certainly give no credence to suggestions that there are genetic differences in intelligence between races. Unlike muscle fibre type, the genetic specification of intellect is extremely complicated, and so could only change very slowly. Once we became upright tool-users, any advance in intelligence would have been a huge advantage to all early human groups wherever they lived, and from this there could be only one possible outcome. All modern races of man must be essentially identical in intellectual capacity and, for that matter, were we to find a late Cro-Magnon in our midst, there would be no reason why he or she could not collect a Nobel prize.

★

Although I have dwelt upon the genetic background that is needed in order to become a world-class sprinter, I do not want to leave the impression that the majority of men and women are poor at it. Because most of us have around 50 per cent fast fibres we can all sprint, and it is only due to our rarely indulging the capability that we think of ourselves as being none too good at it. Successful sprinting is only possible if your natural endowment is developed through training, and the right training will have a spectacular effect on anybody's speed performance.

The most striking change seen with either power or endurance training is an increase in the size of the muscles involved in the activity. It was once thought that this was due to an increase in the number of fibres inside the muscles, but it is now known that this remains unchanged from birth and it is the size of individual fibres that increases, with very specific enlargement of the type of fibres most appropriate for the exercise demands being made. Sprint or power type of work will therefore enlarge the fast fibres, with an increase in the number of contractile fibrils within each one, and simultaneous improvements in the enzyme systems that provide anaerobic energy and the development of greater tolerance to lactic acid build-up. However, these anatomical and biochemical changes are not the only benefits. There are also dramatic changes in neuromuscular co-ordination which have far swifter and greater effects than the changes in muscle size and content.

Every muscle is supplied by nerves running from the spinal cord, each of which controls a group of muscle fibres known as a motor unit. The number of muscle cells in a unit varies with the degree of fine control needed by the brain. In the large muscles used for sprinting, 5,000 muscle fibres may be controlled by a single nerve fibre, whereas in the muscles that control the eye, the number is closer to five. While the brain can exert extremely fine control of eye movement, it has only much coarser control of running action. Even so, it can still regulate the speed and strength of leg contractions pretty accurately.

When you are walking, relatively few motor units – a few tens of thousands of muscle fibres – are being switched on and off in your leg muscles, while when you run, hundreds of thousands may be employed. The faster you try to go, the more units come into action, but if you are untrained you will not activate many that could be of help, and so you will fail to generate the maximum strength and power of which your muscles are capable. With training, the situation changes rapidly. After only a week or two of regular sprinting, many more motor units can be fired simultaneously, as if the muscle has learned to perform the action more efficiently.

I have noticed this phenomenon myself on several occasions while attempting to train for power fitness. By nature, I lean towards endurance events rather than speed, and endurance has been a common feature of my long distance Polar expeditions which have entailed dragging heavy sledges across Arctic or Antarctic ice. However, such journeys do require some high intensity anaerobic type of work, for sledges often become stuck on ice slopes and take brute strength to move them. To cope with this, I have needed to mix some power work with endurance training when preparing for these trips.

Unlike most professional athletes, I have never been good at maintaining levels of fitness in the periods between expeditions. Occasionally I go for a run, but I do no power training at all. I have therefore had the repeated experience of trying to get power fit from a relatively poor base. Instead of sprinting, I have always done this type of training by going up and down flights of stairs with a heavy back-pack to simulate the demands of hoicking a sledge over an ice obstacle. With an expedition approaching, my training programme would start with finding the task impossible. After only one flight, certainly two, I would simply feel too sick to continue. My legs would turn to jelly, filled with lactic acid and painfully useless. But within a couple of weeks, and long before I became more muscular, I would find that I could run up and down the stairs again and again, perhaps repeating the flights twenty times with only thirty second breaks between them. It was always a pleasant surprise, and clearly due to improved motor recruitment rather than significant muscle growth or adaptation.

Even after full training, neither I nor the proper power athletes can recruit *all* the relevant motor units simultaneously. This is a safety mechanism since if all the units within a large, well-trained muscle were to fire at once, the tension could exceed the strength of tendons or even bones. The body has a built-in regulator to prevent such self-injury, although in circumstances of extreme psychological drive, it can be overcome. It is thought that this phenomenon may account for extra-ordinary tales of strength in which individuals have lifted cars to release loved-ones trapped beneath them in accidents. It may also be why the very best athletic performances do not occur during training but await great arenas and big events such as the Olympic Games. It is probably what happened when Bob Beamon broke the long jump record in the Mexico Olympics with a leap that was so much more than he had ever attained before or would ever achieve again.

Psychologically driven full recruitment can also cause less beneficial outcomes. During the 1995 'World's Strongest Man' event, there were a series of contests in which anaerobic muscle power was called upon in excess. During the arm wrestling, one of the finest of the competitors, fired by the pressures of the competition, generated such strong forces in his arm muscles that his humerus, the bone of his upper arm, sheared completely – a horrendous break in a bone much stronger than an average man's. All of us have muscles that are stronger than we know.

★

In addition to training, it is possible to influence sprinting or other power performance through modifications in diet. A great many individuals think that this means consuming large amounts of protein, as well as a host of other specialised substances, either legal or illegal in nature. Body builders are particularly gullible, gulping down handfuls of expensive supplements that are almost certainly entirely useless. If you eat a good mixed diet, your protein and micro-nutrient intake is probably way over your requirements even if you are an exercise fanatic, for the exercise itself will lead to a high dietary intake.

Until recently, the only things that really made a difference were anabolic steroids – illegal and the cause of violent personality disorders as well as liver cancer. In the last four or five years, however, research has identified one dietary supplement that can make a difference. Studies have shown that the short-lasting creatine phosphate stores of fast fibres can be made slightly larger in some people by consuming considerable amounts of creatine itself. In normal circumstances, your body obtains creatine from meat or fish or by manufacturing it. The final amount ending up in your muscles varies quite markedly, with people at the high-end of the normal spectrum having levels 50 per cent higher than those at the low end. If you have naturally high levels, eating additional creatine makes no difference to you, for it seems that there is a natural ceiling on how much your muscles can contain, but if you have naturally low levels, and you eat pure creatine in large quantities, your levels may rise to reach the natural ceiling, and with that rise may come an improvement in your ability to sustain a high work output. The amount of creatine is equivalent to eating about fifteen pounds of steak a day. Vegetarian sprinters or power athletes may particularly benefit from such supplements. They usually rely on their body's manufacturing capacity alone, and so are known to have naturally lower levels of muscle creatine. It probably explains why few successful sprinters have been vegetarians and, incidentally, supports the evidence from the design of our teeth and gut that evolution designed us as omnivores.

The improvement obtained by athletes using creatine or other dietary modification is small, perhaps just a few per cent, but it could be enough to make the difference between no medal and a gold. Nevertheless, nearly all top sprinters now use this legal supplement and it appears to have made little clear difference. Perhaps they all have muscles naturally high in creatine anyway, Hence, although world records do continue to be broken, the increases in speed over the last few decades are more likely to be due to better and more dedicated training than advances in nutritional thinking. A quite different situation exists with endurance sport. For endurance, we now know, nutrition can be everything.

FOUR

★

The Happiness of the
Long Distance Runner

IT IS EARLY April in Greenwich Park and cherry trees laden with blossom provide a pink confetti blown by a stiff wind over a scene of mass madness. Beneath the trees thousands of people are gathering, not to attend a large spring wedding but to enter in one of the world's greatest sporting events. I, like so many others, am about to put my fitness to the test by taking part in the London Marathon.

It is not easy to run a marathon, even when one is well trained, and it has been only a couple of months since I gained a late entry. Such a short time in which to get fit is a far cry from that recommended. According to the running magazines, you should not contemplate starting such a race unless you are regularly covering ten or more miles a day, with a couple of longer runs sometime during the week. Due to lack of motivation and time lost to more pressing engagements, I have been doing less than half that. Nevertheless, throughout my adult life I have always kept up some sort of physical exercise – usually running once or twice a week – and I have ended up very fit indeed after some of my long Polar expeditions. I have persuaded myself that I am not starting from scratch but from a slightly trained base, fortunate enough to have been born with some degree of endurance fitness. My fibre mix must tend towards the slow side. If I do not reckon to go fast, I certainly want to run the whole thing without stopping. I have little confidence that I can do so, but I will do my best. I am nervous, but savour the excitement. As the time for the start approaches, the atmosphere becomes palpable.

It is not hard to see that most of the men and women around me would never make good sprinters. Although events like the London Marathon attract people of all shapes and sizes, none of those who look remotely serious about running could ever be mistaken for fast hundred-metre specialists. These bodies could not boast the body-builder musculature of the sprinter; instead they are men and women with the lean build of the oxygen burner. They are designed to go the distance.

As 9 a.m. approaches the contestants begin to strip down to their running kit and, in the bitter cold, cruise up and down to warm up. Watching them zig-zag through the throngs of people, I realise that they have run considerable distances before heading for the start queue. For me, such an enthusiastic warm-up would be foolish. The last thing I want is to add significantly to the daunting distance ahead. I will simply start slowly and warm up as I go along. I do some stretching exercises and jump up and down a bit. I hope it will be enough to stop tendons pulling. You should never run with muscles too taut from cold.

I need not have bothered. By now the start queue is at least a mile long – running from the gates of Greenwich Park back along the road that crosses Blackheath. Every hundred yards or so, there is a sign indicating an anticipated finish time. The idea is that you should be frank about your chances and join the line beside the time that corresponds to your ability. At the front are the real athletes – the men and women who can complete the twenty-six mile course in under two and a half hours. In the middle are those who have some chance of running the whole distance in four hours or less, and towards the back are those who will run, walk or even crawl to the finishing line some-how, come what may. Unfortunately, the length of the queue means that even if you are setting yourself only halfway back, as I am, it is going to take quite a few minutes from the firing of the starting pistol to crossing the starting line. By the time we are actually running, any warmth will be gone with the wind, and the benefits of stretches lost with it. The queue will also make any time round the course misleading – a disaster for those who want to set a personal best. For some of the runners who feel that their official time is important, the temptation to move up the line is simply too great. You have only to look at the front line group to realise that some of these contestants are in the crawling rather than winning category.

★

Even more than for sprinting, most humans seem to have evolved to cover long distances – maybe not running for mile upon mile but certainly at a walking pace for hour upon hour. The evolutionary need to develop such endurance is obvious. Without settled communities or agriculture, our distant nomadic ancestors went constantly in search of sparse food supplies or tracked animals that themselves roamed widely. No doubt most of them could have run the London Marathon with little difficulty – perhaps not at a modern competitive pace but at a steady lope which would take them through the distance in a perfectly respectable time. But if all our ancestors could perform such a feat, why is it beyond the wildest dream of most adults today? They see that tens of thousands of other men and women can complete a marathon, yet they seem to lack the self-confidence or motivation to do it themselves. Perhaps more significant, most people see no reason for wanting to do so.

We participate in sports from a variety of motives. Some yield rewards through the high levels of skill required, while others generate feelings of camaraderie from working in a team. Yet others give satisfaction through the human instinct to compete, but for most regular runners, none of these motives apply. There is little skill involved in running, even when competing at the top level, and a jog requires no skill at all. Although some men and women do enjoy the comradeship of running in groups, many are just as happy to go alone. Although some people aim for success in competition, most of those who do so have found their ability while running as individuals for pleasure. Overall, few runners are competing with anyone but themselves. So why do so many spend their time jogging round the roads, parks and hills throughout the year, and why are large numbers prepared to train for and suffer through a marathon? Studies of neuro-physiology suggests a possible answer – an answer with implications going way beyond marathons.

★

The starting gun fires exactly on time but, predictably, nothing happens. For some time my section of the line does no more than shuffle forward a few pointless yards, tightening the squash even further. Then, after several minutes trying not to tread on each other's heels, a ripple of action spreads towards us and our part of the gigantic human river begins to move. Slowly, the hesitant ripples become a steady current until, just at the point where we cross the start line itself, we finally break into a lope and move off with the tide. It has taken more than ten minutes since hearing the report of the pistol, but we are now on our way. Ahead lie the historic 26 miles, 385 yards to the finishing line.

★

In addition to electrical transmissions from nerve cell to nerve cell, the brain operates through chemical signals. There are huge numbers of these neuro-transmitters, and it has been shown that general states of mind such as pleasure, sadness, alertness and drowsiness can be dictated by their prevailing concentrations in different parts of the brain. Perhaps the best known example is the compound serotonin which seems to dictate changes in mood. Few of us can claim that we feel as happy through the long dark months of winter as we do in summer's brightness, and for some the winter sees a period of dark depression. This seasonal effect seems to be regulated by serotonin levels in the brain which are higher when we see a lot of the summer sun, and lower during winter's gloom. The changes are thought to relate to more rapid changes in another brain chemical, melatonin, which varies on a day to day basis as part of the careful programming of wakefulness and sleepiness designed to keep us in step with the daylight cycle. A circadian rhythm that evolved to wake us up and feel good when it is light, safe and our efforts can be productive, and to make us feel more drowsy and inactive with the onset of darkness.

As with all biological designs, natural selection would have led to near perfect control systems for these mood and alertness brain chemicals, but they were only perfect for the environment in which they evolved. Now they are unable to cope with the changes that have arrived since civilisation. The wakefulness regulator, triggered by the daylight cycle, still thinks we are hunter-gatherers, living in Africa, where winters are warm enough for us to be outside and the daylight cycle changes little with the seasons. It simply does not understand that many of us now dwell at latitudes where daylight hours vary greatly, and where the cold of winter tends to keep us indoors, even when there is daylight.

A bright day can see light levels outside approaching one hundred thousand candle power while normal indoor lights give out just a few hundred. A few hours of low winter's sun, passing through the windows of our buildings, is simply not enough for our serotonin system to realise that it is daytime and so, for many of us, winter sees a steady depletion of serotonin levels and a consequent lowering of mood. Even resorting to Prozac may not help, for it is only a second-rate means of doing nature's job. It raises serotonin levels to some extent, which is why it has some benefit, but it is nowhere near as effective as a daily dose of sunshine.

Running probably has no direct impact on serotonin levels other than getting you out in daylight. Even so, there are millions who can vouch for its beneficial effects on mood, which probably occur through the stimulation of other brain regulating chemicals, including the endorphins which are incompletely understood. Endorphins are widely spread throughout the brain and appear to grant pleasure and relaxation combined with a surprising clarity of thought. They also increase tolerance to pain and discomfort and may have evolved as part of our pain control mechanism. Clues to their action may be seen in the effects of extraneous substances that mimic them, and these include all the opiate drugs. Morphine and heroin act by artificially stimulating some of the endorphin receptors and so they too grant satisfaction, relaxation, and pain relief. The pleasures we derive from running may therefore be closely related to those sought by the serious drug-user or, to put it the correct way round, the pleasures of opiate use may be mimicking some of the natural pleasures of exercise, albeit more intensely.

Why running should influence endorphin levels is not clear, but a teleological approach suggests that if prolonged exercise was required for survival, and such exercise was painful, the body might well develop a pain reduction system triggered by exertion. Certainly all types of pain lead to elevated endorphin levels and running does produce constant low grade trauma. It is also possible that the increases are related in some way to the rhythm of exercise – either through the regular pattern of breathing or through the regular repetitive movements. All sorts of rhythmic processes can induce relaxation. Yoga is a good example, and either deep breathing or chanting mantras can also lead to a heightened sense of well-being. Music can have similar effects. Psychologists attempt to explain such phenomena by citing hidden memories of maternal heartbeat and unconscious feelings of pre-birth security, while physiologists demonstrate changes in cerebral blood flow with marked endorphin release. Whatever the cause, the root of the pleasures derived from rhythmic activities are complex, but there can be no doubt that the effects are real. Yoga, music, chanting, running – all can definitely promote happiness.

Talk of drug-like parallels suggests to some people that running must be bad for you. This is very unlikely since the unwanted side effects of drugs, especially addiction, probably occur because of inappropriate, poorly targeted and badly regulated dosing. Nevertheless, there is anecdotal evidence that exercise can become rather more than habit-forming, and many runners find that, once they start doing long distances, they cannot easily do without them. If they miss a day, they may become restless and dissatisfied. Certainly I have found this to be so when training hard for expeditions. Whenever a new departure has drawn near, I have gone from my one or two runs of a few miles a week up to long bouts of exercise on most days. Each time I have done so, I seem to cross a threshold beyond which something changes. Instead of having to force myself to go out and run again, I find it difficult not to and, once out, instead of constantly battling with a desire to foreshorten my run and turn for home, I find that I want to go on and on. Even ten miles becomes hopelessly inadequate, and I am only satisfied if I run for a couple of hours at a time. Although such drives might exert unwelcome social pressures, and I am sure there have been examples leading to divorce, there is no evidence that they are physically harmful. Indeed I believe that they may be of considerable benefit.

Once again, a teleological approach provides a possible explanation. If running genuinely tends to be addictive, the compulsion may serve some evolutionary need. Our ancestors had to exercise for much of their lives, and so natural selection is likely to have set them up to be in a good frame of mind while performing physical activity, while aiming to make them relatively soporific in between periods of exercise. The pleasure and freshness of thought that may be experienced during and immediately after exertion could therefore be another hand-down from our evolutionary development. It would certainly have helped our ancestors to become more effective hunter-gatherers.

Since exercise appears to induce a heightened state of thought, there are implications for those who do not exert themselves. It seems to me likely that, with no physical activity, you will inadvertently spend much of your life in a slightly depressed mental state – especially in the winter, when lack of exertion will be exacerbated by your lack of exposure to light. Runners may well benefit from a double whammy as they pound their way round on a winter's weekend. The commentators may pour scorn on them and suggest that exercise is unnecessary, laughable and probably downright harmful, but those critics are invariably talking of the physical side, claiming that the health benefits are unproven. Even in this respect, I would dispute their position, which merely reflects their need to justify inactivity and to support the delusion that they have no problems. But, in fact, this is not the main point. The real benefits of exercise go far beyond a toned up body and a bastion against heart disease. They go to the root of our thinking, and the sceptics simply know not what they miss.

*

For the first part of the race my body reacted in a familiar fashion – with pain and protest. Even though the crowding dictated a slow pace, my legs became tired and I was breathless. I had not completed a couple of miles before I was wondering whether I could keep going. Why was I doing it? Only past experience persuaded me that things would improve. If I stuck with the running, the breathlessness and fatigue would eventually go, washed away by alterations to my breathing and warmth in my muscles. It would not be long before my physiology made adjustments for my needs.

When you first start running, leg muscles are at only 37°C, or even a degree or so less on a cold winter's day. Yet it is at 38°C that they work best. Of course, once you start running, you generate a lot of spare heat and the muscle temperatures rise, but this can take a couple of miles, even on a warm day, and through that distance you will not feel it. The reason for muscles working better when warmer is that their enzymes are set up to operate at the 38°C level, which is an interesting evolutionary choice. If we had evolved to be sprinters in temperate climes, we would have been better off if our muscles worked best at 37°C or even slightly cooler so that they were ready for action at any time. The fact that it takes a considerable distance to bring them up to operating temperature supports the view that our origins were warmer and our usual needs were for endurance rather than speed.

The same sort of argument applies to breathing. When you start to run, your muscles need extra oxygen but your body is not set up to increase the supply immediately. For the first few minutes of a race, you develop an oxygen debt as you use more energy than aerobic systems can supply. It is only when oxygen in the blood has been depleted significantly and levels of carbon dioxide have risen that your brain senses these changes and sends instructions to set things straight. At that point, perhaps after a few hundred yards, you will begin to breathe harder and your heart will pump more strongly. But by then, besides having to meet the demands of your continued movement, you also have to repay the oxygen debt and clear the lactic acid that has accumulated. This takes some time, and so the first couple of miles of any run can be rough. It leads to an odd phenomenon. Most people feel less fatigued after running five or six miles than they do when they have run just one or two. Some inexperienced runners never realise this, and even quite reasonable athletes may believe that distance running is not for them. They have never run far enough to reach equilibrium and comfort and so have never found the capability that evolution bestowed upon nearly all of us.

★

As the discomfort slipped away I settled into truly relaxed running and knew that now I could carry on for many miles. I started to enjoy my surroundings. Moving along in this great throng granted me a special feeling of comradeship. The runners came from all walks of life, and were doing it for all sorts of reasons, but we all had a sense of sharing in something overwhelming that was mixed with those feelings of relaxation and clarity described earlier. The result was a state of mind that most people would not recognise. It was deeply pleasurable.

At the halfway point, however, things began to change. My legs began to weaken and complain once more, and I felt the return of breathlessness. The problems became steadily worse, and by the time I had been running for a couple of hours, and had completed around seventeen or eighteen miles, all pleasure had departed. My mind and body were hurting and my legs had become a strange mixture of rigidity and floppiness. In part, they felt like wooden sticks – sore, stiff, peg-legs. Simultaneously they were the useless, lax limbs of a rag doll. They had become wayward, quite beyond my control and, as I ran, one shoe or the other would brush against the opposite calf, causing deep red scratches to appear. The blood ran down my calves and vivid scarlet stains were added to my white running socks. I was barely able to go on running and kept tripping over my own feet. The tiredness felt similar to that which I had suffered near the beginning of the race, but I knew the cause was entirely different.

★

Fading at somewhere between fifteen and twenty miles is due to a shortage of glucose – the fuel that is favoured by muscles working hard. When sitting, lying or walking, fat is the preferred substrate, but when you start to run, things change. At an easy jogging pace, you are burning about half fat and half glucose, while at full pace you usually switch to glucose entirely. There are two reasons for this change in fuel usage. At low levels of effort, muscles can keep up with demands for ATP by burning the limited fat stores within them but, as you speed up, they need to mobilise fat from elsewhere, and the rate of transport in is limited. More important, however, the change towards glucose burning occurs because fat contains fewer oxygen atoms relative to carbon and hydrogen atoms than glucose, and so it takes more extraneous oxygen to burn it. This oxygen has to come from the air, and for the same rate of work, burning fat needs you to breathe fifteen per cent harder than burning glucose. If already working at levels that make you breathless, this addition is unwelcome and may well force you to slow down.

Most of the glucose needed for a marathon is stored as glycogen. The stores are mainly in the muscles themselves, although some glycogen in the liver helps to maintain blood glucose levels. At the start of a run, people usually have a total of 400 to 500 grammes of glycogen in their muscles which is enough to fuel them for about an hour and a half. After that the supply runs low. The liver then starts to manufacture more glucose, but only enough to maintain reasonable levels in the blood rather than to refuel muscles. As a result, the muscles run out and have no choice but to switch over to fat burning, despite the disadvantages. With the change, you inevitably feel more tired and breathless, and simultaneously, a lowish blood glucose can make you feel nauseated, sweaty and vague. This is an experience familiar to all marathon runners. You have hit the 'wall'.

★

I wasn't helped by the crowds around me. While I was still maintaining a rather unsteady run, the road became packed with people drifting along at a walk. Continuously picking my way round them, I was often forced to stop. It was like attempting to run along Oxford Street at Christmas time, and whenever I did stop, it took an enormous mental effort to break back into a jog. To me, these people were just selfish. Back at the start of the race they were the ones who had not been honest about their capabilities. I had started running at a pace that I thought would see me through the twenty-six miles in a little over three and a half hours – no great achievement, but then I am no great athlete. I stuck to that pace and, at the halfway stage, it even looked as if I might come in under that time. But then the crowd problem, which had started way back in the first couple of miles, became more and more troublesome. Obviously many of the competitors who had no realistic hope of running the race at even my speed started much further forward than I had. They didn't seem to care that, from the word go, they would be obstructing more realistic fellow runners, and by fifteen miles it was clear that many of them were not capable of running the distance at all. I could accept that starting well back in a queue, where it would take more than ten minutes to reach the start line, was not a good basis for an 'official' personal best, but time should not be what an event like this is about. Running a big marathon is about seeking a capability within oneself and sharing the experience with others.

As my legs became worse and my way more tortuous, I found myself mentally cursing my fellow participants, sneering at their pathetic athletic abilities. On later reflection, however, I saw that it was no more their fault that they were slower than my fault that I was not up with the leaders. How swiftly one can run a marathon is mainly down to the genetic hand of fate. At least they were out there, pushing their bodies to individual limits rather than accepting that marathons were only for the athletic elite. I pushed away my anger and turned my thoughts to the finish. It was now just six miles ahead.

Thousands would already have finished, some more than an hour ago. That was more than an hour and a half ahead of my likely finishing time. How did they do it? Despite some expertise in sport's physiology, I could scarcely credit it. I would find it difficult to run 800 metres at the speed they ran the entire race and many, who would not think of themselves as either elderly or infirm, would be unable to match their speed over one hundred metres. What mix of qualities allows men and women to run at the head of a world-class field?

★

The ability to run distance at speed is determined by both oxygen and carbon dioxide transport, although the two go together since they both depend on your circulation. The greatest amount of oxygen your body can take in, pump round, and then combine and burn with fuel is known as your maximal aerobic capacity and it varies with the size of your muscles, the number of their slow fibres, and the capacity of your lungs, heart and circulation. Aerobic capacities are very varied, and good endurance runners may have values of more than twice the normal, while the unfit may be limited to less than half the average. The values also tend to fall with age, although part of that decline is due to decreasing activity and is prevented if you continue to take exercise as the decades pass by.

Just as world-class sprinters may have inherited muscles containing 90 per cent fast fibres, world-class endurance runners may be even more genetically specialised, with muscles containing 95 per cent slow ones. As with sprinting, however, it is not just birthright that grants success. Leading, or even finishing the London Marathon has a lot to do with training, and a great deal to do with eating.

Endurance training improves the capacity for muscles to work without fatigue. It cannot alter the number of your muscle fibres, nor change fast ones into slow, but there are a number of intermediate types of fibre which can be made to become more like slow fibre types, although this may take many years. Along with some of the slow improvements to heart and lungs, it is this slow influence on the nature of intermediate fibres which allows good young 400 and 800 metre runners to become top class 1,500 or even 5,000 metre specialists later in their careers.

The main influence of endurance training on muscles, however, is not through changes to intermediate fibre types, but through improving the slow ones you already have. Just as sprint training makes fast fibres larger, endurance exercise increases the size of slow ones, and these may enlarge by as much as 20 per cent in just a few months. They then contain more fibrils, more glycogen, more fat and more enzymes, and in addition show improvements to their local oxygen delivery systems with an increase in the capillary blood vessels that surround them as well as a rise in the fibre content of myoglobin – a specialised oxygen-carrying molecule similar to haemoglobin in the blood. However, in contrast to sprinting, improvements to endurance performance through training are not so dependant on changes in the muscles alone. The improvements in local fuel supplies and oxygen delivery systems are useless if they are not accompanied by improvements in the shipment of oxygen and carbon dioxide to and from the muscles by the lungs and heart. Genes dictate the potential of lungs and heart in the same way as they do for muscles, but training brings phenomenal benefits.

The muscles employed in breathing are all very specialised, comprising almost 100 per cent *extra slow* fibres which have two or three times more capacity to use oxygen than any other muscle in the body. This is useful since they never need to turn to anaerobic systems which could lead to lactic acid build-up with consequent cramp and fatigue. Just a few fibres in the diaphragm have limited anaerobic capacity, and these can make themselves felt if used; they are the source of the familiar 'stitch'.

While training improves your leg muscles to carry you through the distance, your breathing muscles also become stronger, allowing you to breathe faster, with greater volumes and with less fatigue. The overall size of your lungs – mostly dependant on your body size – actually changes little, but maximum ventilation can increase from perhaps 120 litres of air per minute to more than 200 litres. This will obviously help you to run faster and further with less breathlessness. Yet, strangely, it is not usually deficiencies in your respiration that dictate whether or not you feel breathless. Indeed, it is not even a shortage of oxygen that makes your breathing laboured. That is due to the build-up of carbon dioxide from heavy work making the blood acidic. The acid then stimulates your breathing control to try to get rid of it. The extra breathing does help to clear this carbon dioxide but cannot solve the problem alone. You also need an increase in blood flow to carry the excess from the muscles to the lungs, and so a stronger heart is also required for distance running. Fortunately, the heart is the most trainable organ of all.

One of the most striking changes as your endurance fitness increases is a fall in your resting pulse rate. At first it may drop by one beat per minute for every week of training, and so in just a few weeks it can fall to well under 50 beats per minute. Further reductions, as low as 30 beats per minute, may then occur, although these take much longer. The reason for the slowing is an increase in heart size. Where an average resting heart might pump 60 millilitres of blood per beat, a trained one will pump twice that much, and the improved performance is even more marked during exercise when untrained hearts might reach 120 millilitres per beat but trained ones can exceed 200 millilitres. It means that, overall, the maximum output can go from perhaps 15 litres of blood per minute to more than 30 litres, and with the other improvements in muscle fibres and lungs, maximal aerobic capacity may rise by 20 per cent or more in just six months.

Maximal aerobic capacity, however, is not the be-all and end-all of endurance fitness. Equally important is your anaerobic threshold, or the level of exercise at which some of the work that you are doing needs to be met by anaerobic processes. If you are untrained, it usually occurs at about 50 to 60 per cent of your maximum capacity, and this means that when you work at levels higher than this for any length of time, lactic acid will accumulate. You will feel fatigued and breathless. In the well trained, on the other hand, the anaerobic threshold can become much higher, and lactate will not begin to accumulate until you reach perhaps 70 or even 80 per cent of even your improving aerobic capacity. It is a change that can make much more difference than anything else. Whereas much of your aerobic capacity is down to genes and is unlikely to improve by more than 20 to 25 per cent, the training change in anaerobic threshold from perhaps 50 to 80 per cent and will allow you to work very much harder. Indeed, it has often been exceptional anaerobic thresholds that have made for marathon winning talent. Alberto Salazar, the winner of many world-class marathons in his time, had a maximal aerobic capacity which was scarcely exceptional (70 ml/kg/min), yet he could work at 86 per cent of that without accumulating lactate. He ran the marathon in just two hours and eight minutes.

★

In addition to training, a diet to ensure good carbohydrate supplies is the key to success in marathon running. When you run, your muscles' glycogen stores are depleted, and upon finishing your efforts, mechanisms are switched on to refill them. These seem to become more efficient with repeated use, especially if they are given plenty of carbohydrate to work with. As a result, many endurance athletes have come to be thought of as living on pasta alone, although in reality they eat a wider variety of high carbohydrate foods.

In Britain, the United States and other western nations, most people consume a diet that provides about ten to fifteen per cent of their energy intake as protein, fifty per cent as carbohydrate and the remainder as fat. Since an average adult eats between 2,000 and 3,000 calories a day, this normal diet contains up to 1,500 calories of carbohydrate. An athlete in training often eats more calories overall and so, if sticking to a conventional pattern, might consume about 2,000 calories of carbohydrate. After a long run, however, they may have used up all of their body carbohydrate stores which, in a trained individual, could be more than 700 grammes, or 2,800 calories. A normal diet pattern would therefore be inadequate for their needs. Serious long distance athletes have little choice; they must eat abnormally large quantities of foods containing carbohydrates.

I use the term 'abnormal' with one important proviso. Although athletes consume large quantities of carbohydrate by western standards, their diets are actually quite normal in a worldwide context. What is more, they are almost certainly closer to the intakes of those ancestral hunter-gatherers who would have eaten primarily vegetarian fare. This is interesting from the teleological point of view. If our ancestors had no choice but to eat a high carbohydrate diet, and no choice but to undertake endurance exercise, it is hardly surprising that evolution set us up to achieve optimum endurance performance when on that type of intake. The 70 per cent carbohydrate diet used by marathoners to meet the exceptional needs of their sport is therefore quite natural. Without realising it, they have returned to the food patterns of our past.

Besides eating a lot of carbohydrate in general, endurance athletes also aim specifically to increase glycogen stores immediately after long events. They are able to do this because muscles which are nearly empty of glycogen have such an avid appetite for sugars that if flooded with plenty, they not only restock but actually overfill. This observation led to a dietary practice known as 'carbohydrate loading' which was used extensively in the 1970s and 1980s. At that time, endurance runners would deplete their muscles about a week before an important race by exercising hard over several days while deliberately avoiding all carbohydrate intake and then, with muscles completely empty, would switch to eating virtually nothing but carbohydrate for the remaining time before the event. With no exercise at all in that final pre-race period, their muscles were stuffed full for the start. Nowadays, such formal depletion and then loading with carbohydrate is less favoured, for it has been realised that if a high carbohydrate diet is eaten all the time, all that is needed to attain equally high glycogen levels is to wind down activity a few days before a race.

I did exactly this before my marathon, and during the race itself I tried to stave off the dreadful 'wall' by taking plenty of sports drinks as I went along. Such drinks are another modern accepted tactic to help runners through long distances. The rationale is that by taking drinks containing carbohydrate while on the move, stores will run down more slowly. It sounds logical but there are some problems. It is not possible to absorb carbohydrate as fast as the body uses it, and so you lose ground however much you consume. Furthermore, the carbohydrate in the drinks can have counter-productive effects. Because it leads to more sugar in the bloodstream, muscles are deceived into believing that there is plenty available and are switched towards burning glucose rather than fat. Simultaneously, the higher blood glucose also misinforms the liver about glucose needs and so switches off glucose manufacture. Finally, drinking carbohydrates to excess can make your gut feel pretty uncomfortable. All in all, the benefit from the carbohydrate in sports drinks is less than might be expected.

Even so, sports drinks can be an aid to marathon runners, especially by helping to prevent dehydration. Becoming dried out has a devastating effect on human performance, with as little as a two per cent deficit in body water leading to a ten per cent reduction in endurance capacity. Two per cent is less than two litres and this amount can easily be sweated in the first hour of a run. Over a long race, even when the weather is not hot, the body may lose many litres more. It is therefore vital that you drink as much as possible during the event to avoid this happening.

Even a commitment to drink plenty, and the provision of drinks along a marathon course, do not necessarily keep dehydration at bay. Our thirst mechanism is not one of evolution's greater successes, and we do not necessarily feel very thirsty even when fluid losses are significant. Furthermore, if you do drink a lot, the gut often cannot handle it because exercise has diverted blood away to the working muscles. If over enthusiastic, it is easy to end up feeling bloated, nauseated or in pain. Sports drinks are formulated to overcome these difficulties, but problems arise because they also aim to increase both fluid and carbohydrate absorption – targets that, unfortunately, are incompatible.

All sports drinks contain sugars and salts which are picked up and actively absorbed by the intestine. This helps to maintain sugars for energy and salts lost in sweat, and simultaneously pulls extra water through the wall of the gut by the process of osmosis. Drinking sports formulations can therefore help stave off dehydration for longer than drinking pure water, especially if they are cool and appealing in flavour to help overcome your poor thirst mechanism. Most of the commercial drinks, however, contain about ten per cent carbohydrate, a formula aimed more at the provision of energy than the prevention of dehydration. They are too strong to give best water absorption and, although it is difficult to be sure whether a shortage of glucose or becoming more dehydrated is going to be your greatest problem, I suggest that the latter is more likely to be critical. In my view, the average sports drink should be diluted two-fold.

★

The pain and discomfort continued as I staggered along the Thames Embankment beside the Tower of London. I had to keep gritting my teeth to prevent myself from grinding to a halt. My feet were sore, my legs had seized up and my brain felt fatigued beyond words. I felt sick, and guessed that I had a lowish blood sugar as well as low levels in my muscles. More than anything, I just wanted to stop and walk for a while, but I knew that if I did so I would lose something from the experience. I had entered the phase in which psychology was everything. Whatever I decided mentally, my body would obey. Although it was incredibly difficult to keep on running, the remaining distance was falling and finishing was becoming an ever increasing reality. I just had to keep going.

London's Thames bridges fell away behind me one by one, and as they did so, the distance remaining diminished steadily. Now, within six miles of home, I had entered realms with which I was familiar. Every step brought increasing certainty of success. This lifted my spirits enormously and the cheering crowds gave me more mental strength. The pain could not go on for much longer. I found myself chanting repeatedly an adage I had heard somewhere before: 'when the going gets tough, the tough get going'. It was a mantra that brought a revival in my running.

To be able to go on when it feels so uncomfortable must be the most essential quality for anybody who wishes to run such a race and, although easily belittled, I believe it is a quality with an import beyond running. As we approached Westminster, I and all the thousands with me had taken an opportunity to learn about, and even be impressed by, ourselves.

Past Big Ben, my pace was back up to the rate at which I had started, and on the road towards Buckingham Palace I became faster still. My legs were just as wooden, just as stiff, and just as painful, but we are designed to cope with such hardships when the goal is worthwhile. We can draw upon the very strengths that allowed our ancestors to survive.

The finish was fantastic. For the last two hundred yards I summoned up a blissful sprint for the line just to prove to myself I could do it. Then came the feeling of elation. The enormous satisfaction of achieving such things cannot be expressed, but as I lay down beyond the tape and gazed around at the thousands with me, I knew we all shared something special.

After half an hour or so, I left the Mall and hobbled to the tube. My legs were making slow progress when an old woman, also wearing running kit, drew up beside me. She smiled.

'How was it?' she asked, 'You look as though you felt it.'

She was right. I did, whereas she looked as if she had just come in from a stroll in the park. I grimaced as I made my reply before asking about her race.

'Fine,' she responded. 'Not as good as last year, but still under three hours.'

Three hours, I thought – so much faster yet so much older.

'Do you mind me asking,' I said, 'how old you are?'

Another smile.

'Sixty-eight,' she replied. 'Not bad for a pensioner.'

FIVE

★

Crossing Antarctica

BEYOND MARATHONS, ultra-marathons and other athletic events, the wild places of the Earth offer challenges to human endurance which can take men and women to their very limits. In mountains, jungles or deserts, a whole host of other factors are added to the problem of mere progress. They provide stern tests of the resilience granted by our evolutionary past.

The British have a long history of Polar exploration. Back in the glorious pioneering days of the early 1900s, British expeditions led the way in many attempts to unlock the secrets of Antarctica and, in particular, to become the first to reach the South Pole. This led to the most famous story of them all. In 1911, Captain Scott and four companions reached the Pole after toiling there on foot, dragging sledges behind them. Originally, they had planned to use motor tractors and Welsh mountain ponies, but the former had broken down almost immediately and the latter fared little better. Their companions had helped them by laying out a series of food and fuel depots for the return journey. However, at the Pole, the already weakened men met the cruellest blow of all. A Norwegian, Roald Amundsen, had beaten them to their goal, making an efficient journey with four companions, and with dog teams to pull their sledges. Mentally broken, Scott and his companions turned for home, but unlike Amundsen, who made a safe and rapid return, they were doomed. Two died early on the return journey, while the remaining three were to get more than halfway back to base before dying just eleven miles from comparative safety. They had almost reached the One Ton Depot when they perished in their tent as supplies ran out and a storm prevented them from travelling that last short distance to the replenishment they needed. They wrote down the whole harrowing tale right up to the moment of their deaths, and their bodies and diaries were then found the following year by a search party. They won their place in history for their heroism.

My first trip to Antarctica was made with British Antarctic Survey soon after I graduated as a doctor. I had always intended to use my qualification to take me on expeditions in addition to more usual practice, and in 1980 I spent nine months working on ice breakers and at one of the BAS scientific research bases on the Antarctic peninsula.

During that time, I met two men, Robert Swan and Roger Mear, who were going to change my life. Robert had gone down South with a mission: he planned to repeat Scott's journey – although without the dying part. In Antarctica he teamed up with Roger Mear and the plans changed. Roger, a professional mountaineer who had been involved in the move away from large Sherpa-assisted Himalayan ascents to small Alpine-style assaults, put it to Robert that the South Pole could be approached in the same vein. He saw a lightweight, self-sustained journey to the Pole with men pulling their own loads, but he believed that to return as well in such a mode was impossible. Instead, the walk should be a one way trip, to which Robert agreed. The pair of them, with a Canadian mountaineer, Gareth Wood, were to manhaul from the edge of the Ross Sea ice-shelf to the South Pole where they would be picked up by an expedition aircraft and returned to their start point. It was to be the longest unsupported Polar trip ever made by man.

At that time, the only way for a private expedition to get to Antarctica was by sea, yet the waters around Antarctica are frozen for nine months of the year. A ship can approach the continent only towards the end of the southern summer when the surrounding ice has broken out. For that reason, the expedition had to do the same as Scott's – to go in at the end of the summer and drop off the team who would set up winter quarters in which to live until the following spring. Because they would be so isolated, they felt it was essential to take a doctor with them. Failing to spot that I would be the only one without medical cover, I eagerly volunteered.

To help my career withstand this unconventional break, I decided to undertake some research while I was away in order to return with some benefit. My topic was to be the study of our body weight regulation through the winter and the physiological effects of the walk to the Pole. They were areas of research which, in the end, would influence my entire future career. Indeed, they were the areas that would lead to many of the themes in this book.

In November 1985, after months of winter darkness, the three men of the 'Footsteps of Scott' expedition set off, taking 70 days to reach the South Pole. My studies showed that the three men had each burned up between 5,500 and 6,000 calories per day. At more than twice the energy expenditure of an average adult, it represented hard work. Nevertheless, they completed the walk in good physical condition. It raised the possibility that one could do even more. Could Antarctica be crossed, from coast to coast, without the aid of other men, animals or machines?

★

Each of the 'Footsteps' men had lost around 6.5 kilogrammes (14 pounds) of body weight during their trek, and it was to these figures that I turned when considering the logistics of the much longer Antarctic crossing with Sir Ranulph Fiennes some six years later. How Ran and I met and evolved this Antarctic plan after several unsuccessful attempts together to cross the sea-ice to the North Pole is told in my book *Shadows on the Wasteland*.

Because of the much greater distance than the 'Footsteps' team had accomplished, Ran and I had to manage more hours of sledge hauling each day than they had, and the sledges would need to be heavier. Against the 'Footsteps' consumption of 5,000 calories per day, I therefore considered that our ideal intake should be as much as 6,500 calories each, but to haul a sledge filled with 100 days' rations containing that much energy was not a practical proposition. Instead, I decided that the best compromise was to eat only 5,500 calories and accept the loss of body weight. If it turned out that we made the crossing on an average expenditure of 6,000 calories each per day, we would lose no more than 6 kilogrammes of body weight by the time we reached roughly the halfway point at the Pole. If, on the other hand, the work rate rose to 6,500 calories each, our weight losses at the Pole would be doubled. Either way, it did not seem critical. A normal individual can lose up to 15 per cent of body weight without weakening much, and we intended to get fatter than usual before even starting. Beyond the Pole, with lighter sledges and heading downhill, we would perhaps suffer no further weight loss, especially if we could be carried along by the winds. These tend to blow away from the South Pole and so should help us on our way during the second half of the journey.

The idea of putting on some weight before departure did not go according to plan. For many months, Ran and I followed a policy of dedicated piggery – never missing a sweet, snack or lump of fatty food that came our way. The difficulty lay in our hard training, for it is almost impossible to put on weight while engaged in a lot of regular exercise. In fact, the opposite occurred and both Ran and I became lighter. This was a matter of some concern until we were saved by fate.

During our journey down to the Antarctic a combination of bad weather and a technical fault with our aircraft saw us trapped for ten days in Punta Arenas, a city in southern Chile. There we were unable to do anything but wait, go for the odd run and sample every restaurant in town. The result of this short period of gluttony was a gain for each of us of nearly 6 kilogrammes – a store that was later to prove very useful.

As well as limiting the calories we could take with us, the unsupported nature of our journey dictated the type of food we should eat. In order to gain maximum energy from minimum weight, fat had to be predominant in our diet since it provides 9 calories for every gramme compared to only 4 calories from carbohydrate or protein. Taking this to an extreme would have seen us eating one hundred per cent fat, with our ration of 5,500 calories per day weighing just 610 grammes. Pure fat, however, is not enough to sustain you. The question was: how much protein and carbohydrate were essential to our well-being?

Contrary to many people's belief, a lot of exercise does not necessarily lead to very high protein demands. A normal western diet already contains around twice as much protein as we actually require, and so our very high planned intakes would provide us with plenty even if the percentage of protein in the ration was quite low. How much carbohydrate we would need was a more difficult calculation. It would be beneficial to have high intakes for the same reason that it is helpful to ingest carbohydrate when running marathons – if working hard, the muscles prefer to turn to glucose. Although, like marathon runners, we would feel better if we ate a lot of carbohydrates, the benefits would be countered by their making our sledges heavier if they replaced too much of the fat in our rations. Furthermore, unlike marathon runners who may be working at levels of up to 80 per cent of their maximal aerobic capacity, and so feel terrible when they hit the 'wall' and are forced to turn from burning glucose to burning fat, I thought that we would plod along at levels of more like 40 per cent of our maximal aerobic capacity. At that work rate, muscles should be fairly happy to burn fat for fuel, so I decided that we could get away with a diet containing much less than the 70 per cent carbohydrate of the endurance athlete and opted for a ration containing 57 per cent fat, 35 per cent carbohydrate and 8 per cent protein. Of course, these were percentages of more than double the normal intake, and so the amounts of carbohydrate and protein actually consumed would be very large by normal standards.

Before starting our journey, we had hoped that the first two or three weeks crossing the 250-mile Weddell Sea ice-shelf would be relatively easy, since the surface would be flat and smooth. However, the Antarctic ice-shelves are made of glacial ice that runs down off the continent and floats on the sea, and the ice we encountered was split by deep crevasses. Pulling across such a fractured surface was incredibly difficult. With 100 days' food and fuel on our sledges, we each hauled loads of 220 kilogrammes (484 pounds), despite carrying a bare minimum of clothing and other survival gear. The friction from the ice was also far greater than we had expected. Even though our sledges were equipped with runners of the best non-stick materials, the slipperiness of the ice and snow was minimal. Even at low altitude and far from the Pole, the temperature was often as much as minus 30°C, which is too cold for downward pressure to cause the usual melting that provides a slippery mixture of ice and water under the runners. It was more like dragging huge loads over sand, and we struggled to make about one-and-a-half miles per hour. That meant pulling for between ten and twelve hours a day to make even reasonable progress. Twenty days passed on the ice-shelf before we met the edge of the land itself.

Once we reached land, we began the long climb up towards the Polar plateau, an ascent of more than 10,000 feet. The hard slog rapidly changed to bone-crushing toil. To our dismay, we found that we would haul our loads uphill all day, only to lose all of the height gained in just a few minutes of dropping into some valley that crossed our path. It was very disheartening. Adding to our misery, the winds blew constantly from the high plateau straight into our faces. We battled with white-outs that lasted for up to five days, seeing nothing and being pushed back by the force of the storms. The winds also carved huge dunes of snow and rock-hard ice ridges, known as sastrugi, around and over which we had to wind. As we climbed, it became steadily colder. Our pace grew slower and slower until we could no longer make even one mile per hour.

Mentally it was also difficult. At first I occupied the days of slog with thoughts about the outcome of the expedition, reminiscences of good times past, and all the things I would do when I got back. It was an ideal time for optimistic introspection. After a few weeks, however, I seemed to have thought of everything. I felt as if I had covered every subject so many times that I could think about them no more. Instead I became locked into the desperate tedium and hardship, and my mind was filled with thoughts of defeat.

There is no doubt that when one plans to take on another challenge of this nature, you cannot remember just how hard and painful the last one was. It is only after you have set off that the sheer hell of the work becomes real once more, and you remember the terrible discomfort last time. It is then that you recall why you said you would never do anything like it again, but by that time it is too late. It is not easy to announce to the world shortly after setting out – 'Sorry, I made a mistake.' You let down too many people, as well as your own ego, and so you find yourself stuck in the most difficult situation. In one sense you want to go on, but in another you are also desperate to stop. It leads to great temptation – to find a way out without losing face.

As a doctor, the obvious thing for me to do was to pretend to have an illness. Within a week of starting our crossing, all day had become occupied by my thinking of the best illness or injury to feign. For some reason, I thought that a sub-arachnoid haemorrhage – a kind of stroke – might be a good one, and I walked along wondering when exactly to put on the act. Each day I managed to resist the temptation to act it out there and then, usually making a decision to save it until we had put up the tent that evening. However, once inside the tent, tea and biscuit in hand and a few more miles under our belts, life always seemed rosy.

I am often asked what qualities are needed to make a person successful at such difficult undertakings. My answer is always the same and often surprises people. I firmly believe that one does not need to be particularly brave, strong, or even foolhardy; instead you need a very defective short-term memory. Of course, I am not unique in possessing such a quality, and indeed I would ascribe it as a general phenomenon in humanity. Which woman would voluntarily go through a second childbirth after the experience of a typical first? I believe that we are programmed to be incapable of remembering such pain, and that we cannot recall suffering to anywhere near the degree to which we can recall the times of pleasure. I am sure this is a quality that has been selected by evolution for a purpose. Optimism is needed for survival, and realism must be its worst enemy.

Lulled by my poor short-term recall, I knew that to give up without really needing to do so would mean spending the rest of my life with the knowledge of my failure. Determined not to damn my future, I broached the subject one evening.

'Ran,' I said hesitating. 'Do you ever think of feigning an illness to get out of this?'

He looked over, questioningly.

'No . . .' he said slowly. 'Do you?'

'Well . . . yes.' I was embarrassed. 'Pretty much all the time actually.

I keep thinking that I should pretend to have a stroke. So you should ignore me if I suddenly collapse.'

By telling him, I closed the door upon the option and so forced myself to continue the journey. I was proud of the admission, but there was a frosty silence.

'Are you shocked?' I ventured.

'No, not at all,' he replied. 'I may not have thought of putting anything on myself, but I do keep on hoping you'll be ill.'

★

In addition to fooling me into thinking I was enjoying myself, the relative comforts of the tent brought distraction through being busy. People often ask whether we took books to read upon the journey, but we did not: books are not edible. However, as well as their lack of calorific value, we would also have had no time to read them. At the end of each day's hauling, we had to get the tent up, cut snow blocks, erect a radio antenna, and get inside. We then had to de-ice all our clothing before we could light the stove on which we would make a hot drink, melt water for the next day's vacuum flasks and cook the evening meal. After eating the meal, we would make a radio call, do some navigation, write our diaries and attend to the dressings on our blistered, frostbitten feet. Often we also had to mend our clothing or take various medical samples for my research.

Although not all my scientific studies were to his liking, Ran undertook them with remarkable stoicism. The most frequent samples I needed were small nightly urine specimens which caused neither of us any problems. The little bottles also proved to be useful pieces with which to play the occasional game of chess or draughts. More problematic were the five times during the expedition when I had to obtain 24-hour urine samples and so collected all our output through the day and night into a large bottle. The volume was then measured before we took a small aliquot for later analysis back home. The purpose was to quantify what was being lost from the body over the whole period. These were important studies for my research but collecting the urine in a narrow-necked bottle, in a high wind at minus 50°C, proved to be a tricky business. We observed frost injuries never described previously! Yet as far as Ran was concerned, the difficulties with urine collections paled into insignificance beside the problems that blood sampling caused him. He really cannot tolerate the sight of a needle coming at him and will promptly feint if he is threatened in this way. Needless to say, he got around the problem by simply looking the other way, or by closing his eyes, while I took blood from him. The real test came when he had to take blood samples from me. He would not look then either.

As well as being comforting, conducive to memory loss and busy, the tent was also a place of friendship. It is my experience that to share hardship with another is to forge very strong bonds which will last a lifetime, and of course similar bonds are felt between members of a sports team on a Saturday afternoon. I believe that such camaraderie is also a result of evolution. Within the framework of conventional Darwinian 'survival of the fittest' thinking, it is not easy to see how altruism of this kind can fit. Like the inheritance of longevity, the selection of genes that help others to do well seems impossible, yet it can happen if it favours group success. A good analogy is a bird which gives a warning cry as danger approaches. Logic tells you that, if it stayed quiet as it flew off, it would be less likely to become the next meal and so hand down more of its genes. If the bird remained silent, however, other members of its species around and about, often closely related and so sharing its genes, would get no warning of the impending danger, and the predator would be more likely to take a life. The survival of the group as a whole would be more threatened. It is a process known as kin selection, and although rubbished by Richard Dawkins in his book *The Selfish Gene*, I still believe that it can operate.

Gene programming or not, it has been my experience that the more uncomfortable and dangerous a situation becomes, the more strongly camaraderie is felt. At the same time, I cannot claim that the hardships of our Antarctic journey were always faced with friendship. The pressures took a harsh toll upon our relations, and sometimes, while hauling, bitter disputes made things even more difficult to bear. Most of these quarrels stemmed from the fact that we did not go at the same pace, and often one of us would be left behind. These situations inevitably led to frustrations boiling over, though more usually they were resolved with humour. Seeing the funny side of a situation is another vital asset to survival, and I am sure that it is not just through chance that humour flourishes in times of adversity. Once again, I am confident that it is a witness to our evolution. The selection of the fittest has, since the time of our intellectual growth, always included the selection of the funniest.

★

On Day 67 of our journey we came close to disaster. The wind was blowing briskly when I stepped from the tent and the temperature had fallen below minus 40°C. The sun was still shining and visibility was good, but the weather was obviously changing. I reckoned it could turn really nasty. Earlier, when I had left the tent to answer nature's morning call, it had seemed quite comfortable. The air had been still and silent, and the sun's warmth had poured through my thin clothing so that there was no real need to hurry. Back in the tent, I had added only a cotton windproof and some fleece salopettes to the long underwear in which I had slept, but now, as I was packing up outside, I realised I had made a mistake and might run into trouble. Still, we were about to set off and the hard work should keep me warm.

Our morale was high. The South Pole was only twenty miles away and we were likely to reach it within two days. It helped to have it so close. I was in a better frame of mind than for many weeks past. Just to reach the Pole would be something – only the second time in history that it had been achieved unaided. Besides, we now knew we could go on. A complete crossing of Antarctica might still be far away, but beyond the Pole we would be setting a new world record. It was a satisfying thought and it made all our previous efforts seem worthwhile.

In retrospect, I can see that I should have stopped there and then to put on more protective clothing but, once out of the tent, the aim was to get moving as fast as possible. To stop work and start messing around with kit before even warming up was more likely to prove disastrous than helpful. The weather was deteriorating, but I pressed on, hoping that once we started hauling the heavy sledges, my efforts would rapidly dispel the chill from my limbs. That was how it had been on many previous icy mornings, some of which had begun in much worse weather than this. But this time, when I started moving, I felt much colder than I expected. My hands, in particular, with only gloves and thin mitts on, became increasingly numb. I curled my fingers into my palms in an attempt to restore the circulation but it felt for all the world as if I had been handed a bunch of frozen sausages. My fingers would be frostbitten if I didn't do something about it soon. I decided that I had to stop, despite the risk of getting generally colder, and go back to my sledge to get thicker outer mitts.

They were there on top of the load. Survival in the cold is an art, and anticipating problems is very much the key. But although the mitts were instantly available, I could not get them on my frozen fingers. I couldn't grasp with one hand the cuff of the other to pull it up. To make matters worse, although numb and dead to touch on the surface, my fingers felt unbearably painful inside. I found myself whimpering like an injured dog.

Windchill had now taken the effective temperature down to something nearer minus 80°C, although that was not unusual for the Antarctic plateau. Why had I not felt so cold before? One possibility seemed likely. The enormous efforts we had made were beginning to take their toll. We had lost a lot of weight, despite eating the 5,500 calories a day. It must have added to our vulnerability and meant that our energy expenditure had been enormous.

Ran, who had started out behind me, came up to help, but even with his tugging on the cuffs, I still could not get my inert hands inside. As the minutes slipped by the situation became desperate. We were now both losing the function in our hands and getting rapidly colder. It prompted Ran to make an almost ultimate sacrifice. He took off his own still supple mitts and gave them to me, struggling to get his large hands into my stiff cold ones. Then he urged me to get moving again. It was a wonderful gesture.

Although my hands soon felt better, the stop had taken ten minutes and I was now deeply chilled. I wanted to step up the pace, to get warmer, but could scarcely move my sledge. My muscles had cooled to the point where they no longer functioned properly. Unable to move fast, I could not generate enough heat to match my further losses. I was getting colder and colder, and before long I realised that I had to stop again. I needed more insulation on my body as well in what had become a full Polar storm.

The only other garment I had was a fleece jacket. It would be tricky to get it on, especially as it needed to go under my windproof, before I became truly hypothermic. The dressing would also need some dexterity, which meant taking off my gloves again. Once I removed my gloves it was only moments before my fingers would not obey commands. This time it was the zips on the jackets that defeated them and failed to move. I was dismayed to see the tips of each digit turning chalky white and I could almost feel the ice crystals forming inside them. I began to panic. I was stuck, half-dressed, with literally freezing fingers when Ran caught up once more.

With his help, I managed to get my two jackets on and then to force my pitiful hands back inside the mitts, but the stop took another ten minutes, during which my body cooled further. I had reached a point too deep in a downward spiral, and as we set off again, my thinking began to fade. Although I kept on hauling for another half-hour or so, it was never fast enough to generate much heat. I steadily became colder still and more withdrawn until finally I began to wander from our course. Later Ran told me what happened next.

'When I caught up and called out, you didn't answer, although you passed the compass when I asked for it. Your head was covered and I couldn't see your face, but something struck me as odd. I asked if you were okay but got no reply, and when I pressed you again, you made a sort of unintelligible croak. It was then that it dawned on me you must be hypothermic.

'I tried to get you to help me put up the tent,' he told me, 'but you were just standing around doing nothing, so I put it up by myself and pushed you inside. I wanted you to light the stove but you just knelt there in the middle, completely still and unresponsive. You were in the same position when I came back a couple of minutes later after I had unpacked your sledge and thrown your sleeping bag inside. I shouted at you again to light the stove, and opened the box and gave you the matches. Then I went to get my things from the other sledge. I was another couple of minutes and, when I returned, you had started to light the stove and seemed a little more with it. Perhaps you were better for being out of the wind. Mind you, you weren't exactly compos mentis, and I had to get you into your sleeping bag and make you drink some tea. It was about half an hour before you started making real sense and to appreciate what had happened. When you started to shiver and complain of the cold, I knew you were better.'

In many ways, I was lucky to survive. If Ran had been out in front when I started running into difficulties, I would have been struggling with the gloves and the jacket alone. Inevitably I would have become more deeply hypothermic and probably wandered off and collapsed. Motionless on the Antarctic plateau, my survival time would have been a few minutes at most. By the time Ran realised that I wasn't behind him, his tracks back to mine would have been blown to oblivion.

On Day 68 of our expedition, the day after my hypothermic episode, we reached the Pole and equalled the current world record. We had completed nearly 900 miles of pulling and still had the means to go on, but it was clear that we were weakening and becoming increasingly vulnerable. We sat in the tent and discussed the situation. The Pole was the last place from which we could easily withdraw. Anywhere beyond this point might be difficult for a plane to reach. The fact that I was now so vulnerable to the cold raised a vital question. Was it sensible to go on or should we pull out of our expedition while it was safe?

Ran was seriously concerned. If I could become so weak before arriving at the halfway point of our journey, how would we fare later? I argued that my hypothermia could be attributed to a straightforward misjudgement; now that I realised I could no longer rely on hard work to keep me warm, I would wear more protective clothing whenever the wind put us at risk. Although Ran accepted it, the argument was somewhat flawed. By the time I did become ill, I was actually wearing all the warm clothing I possessed.

★

Many people are surprised to hear that on this walk across the coldest place on Earth we had so little insulation against the intense cold. In retrospect, I too am astonished, but it was not a complete lack of sense that had led us to take with us so little clothing. It is a fact that, while working hard, one can keep warm in almost any environment. At rest, a man produces about as much heat as a single 100 watt light bulb but the output can rise to that of a 2-kilowatt bar fire when working hard. It is truly central heating, and it is a common experience to wrap up well before a winter's walk and then find that, after only a few minutes, you are a sweating wreck; within a few hundred yards of the house or car-park you need to strip off all your spare clothing and carry it for the rest of the day. We didn't want to do that. While planning loads with 100 days of supplies in them, weight was the severest problem. The sledges must contain as much food and fuel as possible and, like books, spare clothes are neither edible nor flammable. We would cope with the cold by working hard or quickly getting inside our tent. It was a policy that worked well while we were of normal weight, but it began to go wrong as we lost more and more. In fairly thin clothing – the equivalent to normal office clothing in Britain – more than half of your insulation comes from the layer of fat beneath your skin. Ours was rapidly disappearing. At the same time, while we towed less, we could never stop moving for long. In each of our twelve-hour days, we stopped only twice for short five-minute breaks.

In any case, I am not convinced that thin garments were the only cause of my hypothermia. Although my symptoms were typical of severe cooling, my recovery in the tent was rather too fast for that. Perhaps I was also suffering from a low blood sugar – a condition known as hypoglycaemia – which can stop the temperature controller in the brain from doing its job properly. During our Antarctic journey, eating the high fat diet and working hard enough to consume our glucose stores, we had good reason to end up with low sugars. The possibility is also supported by some of the samples we took at ten-day intervals during the expedition and later brought back home for analysis. These yielded some extraordinary results, the most striking of which were abnormalities in our blood sugar levels. To put it simply, towards the end of the journey the levels were so low that we should have been dead.

Most of the sugar in the blood is in the form of glucose, with normal levels between 4 and 10 millimoles per litre. Above these figures lies the tendency towards diabetes, whereas below you are moving towards confusion, coma and death. Analysis of our samples showed that our blood glucose had been low from the very first day, in fact at the bottom of the normal range. During the last 30 days of the expedition they appeared to be impossible. On one occasion Ran had a level of 0.2 millimoles while I had one of just 0.3. In ordinary circumstances, these would be fatal, for it is thought that the brain cannot survive without a reasonable amount of glucose for fuel. Indeed, many experts with whom I discussed the results thought that they must be invalid and that we might have left the preservative out of the sample tubes. This seems unlikely. The tubes were loaded with preservative before our departure from Britain, and other results on the same samples were normal. It seems that we really did run these impossibly low levels and must have adapted in some way to cope with the situation.

One possibility would be for the brain to switch to using ketones for fuel in place of sugar. Ketones are chemicals associated with starvation, and it is known that they can be used by the central nervous system to a small degree. Perhaps we had shown that they can be used more extensively, and certainly such a capacity would have served mankind well through evolution. During droughts on the plains of Africa, or out on the Siberian steppes in winter, long searches for food would not always have been successful. The ability to keep on working in the face of prolonged starvation would have been a survival asset, and so perhaps we had merely demonstrated a capacity latent within us all. Yet, if we can all adapt and survive such low sugar levels, why is it common to see patients unconscious in hospital with levels nowhere near as low? The answer might lie in the speed of change. The hospital patients are often diabetics who have taken too much insulin or missed a meal, or patients who are both starving and sick. In such cases, their blood glucose levels will have dropped rapidly, and so their brains would have no time to adapt to an alternative.

★

The couple of weeks before reaching the Pole had seen our strength go into steep decline. This had to be due to our weight losses, although I was unclear how great these were. We could tell from the fit of our clothing that a lot had gone, yet it did not seem possible that we had lost much more than the top of my pre-expedition estimates of ten to twelve kilos. I certainly did not trust the readings on the tiny scales that we carried. All the weight measurements had been taken in the tent, on a floor that tended to be soft snow which allowed the scales to sink in when we stood on them. Still, my belief turned out to be wrong. At the American Scientific Research Station at the South Pole, we weighed ourselves on a piece of plywood board. Much to my dismay, when Ran and I undressed and stood upon the scales, I discovered that our previous readings had been accurate. We had each shed more than 20 kilogrammes (almost 50 pounds). It was a loss far worse than my predictions and way beyond the levels at which strength declines, a loss equivalent to a food deficit of around 120,000 calories. It meant that over the 68 days we had travelled so far, our average energy expenditure had been well over 7,000 calories a day. It did not bode well for our continued journey.

The scale of these weight losses led to more careful analysis of our energy expenditures after we came back to Britain. Our pre-packaged diet on the journey made it easy to obtain accurate estimates of daily energy intake, and from these and the measurements of our body weight losses we could calculate energy expenditure for different segments of the journey. The results were extraordinary. During the period between Day 20 and Day 30, for instance, when we climbed from the ice-shelf to the plateau, we burned up more than 11,000 calories every day. This was 5,500 calories more than we were eating – a deficit equivalent to total starvation while running a couple of marathons a day.

Our energy use far exceeded any measurements previously reported in the scientific literature. For example, levels measured on the Tour de France – adjudged to be one of the world's hardest endurance sporting events – showed that riders use only about 8,000 calories per day. It was difficult to believe that we had been working so much harder, but any doubts about our data were dispelled by a second means of checking the figures. We had also measured our energy expenditure by means of a sophisticated technique employing labelled isotopes of water.

Measuring energy expenditure in men and women as they move around and perform all the tasks of life has always been difficult, due chiefly to the fact that the equipment needed to make the measurements interferes with what the subjects in any study are doing. What is more, in many situations – certainly when crossing Antarctica – you simply cannot take the necessary measuring equipment with you. Recently, however, a technique has been developed for estimating how much carbon dioxide an individual is producing from the difference in the clearance from the body of two labelled isotopes. Subjects drink water that contains very high concentrations of heavy isotopes of hydrogen and oxygen. Fortunately, although difficult and expensive to come by, both are stable and so remain in the heavy form indefinitely without radioactive decay.

When drunk, the heavy hydrogen and oxygen become evenly distributed throughout the water in the body and then, as body water is lost, they also come out – appearing in the urine, in sweat, or on the breath. The heavy hydrogen is lost in this way alone, and so its disposal rate reflects the loss of body water. The heavy oxygen, on the other hand, disappears from the body in both water and carbon dioxide, and so tends to diminish faster. The difference in the disappearance rates of the two isotopes depends upon how much carbon dioxide the body is making. By measuring that difference following a dose of the two isotopes, it is possible to calculate carbon dioxide production which, as the final product of nearly all fuel burned by the body, can be used to estimate total energy expenditure. With no equipment at all, just samples of urine for later analysis, isotope-labelled water allows measurement of an individual's energy expenditure over many days or weeks.

We had performed the isotope study in two halves – taking one dose of isotopes on the day before we started and following their disappearance up to Day 50, and then taking a second dose for the latter part of the trip. During the first 50 days, the results from the isotope studies tallied almost exactly with those calculated from the energy balance data, the isotopes giving average measures of 8,500 calories per day for Ran and 6,700 for me, while the energy balance figures were 9,100 calories in Ran and 6,800 calories in me. Once we had seen that the method worked, it was also easy to take the data and work out the energy we were using up during each ten-day period. These results supported the startling figures from the energy balance data. For Days 20 to 30, when we made the ascent to the plateau, the isotopes gave daily energy expenditures of 10,670 calories in Ran and 11,650 in me. They confirmed the highest maintained energy expenditures ever documented – values that must lie close to the physiological limit.

When we looked at the data for the second part of the expedition, following the dose of isotopes on Day 50, the results looked dubious – not because they gave very high levels of work but because the isotopes that were meant to disappear from our bodies diminished until around Day 80, and then started to rise. Additional heavy hydrogen and oxygen was appearing from nowhere. At first we were at a loss to explain this phenomenon. Then it occurred to me that, after drinking the isotopes, some of them could have been transferred from water into other more complex molecules that were effectively locked away from metabolism. They were then re-released rapidly when, towards the end of the journey, our starvation led to us breaking up our body structure, the sequestered isotopes reappearing in the body water pool faster than we cleared them, so that levels rose rather than fell. Like the blood glucose levels, here was evidence that things were seriously deranged by the end of our journey.

★

Immediately after leaving the Pole we were back to drudgery and, in my case, back to the consideration of those feigned sub-arachnoid haemorrhages. Our weight loss made me wonder how long we could go on. Success looked very distant. There were still more than 200 miles of plateau to cross and then half as much again descending through the Trans-Antarctic mountains to the far side of the continent. Even after that our journey would not be over. The Pacific side of Antarctica is bound by a second ice-shelf, the Ross, and that meant another 400 miles to open water. Only then could we catch the ship that would be calling briefly in mid-February – little more than one month away. However, we now pinned our hopes on the wind to assist us.

The Antarctic is a high cold dome, with the South Pole up at 10,000 feet. Because cold air is denser than warm, winds tend to blow away from the Pole as the air sinks towards the coasts. On our way up to the Pole we had walked into these winds almost constantly; now, as we left it, we could use them to advantage. We carried with us sails adapted from parachutes. We would stand on our skis with the sledges behind us and the sails attached to our harnesses in front. As the wind filled the sails, we were off, rapidly building up high speeds. In theory, one could regulate this acceleration by opening up a central hole in the chute by means of a control line. But the chutes had been designed for wind speeds of up to thirty or forty knots, and down in the Antarctic the winds are often far greater. It had almost spelled disaster earlier in the journey.

We had not expected winds to blow in our favour until we were beyond the Pole, but shortly after we had started, they blew in the right direction. We decided to see if the sails could pull our fully laden sledges. With the wind speed rapidly building to about fifty or sixty knots, the answer was yes – very rapidly. We tried to slow ourselves down by making the central hole bigger, but that only made the chutes unstable. With no appreciable decline in speed, the chutes had begun to swing from side to side, obscuring the way ahead and catching on some local icy protrusion. Still at great speed, skier and sledge then went through the hole, creating a tangled disaster of strings, sledge and ski bindings. There was then the risk of frostbite as we tried to undo the mess with thinly gloved or bare hands.

Sailing on the wind, it was also difficult to discern hidden crevasses. On one occasion I had seen Ran, just ahead of me, fall to the left of a long open crack. Further to the left, the ground looked quite firm, and so I had steered that way and carried on. What I failed to realise was that he had not fallen over by chance. Hidden by a rise, the crevasse actually extended to either side of him, and he had thrown himself over to avoid going into it. He tried to warn me as I approached, but I was going far too fast, and both I and my sledge were pulled into the gaping chasm. I was very lucky to survive. Instead of falling freely, I was partially held up by my sail and, even more fortunately, I went into the crack where it turned a tight bend, just at the point where blowing snow accumulated to build a delicate arch of drift. Instead of plummeting into bottomless darkness, I dropped only twenty feet on to a snow bridge – winded but unhurt. The next moment I was terrified. Crevasse walls rose almost vertically above me and to either side, just a few feet away, blackness beckoned. The bridge was clearly frail, and it took considerable effort to quell my fear and to unload the sledge one ration, one fuel bottle, one piece of equipment at a time. Each of them had to be thrown up to Ran on the edge twenty feet above before finally the sledge was empty. Then, with the aid of a rope, I gratefully climbed out. It had been a narrow escape.

Soon after leaving the Pole, we were suffering even more mal-nourishment and became dreadfully debilitated. It was clear that we needed a boost if we were not to grind to a halt, and there was only one way to get one. We decided to consume an extra quarter ration pack every day. It took our intake up to over 7,000 calories. We would eat this much for however long it took to cross the remainder of the plateau, and then, when we reached the top of the glaciers on the far side, we would cut the food right back and try to survive on just half rations. It meant that as we went down and off the continent, we would be eating less than 3,000 calories per day. If the winds came soon, we might still finish the entire journey, including crossing the second ice-shelf; if not, we might at least succeed in crossing the continent itself.

Eating more, but still very thin, we struggled to maintain our core temperature. The idea that we could keep warm through our work alone now proved to be even more misguided. Day after windless day went by. It seemed so unfair. In the ten days after leaving the Pole, we managed to put less than 150 miles behind us. Eventually, on two brief occasions, we got some wind from the south and made good distances, although each time it became too strong for safety. We found ourselves falling and being dragged over rough ground at tremendous speeds. Our sledges turned over and were ripped along upside down. After becoming tangled in the rigging, we were numb with cold and my fingers became frostbitten, blistered and were weeping pus. In addition, I was hobbled by a fractured ankle. Ran too had problems with his feet. A graft from a previous frost injury had broken down early in the journey and the pain he had suffered ever since hurt him more than ever now. The area could well become infected and, short of antibiotics, it looked likely that we would be forced to stop.

Even so, the wind had moved us on. Early on Day 82, as the storm cleared, we could see the mountains ahead of us. It was a glorious sight that lifted our morale and we reached them two days later. Here our route began to slope. It was as if we had reached a gigantic weir and we followed the ice as it poured from the plateau into the huge Beardmore glacier. With the move into the mountains, the surfaces rapidly changed from wind-blown snow to a sea of gleaming ice-glass on which we frequently fell. Without the crampons that had seemed like a luxury we could not afford, we resorted to cutting up used fuel bottles to make grips, but the metal proved too soft. Grips made from stitched rope were little better, so we were forced to travel at a teetering run, moving faster than we would have otherwise dared.

As we descended the Beardmore Glacier, the gradients lessened and the ice became covered with soft snow once more. The sledges began dragging again, and we were back to a slog. With the extra work came more hunger. Half rations seemed pitifully inadequate. A few spoonfuls each and the pot was empty, leaving stomachs aching. During the day, we now ate nothing but two small chocolate bars and shared one thermos of soup. Often drifting through a haze of unreality, I was not surprised to learn later that our blood samples showed those dreadfully low glucose levels.

We became steadily weaker. Although the sledges now weighed less than 90 kilos, we could move them only slowly. Well over a thousand miles lay behind us, but our bodies had paid the price. Even the magnificent mountains could not lift us as we approached the last difficulty of our journey. Ahead, the glacier ran through a narrow channel between two high peaks. Choked with crevasses caused by the constriction, our path was blocked by an enormous hump-back of fragmented, disrupted and tortuous ice that extended as far as the eye could see. A diversion to avoid it would have taken days, so we had little choice but to take it on.

At first the slope rose steadily, broken by huge parallel cracks, but we found places where they were bridged and slowly we climbed. At length we reached a region of total chaos, where the ice was split in all directions and more of the area was black voids than white surface. It was terrifying. The sledges had minds of their own, slewing sideways, trying to drag us to destruction in gaping mouths. We made our way through like men in a maze, wandering back and forth, seeking bridges strong enough to bear our weight. It took four hours to move less than a mile, but finally we descended the far side. Ahead of us the huge ice river smoothed out as it wound through the last of the mountains towards the five-thousand-foot Gothic spired summit of Mt Kyffin. It marked the far side of the continent.

We were now close to completing our continental crossing and just days from our first big success. Only Ran's frostbitten foot could stop us, and it would have to be a raging infection to turn us from that goal. On Day 90, close on Mt Kyffin, we became determined to finish in one haul. We set off at 6 a.m. as usual but did not stop at 6 p.m. In the event, we kept on walking until two in the morning – passing through the 'Gate' at the foot of the Beardmore and off the Antarctic continent. I cannot describe how it felt as we set up camp on the Pacific ice-shelf for the first time. After three months of hell, we had achieved something special.

That night we discussed our situation. We had only eight food packs left, sixteen days of half rations if we could keep going on so little. If we could sail, we might manage to cover long distances, but we were far too weak to haul the sledges much more. It was difficult to know if we had any real chance of reaching the far side of the ice-shelf four hundred miles away, but neither of us could say it was impossible. Just four hours after our stop, we set out once more.

Progress was slow. The wind did not come and our legs were giving way. Along with our fat, we had lost much of our muscle, and our bodies were mere sticks with clothes hanging half empty. Whatever our determination, we could not win by manhauling; we needed the wind. Surely it would not forsake us.

It did. For five days the weather stayed bright, cold and still. The surfaces were smooth and stronger men would have made good distance. Yet, in ten hours – now all we could manage – we pulled our sledges no more than fifteen miles. We were crushed by the vastness of our journey, sapped by the weight of our loads, and our muscles were poisoned by the ice. On Day 95, with no wind at all, we gave up. We could do no more.

While Ran made the radio call for our pick-up, I went and stood outside. Our tent was pitched in the middle of a huge white plain and the sun was shining. To the south, a thin line ran back from where I stood to disappear beyond the horizon, towards mountains and wind-sluiced valleys. There it ran back up the glacier and then due south to the Pole. It continued on – straight for the rest of the plateau, and dropped tortuously through valleys, dunes and sastrugi to the ice-shelf on the far side and so to the Atlantic coast. It was the longest, unbroken track that any man had ever made.

★

Ten days later, we were back in a laboratory in England, facing a bank of tests that would give a fuller picture of the effects of our journey. Most of the studies were geared to look at two inter-related areas. How had our metabolism reacted to massive weight loss induced by exercise, and what had happened to our strength and fitness? The results demonstrated that it was more than just blood glucose levels that were pushed beyond accepted limits.

Despite our huge food intakes, Ran ended up losing 25 kilogrammes (55 pounds) in weight while I lost about 22 kilos (48 pounds). In order to find out how much of these losses were fat and how much were muscle, we measured our density – not in the intellectual sense, but as measured by Archimedes in his bath. Because fat is less dense than muscle, the fatter you are the more prone you are to float. Comparing your weight in air with your weight when immersed in water will therefore tell you how much of your body is fat and how much is lean. We performed these underwater weighings before and after the journey and discovered that our bodies had dropped from a fairly normal 16 to 17 per cent fat before starting to dangerously low levels of just two or three per cent after the journey. As the measurements after the walk were made when we had returned home and already eaten an enormous amount of food, it seemed likely that at the end of the journey itself we had been around zero per cent fat. It was surprising we were still alive, and little wonder that we had felt the cold so badly.

If you stop eating and just have water to drink, you usually die after about two months, even if you lie down and use as little energy as possible. This is known from famine victims and political hunger strikers in whom deaths usually occur when they have lost about one third of their body weight. It is also known that for the last few weeks of their lives, such starvation victims are unable to move much, due to sheer muscle weakness, and that at the time of death they still have some body fat. Why then were we able to continue, albeit slowly, when we had lost similar amounts of weight and had pushed our fat levels down to hitherto unreported extents?

Possibly the answer lies in the through-put of food in our unusual situation. Although our daily weight loss was often much more than would be seen in a person on a hunger strike, we still had a normal to high intake of protein rather than low levels or none at all. Our bodies therefore had enough to maintain essential organs, so that our hearts, livers and kidneys did not cease to function as they would in straight dietary malnutrition. Indeed, some of our measurements proved this directly. During our 24-hour urine collections, we had taken another isotope label – a heavy nitrogen atom attached to an amino acid. Amino acids are the building bricks of proteins, and by measuring the rate of reappearance of the heavy nitrogen in our urine, and knowing how much protein we were eating overall, it was possible to estimate the rate at which our bodies were building protein structures. Even as we starved, these levels were high, the opposite of what is seen if you are starving 'normally'. We had been able to maintain our functions in the face of weight losses that as a rule would be fatal.

The measurements of fitness and strength might have been expected to show that, with all our exercise, we became fitter. Instead our maximal aerobic capacity went down by about 20 per cent, and the strength of individual muscle groups dropped even further. This was noticeable. After I got back home, I could not walk upstairs without pulling on the bannisters, or lift my children, who were only very small. It was tempting to think that the weakness was due to the clearly visible loss of our muscle mass, but the causes turned out to be more extreme.

Before and after the expedition, we had specimens of muscle taken directly from the top of our legs with large cutting needles. These biopsy specimens were analysed to measure the activity of the enzymes in the muscles – the molecules that run the entire system within the cells. The analysis showed that, after the expedition, enzyme activity had declined by more than 50 per cent in both of us, whereas under training with normal nutrition they would be expected to rise by as much as 80 per cent. The internal structure and mechanisms of our muscles had literally been burned in order to survive.

Some of the other findings will be mentioned later in this book, but one other warrants mention here. The levels of our testosterone – the male sex hormone – declined progressively from the moment we started until the time we finished, and ended up so low that they were virtually undetectable. The explanation for this may be two-fold. It is recognised that testosterone levels fall with extreme physical stress, and certainly we could claim to have been through that. Testosterone levels, however, also fall in men who are isolated from women. In sub-mariners, for example, levels have been shown to fall steadily through their underwater voyages and return to normal just before they reach port. Not necessarily home, but any port – the anticipation of meeting any member of the fairer sex is enough to correct the abnormality. The effect that this drop in testosterone had on Ran and me is not clear, yet it may have been useful. Throughout our long trek, we never did fancy one another.

SIX

★

Marathon of the Sands

Darkness was falling as we came down from the high Atlas and the
unheated and draughty bus became distinctly chilly. It had been
cold all afternoon. Climbing up into the mountains south of Marakesh,
we had left the warm sunshine of the Moroccan spring to move under
wreaths of dark grey cloud. At the top of the pass we had stopped for
some sweet black tea at a roadside stall. Strong winds had made it feel
bitter and there had been a hint of sleet in the air, ice blowing off the
persisting winter snowfields. Yet, despite the overcast skies and the
cold, it was a spectacular journey, both beautiful and nerve-wracking.
Through intermittent breaks in the cloud we glimpsed peak, cliff and
distant plain, while our young local driver hurled his machine around
the multiple hairpin bends, totally ignoring the precipitous drops. He
saw them as just a part of his everyday world, while we could only
experience the irrational fear that large drops exert – a magnetic
attraction upon your vehicle, willing it to leave the road and plunge to
oblivion. Each bend had brought an unwelcome reminder of our
mortality.

Now, with the onset of darkness, we began to sense that the road was
becoming less steep. Outcrops of rock that flashed by in the headlights
became less obvious and instead the beams swung from tarmac across
hints of desert sands. As the moon rose it illuminated an almost alien
world. Around us, stretching away on all sides, were the dunes – their
silvery, iridescent crescents echoing the source of illumination.
Conversation that had lapsed during the descent from the mountains
started up once more and the bus filled with the sounds of many
languages. It also filled with an air of expectation. We had entered the
Sahara at last and were now close to the test for which we had long been
preparing – the Marathon of the Sands, one of the toughest foot races
on Earth.

The bus drew into a sleeping village square that could have come straight from Spain or Italy, but as we stopped and the inhabitants emerged to greet us, it was clear that they were not of southern Europe. Their clothes, manner and complexion were of northern Africa. They were the people of the world's greatest desert.

The Tuareg are a proud race, the masters of the southern Moroccan Sahara – great travellers, story-tellers and survivors. Many are still nomadic, but others have chosen to settle, and their young children now swarmed up to the doors of the bus, anxious to see and shake the hands of these strange visitors. They also wanted to beg, not for money or food but for pens and pencils. It appeared that the local village school had no materials, and soon we had all given up our writing things. The children could have come from anywhere, though their elders were of striking appearance. All looked prematurely aged by their dwelling place, as if more than one generation separated them from their off-spring. Some appeared to hail from older centuries and, most strikingly, many had dyed their faces with pigment to a deep wizened blue.

After a long wait in the square, shivering in the desert darkness, a row of distant headlights could be seen approaching us across the sands. The vehicles were presumably on a track, though behind each one a huge plume of sand was rising. When the jeeps arrived, we transferred our belongings and packed into them. Engines revved as we headed out from the square on what would be an uncomfortable two-hour journey before we reached our start camp. The Tuareg children waved us away and a strange murmur arose from the blue-stained elders. They all had an air of deep mysteriousness, of knowledge beyond the imagination. They knew that the desert would test our strength, yet not in ways that they would ever seek.

★

Most people, when they hear about 'Le Marathon des Sables', are surprised that anybody should be so stupid as to attempt to run such a distance in the Sahara. Then, when they realise that it is not a mere marathon but a true ultra-distance race, they decide that the word 'stupid' is far too mild. Yet men and women can run much farther than is commonly thought, and they can do it in conditions of heat that many imagine would render us immobile.

The race is held annually, and each year it attracts increasing numbers of competitors. They come from all over the world, keen to pit their strength against one of the most difficult of environments, covering in one week the equivalent of five marathons – not on smooth London tarmac but over rocks, plains and shifting sand. To make matters more difficult, competitors must each carry a backpack containing all food for the entire week as well as a stove, sleeping bag, flares, torches, warm clothing and navigation equipment. On top of all that must go a supply of water although, in that case, it would be impossible to carry enough fluid for one day, let alone the entire week. Runners are therefore re-supplied at checkpoints on the route. At any one time, you only carry enough water for the next ten to fifteen miles. With a fairly heavy rucksack on the back, rough going underfoot, and the fierce desert sun, it is not a venture for the faint-hearted and many find it difficult to understand how such a race can be run. The answer lies in our remarkable ability to cope with heat.

Running through our bare skin is a network of blood vessels that dissipate heat through the processes of convection, conduction, and radiation, and if that is not enough, we also have tens of thousands of sweat glands which can automatically wet the skin to add the power of evaporation to our cooling. This ability to cope with heat stems from the millions of years of our early development in the hot cradle of Africa where our primate ancestors met a climate which would have cut down those who could not keep strong, fit, and above all fertile through the worst of the blistering summers. Natural selection melded early humans to make them extremely heat tolerant and when, 100,000 years ago, they started to spread to cooler parts of the globe, they took this potential with them. It has altered little since. We all possess the heat resistance of our early African forebears, and all modern races are equally capable of surviving, living, working, or even running across the hottest parts of the planet. We are each, after all, the naked ape.

★

We awoke early, the harsh rays from the east pouring into our wall-less Bedouin shelters and rapidly moving us from being cold to overheating as we lay inside. The stories of freezing desert nights had proved to be true, and our ultra-lightweight sleeping bags had been inadequate to prevent us becoming chilled. I had slept badly and now, at only 6 a.m., I was reluctant to rise until one of my companions opened his eyes and became wakeful enough to engage his brain. His sharp cry of warning brought us all to our feet. Around us, scuttling across the rugs, was a host of red and white scorpions. The thought of being joined in my bag by one of those barbed creatures was enough to cast off all remnants of drowsiness. Fortunately, the scorpions were as reluctant to be sociable as we were and, as if by magic, they vanished under the stones where they would spend the day avoiding the relentless sun.

We too would spend much of this first day in shelter, registering and preparing for the race. There were six of us from Britain – three were colleagues of mine whom I had persuaded to come and enjoy this mad challenge, and two others decided to enter the race when they heard that nobody from the United Kingdom had ever done so before. For me, the race provided an opportunity to get back to reasonable levels of activity and fitness, as well as another chance to visit an exciting and wild part of the planet. The heat would also make an interesting contrast to the conditions in which Ran Fiennes and I had crossed Antarctica.

The six of us were joining about one hundred and twenty other runners from a total of eighteen different nations, although French and Americans made up a considerable proportion of those we could see emerging from their shelters. It was our first proper view of the camp at which we had arrived late the previous night. Lines of Bedouin tents, each made of a large piece of dark cloth supported on wooden poles, filled about half the head of a broad stony valley that ran away to the south. The valley had a gentle slope up to the east and a vertical high escarpment of purplish banded rock to the west. The floor, of sharp-edged stones, was cut in places by deep, steep-walled wadis, presumably filled at times with water. At intervals down the valley stood clumps of palms or thorny bushes, their vibrant, rippling greens providing a striking relief from the otherwise earthen hues.

Each tent was carpeted with a threadbare coloured rug on which we had done our best to sleep through the chilly night, waking from time to time to rediscover that hips have no soft tissue cover at all. I had assumed that in the Sahara we would always sleep on soft sand, but it seemed that the Bedu shelters could only be well anchored on harder ground. The shelters were laid out in three sides of a square, with the fourth side made up of large administrative tents, one of which bore a big red cross. Glancing inside, I was reminded of a television portrayal of a MASH receiving station. Then we were summoned to a different larger tent by the aroma of baguettes, jam and coffee drifting across the camp. The organisers of the event were French and a continental breakfast started the day.

Nobody could enter the race without showing all of their safety equipment as well as their electro-cardiograms and medical certificates. Then their packs were weighed. Mine came in at a disappointingly heavy 12 kilogrammes (more than 26 pounds), and that was without any fluid on board. It was a weight somewhere in the middle of the range of other runners. Those who had taken part in the race before had particularly light packs, some achieving this by carrying no food at all except sugar and electrolyte powders to mix with their drinks.

Even more than in a single marathon, I had considered that eating and drinking on the move would be essential, as would eating well at the end of each day's race. My plan was to drink up to two litres per hour of rehydration drink, which contained some calories, and then to consume nearly 3,000 calories of carbohydrate across each evening and breakfast meals so that my muscle and liver stores of sugars would be re-stocked. As a result my pack weighed more than I wished, and encouraged by the voices of experience that were saying one could run on burning body fat alone, I decided I would cut back to a reasonable compromise. Less well trained than the true professionals, my body would not burn fat quite so happily, but I managed to reduce the weight of my pack down to 9.5 kilos (21 pounds) without any water.

Registration also provided us with guides to our route that would see us across the desert. Each day was divided into stages, with water at the end of each stage. The route appeared to be variably defined, for in places guidance was precise – 'follow the flagged tracks' or 'run along the edge of the wadi' – whereas in others there were more indefinite instructions. On Day 3, for instance, we were exhorted to

> . . . enter the band of dunes following a bearing of 176 degrees and maintain this for 18 kilometres. At the top of the final large dune, you will see the end of the stage on a bearing of 145 degrees.

It sounded to me as if this might be easier said than done. Following bearings across even rolling country is hard enough, let alone across steep sand dunes that all look similar and among which one cannot see far. Still, since I would never be up with the leaders, I also guessed that I would always have footprints to follow.

The total distance for the seven days was as expected, but daily distances were more variable than I had anticipated. Tomorrow, the first day, was to be a truly modest 15 miles, but Day 4 provided a mammoth 50 – almost a double marathon. A full single marathon was signalled on Day 6. Since, after the first couple of days, I would have run further than I had ever done in my life, I did begin to wonder if it would be possible for me to finish.

The heat added to my consternation. It became exceedingly warm, far worse than I had thought possible in April. By midday we were positively roasting, even lying shaded by the canopies. When the wind blew, it somehow made things worse, engulfing us in a wave of crushing heat rather than providing the relief one might have expected. Despite knowing I was well acclimatised, and that people had completed similar races before, it struck me that the climate might stop me before the distance did. When dusk fell and we settled down to sleep, these anxieties – added to the cold, the hard ground and fears of attentive scorpions – conspired to make it another restless night.

★

Unlike the earliest of our ancestors, and indeed the fish and reptiles of today, mammals and birds regulate their core temperatures very closely. This allows every cell to operate under stable conditions, using metabolic systems set up around enzymes that work best within narrow temperature limits. The human core temperature is set at around 37°C, although this varies with diurnal rhythms from about 36°C during the latter part of the night up to around 37.5°C in the late afternoon. These are also the resting values, and they rise considerably during exercise, reaching levels of 39°C or more if work is hard and the conditions hot. It seems that we are designed to cope with this for the optimal temperature for working muscle is well above the 37°C norm. Hence the value of the pre-race warm-up.

Theoretically, evolution could have led to any temperature at which to run our normal core level, with enzymes adjusted through eons to be at their most efficient at the temperature chosen. Yet nearly all warm-blooded creatures have similar resting core temperatures. Obviously, the higher a species sets its normal operating temperature, the more heat production is needed to maintain it, while the cooler the normal operating temperature, the more cooling capacity will be required. Perhaps our 37°C temperature reflects an evolutionary compromise with climate during the periods through which our ancestors lived. But if that were the only reason, one might have expected more variation in the core temperatures selected by different warm-blooded species that evolved around the world under very varied ambient conditions. It could be argued that the similarity is due to all warm-blooded creatures being genetically close enough to have had a common warm-blooded ancestor and that the basic enzyme blocks of life, and the temperature regulation system that goes with them, evolved very early.

There is another intriguing suggestion. The unit of energy used most widely in biology is the calorie – the energy required to heat up one gramme of water by 1°C – but its value actually varies according to the temperature of the water, even when insulation is total so that no extra heat is lost. The strict definition of the calorie is therefore the amount of energy needed to heat up one gramme of water from 15°C to 16°C, for above this temperature less energy is needed to make a one degree change. At least, less energy is required until you reach a nadir at around 35°C, beyond which the amount of energy needed to change water increases again. Regulation of core temperature close to this point allows changes in temperature to be made with minimum energy cost. Could this be the reason why evolution ended up dictating core temperatures of around this value?

★

By the time we reached the end of the first day my anxieties over failure were turning into certainties. Most of us were shattered. All of us had trained over greater distances than the 15 miles we had just run, but the rough ground had made it very hard and I felt worse than after the full London Marathon. Although we had not come across any really sandy areas, it had still been mostly soft underfoot, and where it wasn't, it had been viciously stony. Over soft ground, feet felt like lead and thigh muscles burned, while on the stones, toes were stubbed and it was impossible to attain any rhythm. Ankles also constantly rolled from side to side, threatening to go right over. I wasn't sure which was worse, sinking in or stumbling.

The start had been fun, with a carnival atmosphere overcoming our many apprehensions. We set off at 9 a.m., and although the sun could already be felt, it was quite pleasant being bathed in its rays. The air was fresh, with no wind, and the purple scarp of the west side of the valley could be seen heading away with almost undiminished clarity. About six miles on, it turned to the west, and there too we would turn before following the rocky wall down towards a plain. I heard various quips about rounding this huge bend on the inside line and of not getting boxed in at the start. The gun sounded and we were away, and with it our heat production rose. Unfortunately, the sun followed suit.

For the first hour I almost despaired of my pack. I tried to fix my mind upon the next segment of the run and to ignore both the weight and the pain in my neck and shoulders. It worked for a while, but the target I fixed on would come no closer. I soon felt miserable and uncomfortable once more. Around me were a very mixed bunch. Some were international ultra-distance athletes, running hard for a purse of thousands of dollars and the title of champion of the 'Marathon of the Sands'. These quickly disappeared into the distance ahead. Others clearly reckoned to walk for much of the time, and I assumed that they planned to cover the distances only slowly in the hope that such self-control would permit them to complete each stage and push on through each day. They soon disappeared behind. The majority, however, were like me, average runners who wished to test themselves to their own modest limits. We planned to run for as much of each day as we were capable, without pushing so hard that we would end up pulling out. Of course, whether fast, slow or medium, each runner's aim was to complete the task, to exert their strength and will in order to finish all seven days.

Among the more bizarre runners were the two oldest competitors who had both run in all nine previous races. I ran for a while with Ibrahim, from Morocco, who was affectionately known as the Butcher of Fez. Although he spoke little English, we managed to exchange a few words on homes and family. At seventy, he was still active as a butcher, had twelve grown-up children and innumerable grandchildren. He was another prime example that age does not necessarily bring inactivity. The other elderly runner was one year younger and known as L'homme bleu. Every year that he had entered, he had adopted Tuareg dyes for the period of the race. Deep blue faced and grey bearded, he made a rare sight, and it was difficult to imagine him as his more normal self – a merchant banker from Paris.

After the first checkpoint at 7 miles things began to improve. Between there and the second checkpoint at 12 miles it was almost comfortable, and I was reminded once more of how the body will accept anything as reasonable for a while if asked to do so. After that, it was miserable once more, and the heat was beginning to tell. I grew hotter and hotter as midday approached and we headed out across an open plain, devoid of vegetation and of unknown size. Heat rising from the surface of dark stones made the landscape shimmer like rippled water. It seemed for all the world as if we were crossing a huge lake with the far side beyond the horizon.

I began to get cramps in my legs, and slowly I was forced to reduce my pace to a walk. Many others did the same, and we became a weary, ragged line following the trail of those who had preceded us. The sun beat down upon our heads and, despite wearing caps or hats, we began to suffer from headaches. The hats had cloth squares sewn on the backs of them to shade our necks, and we looked like members of the French Foreign Legion returning from a rout. Yet our enemy was the desert itself rather than some human foe. Eventually we all made it, entering the camp and feeling almost instantly better as our spirits were lifted by our first achievement. But how would we feel on the days following? We quickly found our water allowance and made for the shade. Tired, dry and overheating, it was time to drink plenty and eat what food we had. It wasn't all that long before we would have to start again.

★

To maintain our core temperature at 37°C, and certainly below 40°C which is dangerous, the heat we lose to our surroundings must be at least as much as heat we generate along with that which we absorb from the environment. This means that, when working hard, we must lose one or even two kilowatts through conduction, convection, radiation and sweating.

Conduction involves the direct transfer of heat through contact. Thus heat generated within the cells of the body is conducted outwards to the skin, passing through any intervening organs, muscles or fat. However, if conduction to the surface were the only means whereby spare heat could be removed, it would be both inefficient and uncontrollable. Deep inside the working muscles, the enormous heat production would make things far too hot. At the same time, far from working muscles, other parts of the body not in current use or too near to the surface might have too little heat to operate. What is needed, therefore, is a mechanism to transport heat from working areas where there is too much to other areas that need it, or to the skin to get rid of excess. This job is performed by the circulation, for as well as transporting oxygen, food and the chemical waste-products of metabolism, the heart, blood and skin act as a sophisticated version of the car water pump and radiator.

Once spare heat is carried to the surface, it is conducted away by whatever is in contact with our skin. Normally this is air and clothing, but sometimes we may be touching water, wood, or metal. Air is a poor conductor of heat, and so direct conductive losses tend to be quite small, especially if a layer of warm air is trapped against the skin by clothing. Water, on the other hand, is a good conductor, able to transfer heat about twenty-five times more quickly than air. The result is that when immersed in water, our heat losses are much greater than usual. A 25°C day in the garden is shorts, T-shirt and sweating weather, whereas 25°C in the swimming pool is only warm enough if you keep moving around. In fact, if you remain still in water at 30°C, you soon become hypothermic. Heat always moves from higher to lower temperatures. With water's good conduction, the 7°C difference between the swimming pool and your body will suck heat away at a rate exceeding the 100 watts your resting body will be producing.

Metals are even better conductors than water, and they can remove heat from our skin very speedily indeed. This is why metal feels so cool to the touch compared with wood. As long as the metal is at a lower temperature than your skin, it will conduct heat away from the area touching it and the nerves will tell the brain that the object is cold.

Conversely, when the metal is hotter than your skin, it will conduct heat in easily. A piece of metal lying in the sun, soaking up radiant heat, may literally be burning to touch, whereas wood in the same situation is tolerable. Conduction, as with all the heat transfer mechanisms, works both ways.

In reality, conduction rarely operates alone. Mediums such as air and water are fluid and so convection also comes in play. This is the removal of heat by the flow of one substance over another, and the air in contact with our skin almost always moves even when we are stationary. The movement occurs for many reasons. Even when we are lying still, the fact that warmed air is less dense means that air in contact with our bodies rises away from the skin to be replaced by cooler air. Most of the time we are not still anyway or, even if we are, the air around us is moving with the breeze. Convection and conduction take place constantly and in most environments we can easily lose more heat than we would wish. For that reason, we normally wear clothes to trap air against our skin and so limit conduction and convection to levels which are comfortable.

At rest in air of about 20°C, conduction and convection account for only ten to twenty per cent of our heat loss, with most of the rest taking place through radiation from every part of our body. Of course, we also receive incoming radiation from warm objects nearby and, more significantly, the sun. The sun transmits both light and infra-red heat. We absorb almost all the infra-red, whatever our skin colour or clothing, but the light energy is absorbed variably by different colours – light skin and clothing reflect most of it, while dark colours absorb it.

Even on the coldest of winter days the sun can provide enormous heating, and in places like the Sahara its effect can be immense. Worse, when out in the desert, the whole environment of air, sand and rock can all be at a higher temperature than your body surface and so will radiate heat into you. Indeed, once the environmental temperature exceeds 35°C, heat cannot be lost through convection, conduction or radiation and the input can be enormous. To run with a pack in such a climate might seem impossible since heat production will be well up on the bar-fire scale. How then do the competitors in the Marathon of the Sands survive? The answer lies in our heat loss through sweating, a mechanism that allows us to withstand extraordinary conditions.

★

After yet another uncomfortable night we were disturbed early, not by scorpions but by a sudden change in the weather. Half an hour before we were due to get up at 5.30 a.m., the air began to stir and a light breeze started to flap our shelters. Within minutes, the breeze rose to become a stiff wind and then a full-blown gale. It was an extraordinarily rapid change as the stars that had been gleaming with unearthly intensity disappeared behind the blowing sand. Indeed all was lost in an increasing, whirling stream of violence.

Under the shelters, the sand filled everything, and all but the heaviest of our belongings threatened to blow away. It led to another hurried exodus, after which there were pathetic attempts to eat without consuming too much of the streaming dust. There was then a long wait for the start at 8 a.m., but with a full sandstorm now raging, it kept being put back while the camp attendants wrestled with the Bedouin shelters. To get them down and packed up was difficult and not altogether successful. At least one was lost for I saw it – momentarily – whirling into the sky before it disappeared in a veil of dust. Even after the camp was dismantled the organisers were reluctant to allow us to set off, fearing that in such poor visibility runners could easily go astray.

We all stood around the vehicles, vainly seeking shelter from the painful blast of stinging crystals. Eventually it was decided that we should go ahead, although the route was changed to follow flagged desert tracks. It would extend the day's distance from 17 to 21 miles, but it would be much safer than running on a bearing when visibility was less than fifty yards.

At around 11.15 we departed, running down a trail directly into the storm. It was a bizarre experience. The sand was everywhere, and despite scarves around mouths and noses, we inhaled it, ate it, coughed it. Whatever we did with the rest of our clothing, the sand also got to our skin, and soon all moving parts became sore and chafed. It was desperately hard to run against the constant battering, and even when we changed direction the wind still seemed to blow against us. I had never run in such strong winds before, and reckoned that it doubled the work-load. It felt as though I was running steeply uphill, and nobody could sustain 21 miles of such work. There was one way out – we needed to help one another.

It was obvious really. Run behind another competitor, take turns to be out in front, and the load was cut tremendously. It was an example of slip-streaming taken to extremes and, even better, runners could form into long lines of single file, with only one front-man bearing the brunt of the gale. Every now and then that front runner could pull over and drop to the back of the line to enjoy some hard-earned rest. It worked brilliantly, and it brought a tremendous spirit of unity to the race – a feeling that was to remain for the rest of the event. The Marathon of the Sands is not so much a competition as an opportunity to challenge oneself. Everyone was in the same boat, and we knew it. If a lone runner was spotted through the gloom, the whole line would deviate from the course to go and collect him. Slowly each moving snake of runners grew longer, meandering across the desert and picking up new segments.

I also found it helpful to have no option but to stick with the pace. To drop out of the line would mean falling back completely, for it would be impossible to catch up again. There was therefore no choice of walking, even in the face of discomfort from a full bladder. It made progress remarkably steady until, after a couple of hours, the wind began to drop and the moving sands returned to the ground. This was of course welcome, but it now revealed a high sun and soon the stupefying heat of the previous day returned. With still more than half the distance to go, we came to a region of low scrub. It had trapped both today's and previous blowing sands and so had formed an area of small dunes. Although they were tiny by Saharan standards, we toiled through these little undulations as if carrying a hundred kilos, and with the hard exercise we cooked from within and without.

Wherever we looked, the harsh sun shone, casting black shadows on the rippled sand. Its heat was reflected from every surface, burning the face despite the dust that now coated every one of us. Although only the second day, our lips were already dry, cracked and swollen. I sipped at my drinks continually, my throat rasping from my laboured breathing. Slowly, as on the day before, I developed a thumping headache and I knew that my core temperature would be running higher than was safe. Once again, I also suffered cramps and nausea as I became increasingly dehydrated. My body was sweating at a rate that my drinking could not match, and it was with enormous relief that I finally saw the finishing flags fluttering up ahead.

★

Unlike convection, conduction and radiation, sweating can lose body heat at any environmental temperature. To evaporate, water requires an enormous amount of energy, energy that will raise the individual molecules from their relatively placid liquid activity state to the frenzied movement of molecules in a free gas. In comparison with the one calorie of heat that it takes to raise a gramme of water by one degree, the evaporation of a gramme of water takes 585 calories, and it is for this reason that wet skin is so wonderfully effective at cooling us. Most of us are familiar with such cooling when emerging from the sea or the swimming pool to stand in a breeze with wet skin. Whatever the heat of the day, you suddenly feel cool, or even cold, and remain so until the water runs out and once more your skin is dry. Then you have to return to reliance on your built-in wetting equipment – your sweat mechanism – to keep you cool, though not at the level you have just experienced. The sweat control mechanisms are set up to prevent too big a rise in core temperature rather than stopping any rise at all. We barely sweat at all until we are on the hot side of comfortable.

Although evaporation can extract an enormous amount of heat, it can only do so if the surrounding air is dry, at least, dry enough to have the capacity to take up further water vapour. Fortunately, the hotter the conditions, the more water the air can accommodate, and in such desert regions as the Sahara, the very hot dry air has an enormous capacity for vapour uptake. This is not true for all hot parts of the world, for in some regions it is hot and humid. If the temperature is over 35°C and the relative humidity is approaching 100 per cent, there are no possible routes by which to lose heat. Even if you do nothing you will inexorably heat up and eventually die. In such climates, man can only survive when nighttime conditions are cool enough for the heat gained during the day to be lost before the next one begins. Life is also made more miserable when sweat will not evaporate and instead runs down face and body in those dreadful salty rivulets. There is no doubt that humidity has far more effect on comfort than the temperature itself. You are more likely to succeed in completing multi-marathons in the ultra-dry Sahara at 40°C than you might were you to do the same in Britain at the height of summer.

Sweating can occur efficiently only if the body has enough fluid to spare, and in hot conditions, with high sweat rates, we will slowly dehydrate. As we do so, the rate at which we can sweat will fall and our tolerance to the heat goes down. In the Sahara, we were aware of this potential problem before we started. Some runners simply put the bottles of water supplied straight into their packs, stopping occasionally to fish them out and have a good swig. However, there was a better way. To maximise the amount of fluid that one can absorb, you need to be able to get at it constantly. Many of us therefore had drinking systems from which we could take small quantities even as we ran.

The system I used consisted of two 1.5-litre panniers built into the sides of my rucksack, with long flexible straws coming over each shoulder attached to the straps. The free ends of the straws were close to my mouth, with a bite valve to allow me to suck when I wished but remaining sealed at other times. As a result, I could take little sips often and I drank some fluid every minute or so throughout a day's run.

Unfortunately, even with such a system and a commitment to drink as much as possible, it is still difficult to get enough fluid in. Even when fluid losses are on a scale that will severely impair heat tolerance and performance, we do not necessarily feel thirsty enough to drink large volumes, and if we do, the gut finds it difficult to absorb. This is especially true when we are exercising and blood is diverted away from the intestine to working muscles and heat-losing skin. You can easily end up feeling bloated and sick. In fact, in circumstances such as those in the Sahara race, it is virtually impossible for the body to absorb as much as it can sweat.

Sports drinks can help to maximise the absorption rates of fluid from the gut. If they are cool and appealing in flavour, they also encourage you to drink plenty of them. They contain both dissolved sugars and salts which are picked up by the cells in the wall of the intestine. The absorption of these solutes then pulls water through the wall at a faster rate than usual due to the effect of osmosis. Such drinks can promote the intake of everything one needs – sugars for energy, salts to replace those lost in the sweat, and water to match your high sweat rates.

If you are going to use sports drinks during a run, it is important that you get used to them for it is easy to give yourself stomach cramps. My colleagues and I trained for some weeks before the Marathon of the Sands with our special backpacks, going for long runs with the panniers fully loaded. We were using a drink that came in a white powder form which we mixed with water to fill our packs. The habit led to some rather odd looks in the changing rooms from runners who felt that we must be bold drug-users. It also led to an amusing incident after I first received my specially adapted backpack. Running through some woods on an early training outing with my colleagues, who still did not have their drinking system, I stopped for a pee. While I emptied a very full bladder, back turned to my friends, I was also drinking from the backpack straw. Without thinking, I asked the others if they wanted a drink.

'No thanks,' came the reply from Richard. 'I have never had much liking for urine.'

★

The third day was to bring us into contact with full-scale Saharan dunes, and from the start at 8 a.m., we could see them lying in line across our path. After trying to run through the tiny dunes the day before, we knew that this would entail work in a different league, and we were already suffering from cumulative fatigue. Strange gaits also suggested that many of the competitors were carrying sore and blistered feet, and certainly my own were beginning to suffer badly. With the heat and the constant pounding, my feet had swelled up, and the shoes I wore were now tight. Several toe-nails were bruised and, indeed, were later to fall off. The soles were also blistering. It was not much fun until I had got moving, although once under way, it was remarkable how the pain disappeared. Evolution has set us up to ignore repetitive pain signals – a useful coping mechanism for the inevitable injuries of life.

As we ran, hobbled or walked through the first six miles of the day, the undulating obstacles came steadily closer and I began to be astonished by their apparent size. Although I had read of how large dunes could be, I nevertheless thought that it must be an illusion. Yet I was still quite a distance away when the lead runners reached them, and their tiny black-dot figures gave the dunes real scale. Some were hundreds of feet high.

At the checkpoint, I picked up four and a half litres of water which I added to more than one litre which I had left from the previous evening's allowance. Now I felt that I had plenty to cover the remaining distance, sand or no sand. I reckoned I could cross the 11 miles of dunes in about three hours, but on reaching the first dune, I began to recalculate. I had never seen anything like it, and certainly never experienced such a formidable obstacle to progress. Like all the others, it was a magnificent crescent of sand, probably a quarter of a mile wide and rising more than three hundred feet. It lay with the concavity towards me, and both the compass bearing and earlier runners' tracks took me straight towards the middle of it. I found myself entering a huge steep-walled bowl between the two curving arms of the crescent. Intending to maintain the shortest and straightest line, I continued right into the bowl and then attempted to scale the dune directly. Within a few steps, I realised it was hopeless. The slope was steep and so loose that I could only make progress if I tried to power up it using arms and legs simultaneously. But it was impossible to maintain the level of work needed to go on up and, as soon as I slowed, I found myself descending backwards on a fast down escalator. It was like skiing down through soft powder snow. Another approach was needed.

Glancing round, I noticed that the tracks I had followed dispersed to either right or left. Others had tried to scrabble at the face but their marks in the soft sand were nearly filled in. Since I was right in the middle of the bowl, it made little difference which direction I chose, so

I moved left, assuming that the best option must be to thread my way between the dunes. No such luck. The dunes lay in a carefully defined pattern to impede progress. The high points of each line seemed to be staggered so that the distances involved in skirting them were not feasible, and the sand in the gullies between was all very soft. The only way to get through was to ascend one or other arm of the dune crescent, making virtually no ground in the direction I wished to go, before dropping off the top of the dune at the point that would bring me down to the next crescent ridge. It entailed enormous additional distance.

This tactic for dealing with trans-dune travel only became evident when I reached the top of the first dune and looked ahead. There before me lay the desert of the movies – rank upon rank of these huge sand hills extending to the horizon. Some were small, some big, and some gigantic – far bigger than the one upon which I stood. And all were beautiful, having a tremendous purity of line and near perfect sculpted form. In among them I saw the figures of those in front of me, and from what I could see of their tracks, they appeared to be diverging like the branches of a tree. Not only could you not achieve a straight line, but at each dune ascent there was often a choice of 'best route'. Runners were making different choices and the race was rapidly spreading out. Of course, one might have thought that if everybody were following the same compass bearing, the net divergence would be small. But unlike the usual situation of travel on a bearing, when you can set your sights on something distant and then head for it, here there was no distant point on which to focus. The horizon looked the same in all directions, and although dunes varied in size, they all had the same shape and orientation. It was impossible to tell them apart with any certainty. As a result one could only follow the bearing by guesswork and intuition. Some competitors were clearly more intuitive than others.

At first I followed the ill-defined trails of some of the fifty or so runners ahead of me, but it only took a few branchings to discover that I was following what was probably a single set of tracks. Before much further, I disagreed with his or her choice of route and found myself climbing one of these grand dunes without any sign of previous passing. It brought about a strange pleasure, a sense of adventure, although when I descended into the quiet untouched bowl beyond, I also had strong feelings of apprehensive isolation. As I went on, the same anxiety was to recur whenever I was down deep within the sands, but every ten minutes or so I would reach the top of another dune and, if I waited a moment and watched, other figures could be seen emerging between the waves of sand.

A couple of hours passed and it became dreadfully hot. The sun was high enough to eradicate all areas of shade and the sand surfaces lapped up its radiation and then re-transmitted it. Within the still and silent bowls, temperatures were unbelievable. As I descended into them, I could feel the heat scorching my face like the blast from an open oven door. The sand was so hot that it was burning to touch, and this became a serious problem when going uphill, for I still had to use my hands on the steeper parts of some slopes. It was only on the dune summits that there was any relief from the baking conditions.

After a couple of hours I reached a very high dune and spent five minutes looking about. I had no idea how far I had come, nor how much longer I would need to go on. Even taking it easy with my water supply, I had drunk nearly three litres, so I would be hard pushed if the camp was more than two hours away. I could see quite a distance to the south – the rough direction in which we were heading – and even further east and west. Dots of people were everywhere, and I got the impression that many of the figures were standing watching others – as I was – probably trying to answer the same question. There was no information. I appeared to be about in the middle of the spread of competitors on an east/west axis, and was somewhat reassured. As far as distance to go was concerned, I could make no judgement – the dunes went on as far as I could see. If the journey took all afternoon and I arrived beyond the dunes at dusk, it would be hard to find the camp. Even more disconcerting, if it took that long, I would definitely run out of water.

I went on again, and a little later came down to a wider gap between the dunes that I hadn't noticed from my earlier viewpoint. It appeared to be a dry wadi that was running south-west. Although not precisely in the direction I wished to go, it would be much easier to make good speed that way. I could head back east once I exited the dunes. Out in the middle of what was essentially a small valley, I found several sets of fresh footprints and it was obvious that others had had the same idea. I followed them, feeling more optimistic, and even broke into a trot, but the heat was too hot to keep it up for long. The little valley began to wind and then turned further west. I would have to leave it to head south again. I was about to do so when I rounded a bend and came upon an extraordinary sight. Ahead of me, standing in the middle of a small flat area, was a single large tree. In it were two blue-winged rollers.

I was stunned. Since entering the dunes, I had seen no signs of life at all – neither plants nor animals. On previous days we had come across some scrubby bushes and the occasional palm, but this seemed to be some sort of conifer – perhaps a juniper, although I had never seen one so large. It seemed so odd to find it here in the heart of the most inhospitable area we had crossed. The birds were also the first I had seen other than a few scavenging vultures that had turned up at the camp last night.

In the shade of the tree there was other life. Five or six of my fellow runners were resting there, emptying sand from their shoes and redressing their blistered feet. I joined them and we conferred about directions, distances, and water supplies. None of them had any more ideas than I had. During the conversation I noticed something else. Many of the tree's roots were above the ground and, on one side of the tree, there was a deep hollow beneath them. Looking more closely, I realised that a small chamber had been deliberately excavated and in it were a couple of pots and a few bits of cloth and bedding. It was difficult to imagine who would shelter there, but some goat droppings around the tree suggested that some of the Tuareg must come here. Quite where they might be coming from and where they might be going was beyond me.

I set off again after ten minutes. The others had decided to wait through the hottest part of the day before they continued, but I was concerned that nightfall was the real threat. I walked on down the wadi, resisting any further temptation to run. Once back in the dunes, I would be struggling again, and I didn't want to overheat before that. As I suspected, it was only a few hundred yards before the valley turned even further west. I had to take the plunge and began to cross the dunes again.

I was alone in the sands once more. It was 2 p.m. and the sun was as strong as ever. I was down to my last litre of water and, even so, I had not been drinking enough. For the third day my head pounded, and now my vision swam. My mind seemed to wander from unrealistic pleasure in my situation to all the anxieties that could be imagined arising from it. I was not slowing down but crossing the dunes was so much further than I had expected. I felt a strange detachment from reality. Even when my water finally ran out, my intellect seemed to be a disconnected observer of events that were happening to my body. As I became drier and slower, this feeling of detachment became more and more marked.

It was 5 p.m. when I finally struggled to the top of what proved to be the last sand bastion. It was the highest dune of all, and from the summit I saw the huge hill fall away to reach a broad earthen plain. As

I had feared, the camp was not visible. I could not see far to either side for dunes, almost as large as the one on which I stood, stuck out farther on to the plain. Looking right and left as far as I could see and trying to gather where others were heading provided no help. Several figures could be seen as dots on other dunes to the east and west. That did seem to place me somewhere in the middle and suggested that the camp could not be too far.

Looking ahead again, I suddenly spotted that there was a track running parallel with the dunes about half a mile out on the plain. The camp would have to be close to this since the vehicles must have come in this way. Then I noticed figures moving slowly along the track in opposite directions. They would soon meet and pool their information, which might allow them a rational decision. Just then, a better clue presented itself straight from the story books. Over to the right, about a mile away, three large birds were circling. Even if nobody had died, the vultures were surely over the camp. I headed down to the track.

★

The regulation of core temperature in the heat is mainly sub-conscious and under the control of the cells in the base of the brain known as the hypothalamus. The cells receive information about body and environmental temperatures from specialised heat-sensitive nerves in both the skin and selected points deeper in the body, such as the walls of large blood vessels. These nerves are primarily responsive to temperature change rather than the temperature level, firing signals more rapidly when the temperature is climbing quickly. The sensation that the skin is hot is therefore much greater when it is changing between say 20° and 30°C than if it is static at 40°C.

This is strikingly demonstrated by the experience of getting into newly drawn bath water, which almost invariably feels very hot, whatever you have done with the mixer tap. As you touch the surface – probably at around 40°C – the skin in contact with the water rises in temperature quickly, particularly on the toes, which tend to be cooler than any other parts of the body. The toe receptors undergo a huge temperature rise and fire rapidly however tentatively you step in. The brain receives impulses so fast that they are interpreted as: 'WARNING – very hot water'. In reality, of course, after a few moments of forbearance, the water feels perfect or even tepid. Then, as you gingerly venture further, each newly heated region repeats the warning message loudly, inhibiting faster progress.

The cells of the hypothalamus integrate temperature information from the body with their own temperature which is representative of the brain. If they detect that temperatures are on the high side of normal, they trigger physiological responses to increase the rate of heat loss and inform consciousness that it is getting uncomfortably warm. Appropriate behavioural decisions can then be made, such as removing clothing.

The physiological responses to heat include alterations in the blood flow to the skin and the triggering of the sweating mechanism. In the heat, nerves convey signals to open valves at the input end of the skin blood vessel network so that surface blood flow increases. This leads to a corresponding increase in heat that can be lost by convection, conduction and radiation. There are also increases in blood flow to deeper layers of muscle and fat which make them less insulating. More heat is then conducted from the core.

An overheating hypothalamus also triggers sweating via nerves running directly to each gland. In a non-acclimatised individual, the mechanism is activated when core temperature reaches about 38°C, or 1°C above normal. It is very effective as long as the air is not too humid, and with frequent exposure becomes more effective still.

Acclimatisation involves improvements in skin blood flow, the vessels open wider and sooner, but most of the benefit comes from increased sweating. On average, a non-acclimatised person sweats about half a litre per hour, and this can rise to as much as two litres or more in someone fully heat adapted. With the skin changes, heat losses may increase to nearly twenty times normal, resulting in a tremendous change in our capacity to live and work in hot conditions. It goes without saying that such sweating levels put one at risk of dehydration and, on this scale, losses can exceed the capacity for the gut to absorb water to replace them.

High sweat rates cannot be maintained for hours on end, not only due to fluid losses but because of the risk of salt depletion. Acclimatisation, however, does include changes that try to maintain salt balance. Sweat glands consist of groups of cells in the skin which can draw salt and water from the blood and pump it into the hollow centre of the gland from where a duct leads to the surface. The fluid formed by this process is at first extremely salty, similar to the blood itself, but as it passes up the duct, much of the salt is reabsorbed. When sweat emerges on the skin, the salt is much less concentrated. Still, problems can arise if an unacclimatised person is sweating freely. The speed at which sweat flows up the duct can be too fast for effective salt reabsorption and, as a result, when you first enter hot climates, you are at great risk of becoming salt depleted. After acclimatisation, however, the duct cells become far better at picking up salt, and even though the sweat flow may increase to four times normal, the cells still absorb most of the salt that is passing. The kidneys also improve their ability to hold on to salt so that, in the end, a fully acclimatised person in the heat needs no more salt than an unacclimatised individual in temperate conditions.

Acclimatisation can develop changes surprisingly swiftly, especially if exercise accompanies the hot climate, so raising core temperature from within as well as without. High levels of heat tolerance are acquired in just a week or ten days if core temperature rises by more than one degree for at least an hour during that time. Beyond that, further changes tend to be small, although there are some poorly understood long-term adaptations, taking years to develop, whereby the body becomes better at not over-sweating. It is interesting to find that all races have a similar capacity for both short and longer term adaptive changes, so emphasising the closeness of our genetic inheritance. There has simply been too little time for any of us to move far from our heat adapted African ancestors. After leaving a hot environment, the benefits of acclimatisation are lost progressively over just twenty to forty days. It is therefore entirely misleading to hear people speak of their heat tolerance as something that they 'got used to during the war'.

The key to managing the Sahara race was to become fully acclimatised before the event. All the competitors tried to do this, but I and my colleagues had the advantage of working in a research establishment with experimental chambers capable of generating any climate on earth. We ran on a treadmill in conditions similar to those we would meet during the desert race to come, and I topped this off by deliberately wearing more clothing than usual when I went for training runs and by having a hot bath each evening. In fact, the latter manoeuvre can achieve a considerable degree of heat adaptation on its own. It is a trick worth pursuing when going on holiday to a hot region if you want to be affected as little as possible from day one.

The very fact that we could all run in the Marathon of the Sands illustrates the effectiveness of human heat adaptation, but it is perhaps a less dramatic demonstration than that performed by an early Victorian scientist and showman. He used to stun his audience by entering an oven on stage at a temperature of something approaching 150°C. Scantily clad, he then remained inside for more than a quarter of an hour accompanied by a steak that went in with him raw. On his triumphant emergence, he would consume the steak, done to a turn, before his enthusiastic audience.

★

There was no avoiding it – over enthusiastic re-hydration from the night before was triggering urgent bladder signals and they just could not be ignored. Reluctantly I struggled out of my cocoon and winced as I stood up. My feet were sore and degrading badly. I limped over to the latrine. It was cold in the early morning light and there were beads of dew glinting on the outside of our shelter. I glanced at my watch and saw that it was nearly 5 a.m. It was still quite dark, and above the eastern horizon a star, perhaps a planet, was just visible on blue velvet. Beneath, the sky was already changing to a fierce orange glow. It was a harbinger of the heat to come.

Around the camp, other runners were struggling into wakefulness, many clearly suffering with their feet as much as I was. The miles we had already covered had taken their toll and some competitors looked pitiful as they hobbled around. Day 4 was to be the long one, the day that we had all dreaded since signing up for this crazy race. If we feared it before we left home, we feared it even more now. We were in poor shape to run two full marathons.

It had taken me more than seven hours to cross the dunes the day before and I was surprised to find, on finishing, that I remained well up in the field. It turned out that most competitors had run out of water, and many had been out long after dusk. From my sleeping bag, waking frequently to turn sore hips, I had watched anxious officials shining lights into the darkness, and occasionally a flash would be returned if someone was near. As they each approached, a ripple would go round the camp, and people got up to greet them. All the later runners, some of them arriving after midnight, had found themselves crossing a line to loud applause from their weary fellow competitors. I was astonished by this self-sacrificing tradition, but each arrival brought pleasure to everyone. Overcoming hardship together brings closeness, and it was accepted by all that these slower contestants, who had walked for much of the way, were having the toughest race. Those of us with some of the afternoon or evening to add to our night's recovery were lucky. These poor souls would have to be up again and making ready less than six hours after they had arrived.

The first stage of Day 4 was over a flat plain of baked earth and large rocks. I reached the first checkpoint at 9 miles after only an hour and a quarter, but the heat was already beginning to build. While less than confident that my body was back in proper fluid balance after the previous day's events, it was pleasant to find the second stage going so smoothly, and I was comfortable when I reached the second checkpoint at around 17 miles. From there I entered another broad belt of small

dunes. While negotiating these – sometimes running, sometimes walking – I noticed the weather begin to change. A slight breeze increased
swiftly and within minutes I was engulfed in another full-blown sandstorm. The field was already spread out and there were no companions
nearby to form a mutually supporting snake as we had done two days
before. My footsteps were blown into oblivion as soon as I made them.
Fortunately, one of my team fellows, Mike Lean, emerged from the
gloom and we kept close together, following the compass meticulously
and hoping that we would spot the flags which extended on each side
of all the checkpoints to lead contestants in. At last, a little over six hours
into the day, we arrived at the 29-mile mark, checkpoint 3.

Here the route took a turn and, as we left, the wind blew more across
our path rather than against us. Concerns about getting lost also
diminished since we were now following a track which should take us
through the whole of the next 11 miles. We were doing a slow jog,
keeping fatigue at bay by moderating our pace. At one point the track
passed a few buildings which comprised a ramshackle village. They
were made of mud and straw and looked derelict. Although I saw no
people, there were some goats and even one dog around. It seemed that
we were in a generally more hospitable part of the desert. The way
became scrubby, and through the blowing sand I began to see that we
were passing trees of some size. Many of them were similar to the lone
tree of the dunes, and occasionally one stood close to the track. Passing
such a tree on the lee side produced an interesting effect. Since the wind
had started and obliterated the sun, the heat had subsided and we felt
quite comfortable. Out of the wind behind the trees, however, my skin
temperature shot up, and although it only lasted a moment as I jogged
past, it felt like jumping into a sauna. The air was not cool at all – it was
simply dry enough, and the wind strong enough, to extract all the extra
heat I was generating.

Halfway along the 11-mile track, Mike Lean, who was a better
runner than I, disappeared ahead and I was left jogging slowly and alone
through the still dust-filled afternoon. The track was vehicle wide,
slightly sunken, and identified by the absence of scrub or trees. It took
wild swings every now and then to avoid more established dunes, for
the people of the desert have long accepted that you never fight with
the sands. They simply move their roads, their fields and their buildings
whenever the shifting desert dictates. One had to be attentive to spot
where the track would go next. By mid-afternoon, my vigilance had
lapsed.

I was trotting slowly in a daydream, following the vague outline of the route on automatic pilot. My mind was far from the desert, trying to banish the pain and fatigue as I made my way. It was a trick I had adopted in Antarctica, though there I had had less need to concentrate. Suddenly I came back to the present to find my way ahead blocked. In front of me stood a clump of scrub and trees covering the broad side of a large dune and, to my surprise, there was no sign of the track continuing. But I wasn't concerned: obviously the road went round the hill and either way I could just back-track and pick it up. I turned round, and rapidly turned back. The wind had built up and was suddenly blowing in my face. Even with sunglasses on, I found sand poured into my eyes. I knew from my compass that I should now be heading almost with the wind and so, if I cast to right or left along the side of the dune I had to pick up the route again. I decided to head right.

It was a struggle to walk in that direction, for the wind must have been up to something like forty knots and the whipping sand was painful as it scoured my legs and arms. I worked my way sideways, head bent against the blast, looking for the route. After ten minutes, I stopped. The dune was no longer in the way and, as far as I could see – not very far actually – the land in the direction that I should be going was flat. Seeing no track, however, I thought that I must have opted for the wrong side of the dune. I began to retrace my steps – or I would have done had they still been there. The wind had carried away all trace of my passing just moments before.

I was now walking with the wind on my left and the dune to my right back to where I started. Ten minutes beyond, I seemed to be well past the other end of the dune but still with no sign of the track. I began to grow anxious. I had only a couple of litres of water and could not afford to mess about like this. What should I do now? Should I stop and wait for the wind to drop and visibility to improve, or should I go on casting about, using up energy and water? I wasn't sure.

I would first go on a little bit further to the left, but just as I was going to restart my traverse, a flash of red caught my eye. About thirty yards away, back to the right, a runner had just disappeared behind a tree and, as I whipped round, he reappeared heading with the wind. He was either on route or very confident, and I raced off in pursuit. Within thirty seconds, I was on line behind him, and I was also clearly back on the track. I had been standing, pondering my best survival tactic just yards from where I should have been going. It brought home to me how easy it would be to get lost completely. I had been lucky, but another competitor was less fortunate.

We got the news of the lost Italian when we reached checkpoint 4. From there, the finish should have been another 10 miles but as I arrived, I noticed that the administration tent was surrounded by runners, either lying on the ground or standing, stretching and putting on extra clothing. Mike Lean was there along with another of my colleagues, René Nevola. René, the best runner of us all, must have reached here long ago. It made no sense, but he saw me coming.

'Too dangerous to go on,' he replied to my unspoken question. 'We're waiting here for transport.'

I sat down with him. Although the announcement was one of the most welcome things I had ever heard, I was very surprised. There were still a few hours until darkness, and in the race book it was clear that the route onwards continued to follow tracks. It was not like the organisers to let us off so lightly, but I for one was grateful and settled down to wait. As the afternoon wore on towards evening, it began to feel cold sitting in the wind. More competitors trickled in, and the checkpoint took on the appearance of an impromptu campsite, with runners huddled in sleeping bags. Towards dusk the rumours began to fly.

It appeared that a runner was missing. Someone had left checkpoint 3 but had not arrived here at checkpoint 4. He was a first class athlete, and the organisers were most concerned. Unknown to us all, they had been operating a safety system of notifying by radio the next checkpoint as runners departed the one before. In that way they knew the rough whereabouts of every runner on the course, as well as when he or she might be expected at the next checkpoint. Someone had failed to appear, and it was then that they had decided to stop the race.

Rumours began to harden into facts. The mystery runner was Mauro Prosperi, an Italian ex-Olympics Pentathlon champion. René, who was also of Italian parenthood and who spoke the language, had met him and this made it more real for him. You could see his concern. I too felt close to Mauro's situation since I believed I knew exactly what had happened. Nevertheless, although he was probably in for a rough time, I felt sure he would be all right. He might not have enough water, but he would be sensible and simply wait the storm out. All the runners carried flares and a signal mirror, and once the storm had settled, he would be found. The organisers even had a helicopter for just such an emergency. In any case, the race had been designed for safety. The course was out in the open desert but it was deliberately set to run parallel with roads that were never far away. We had been told just the day before that, if we became lost, we should just head south.

There was no possibility that we could miss the southern Moroccan east-west main trunk road.

★

We are designed to starve. Back in the evolutionary past, we could not rely on a constant food supply and so we put on fat in the good times and were able to conserve energy in bad. This propensity is unwelcome for those who wish to remain fashionably slim, but it is a useful adaptation for many peoples of the world and essential for the few that get stuck in a survival situation. Mauro could last for weeks without food, though not without water. Since the time when we were unicellular marine organisms, all processes of life have been geared to operate in a very specific set of liquid-based conditions, and the basic processes have not changed. Adapting to operate in drier circumstances, our evolutionary solution was to carry the right liquid – essentially the same as those primaeval seas – within our cells. To assist in this, we developed regulatory systems to maintain cell constancy, particularly using our kidneys to either excrete or hold on to water in order to match overall input and output, and to vary the loss of salts. Nevertheless, the kidneys cannot excrete no water at all. If you don't have about 700 millilitres a day to spare, you begin to accumulate the poisons of your own metabolism.

Getting lost in the desert is not a good idea, for there the daily water requirement is immense even if you are not running. The 700 mls needed for urine is combined with several litres needed for sweating to stop you overheating. Being acclimatised is only of value if you have plenty of liquid to spare. Mauro had left the last checkpoint with just three litres, and had presumably travelled and drunk part of this before becoming lost. He was therefore at no advantage over anybody plucked from the streets of London and flung into isolation in the southern Sahara without water. Of course his body would fight: it would reduce both urine and sweat rates to a minimum, but as soon as he began to become dehydrated he would risk becoming overheated, especially if he did not act carefully. He needed to combine his intelligence with his physiology in order to survive.

By and large, the human animal prefers behavioural means to limit heat exposure. When the hypothalamus detects a rise in body temperature, it informs consciousness and we begin to feel the discomfort we call getting hot. We then change our behaviour so as to limit heat production, decrease heat absorption from our surroundings and promote faster heat loss. When it is hot, we usually choose to do as little work as possible, and may even lie down. This minimises all muscle efforts, including those that normally work just to keep us upright, and this keeps the levels of heat production below even that of the 100 watt bulb. Lying down also maximises heat loss through convection and evaporation. Air passing over the lower body then becomes warm and humid and no longer rises past the upper body. Any advantage obtained from the horizontal position, however, may be offset if you are lying on hot ground or a mattress that insulates your back. Tropical dwellers found the solution years ago by inventing the hammock, which is a most efficient way of minimising heat production and maximising convective and conductive heat loss. It can even be hung in the shade, with the added touches of a fan and cooled drink.

Most behavioural adjustments to our heat balance are made through adjustments to clothing. Here a series of very complex interactions have some influence, and the choice of best clothing for the heat is not as straightforward as it might appear. You can dress up or dress down.

Dressing up involves protecting oneself from the hot environment through the use of plenty of material to cover the body, providing it with shade from the sun's radiant input, and insulating it from hot surroundings. This is the traditional choice of desert peoples, and so one might think that we would have been wise to follow their example by setting out on our Saharan run swathed from head to foot. Unfortunately, it is only the good way for people who plan to rest through the hottest part of the day, for the approach retains any self generated heat, severely limiting heat losses through sweating.

Dressing down means wearing as little as possible. As a result, you expose yourself to high heat input from the environment but allow sweat and the other heat loss mechanisms to work efficiently. It is fine as long as you have plenty of water, and effective in a modern desert run so long as you are not profligate. For traditional desert dwellers, who must minimise water use at all times, it would be useless. While he was lost without water, it was also inappropriate dress for Mauro Prosperi.

★

The fifth day was free for rest. We all needed to recover, and it would allow time for Mauro to be rescued. It dawned clear and still, and at 6 a.m. the desert quiet was shattered by the whine of helicopter turbines. We were confident he would be back within an hour or two.

People took a leisurely breakfast and after that strolled up to the top of some nearby high dunes to look out over the beauties of the Sahara without the pressure of having to move on. The day passed slowly, with just occasional excitement as the helicopter returned and everyone gathered to greet the lost one. Each time our hopes were raised, they were dashed again – it was only a return for refuelling. Finding Mauro was not as easy as had been thought. The jeeps were despatched to help search on the ground, with some going down to stake out the southern road. Wherever he was, he would surely hear the search helicopter and use his flares or signal mirror when it came near.

At about midday the organisers called for me. They knew I was involved in survival research and were clearly becoming concerned about how long it might take to find him. They wanted to know how long he might last in the desert, but so many factors were involved – the greatest of which was what he decided to do – that I found his chances unpredictable. If he was stupid and made great efforts to travel far and fast by day, his water losses would be immense, his heat stress appalling, and without fluid replacement he would collapse within a few hours. In a couple of days he would be dead. If he was wise, and waited in the shade, covering his body, he could last some days, although he would probably be unconscious after four or five, and then the chances of finding him would be remote. There was also another possibility. The fact that he had not been found on a clear day, lost from a route well marked and with a road running just south of it, was not a good sign. He might have pressed on to the point of collapse and then fallen, only to be buried in the moving sands. If that were the case, he would be dead already.

The question was academic and the organisers had no choice. They must search for at least a week, if not longer, and since they hadn't found him quickly, they needed assistance. The Moroccan authorities were notified, and the Army and Air Force moved in. The local tribes were also told, and many of them began to scour the country on foot. There were nearby, I understood, several small villages of the type that I had noticed the previous day. Unless Mauro had gone on to the point of collapse, there was no reasonable doubt that he would be found, or that he would find his own way to the road or a village. Everybody remained confident.

By evening it was becoming difficult to remain optimistic. The mood in the camp began to sink tangibly low, and by the time everyone tried to settle down for the night, anxiety was added to the many other reasons for sleeping badly. In the middle of the night, darkness was broken by the sounds of weeping. The Italian camp was going into mourning.

Next morning, Mauro had still not been found, but the race was to go on. Not all of us felt easy with that decision, but we all convinced ourselves that it was reasonable. There was nothing of practical use for us to do to help him. We could not all head off into the desert to look without the risk of others being lost. It was better to let the vehicles go on with the search, rather than tying them up with transporting us out of the sands. We would continue on foot towards our final destination. There was only a marathon on Day 6, and a 10-mile run on Day 7. After beating the long day, it felt as if little were left to do.

The reality was different. A marathon should never be taken lightly, and in the desert it is a very long way indeed. The fact that it was a familiar distance was also to have some surprising consequences. Nearly all of us had run marathons and so each man and woman knew roughly how long it should take them in cooler conditions on normal surfaces. That self-knowledge tended to set time targets which threatened self-control, and pushing hard could be dangerous especially as, for the sake of variation, the organisers had included a new physical test. That day, instead of dunes, our run would take us over a range of rocky mountains.

The race started well enough – 10 miles across a boulder-strewn plain, with an occasional wadi to break our rhythm. Yes, there was great pain as raw nerve endings screamed from blistered toes, heels and soles, but our brains were soon quite blind to their message for we had all made the conscious decision to ignore them. Even our general fatigue and stiffness dispersed as muscles warmed up, and soon we had all settled into a tiring but comfortable and familiar action.

As the rocky slopes of the mountains grew nearer, they looked more and more impenetrable. Huge cliffs of red and ochre sandstone merged with scarcely less steep, deeply-fissured slabs above. I wondered how we could climb, let alone run, over such an obstacle, but at the base of the rocks, beyond the first checkpoint and restocking with water, the route turned beneath the cliffs, rounded an outcrop, and entered a steep-walled canyon. It was a strange and wonderful place. Scarce rains from perhaps a million years had carved a way down from the mountains which now allowed us to go up into them. High walls with glowing bands of colour rose on either side, water-carved into Henry Moore sculptures. We followed the floor of the canyon, part sand and part rock, as it twisted and turned, scrambling over boulders that blocked our way. Inside the dark walls it was cool, and I was grateful for this as my body churned out more heat from the effort of ascent. But as I climbed, so did the sun. By the time the canyon widened and I emerged on the upper slopes of the mountains, it was beating down on me ferociously. The last thousand feet to the col were desperate.

I reached the col shortly before midday and found others who had got there before me crouched beneath a small rock overhang, trying to escape the fierce radiation while they drank some water. Few of us had much left. At the checkpoint before the canyon we had been given only three litres and, despite the climb taking less than an hour, our sweat rates had been immense. The ascent had cost me two litres of that water and I could expect no more until well beyond the mountains. It looked like a steep and difficult descent, and none of us could be wasteful.

The view from the top was spectacular. Behind was the undulating land we had covered in the last few days, while ahead was a flat plain with the next checkpoint at around 19 miles well visible. Far beyond, we could even see the finishing camp, shimmering through heat haze. But what really drew the eye lay further still. On the far side of the plain, a slash of emerald green divided it from a further range of arid mountains. Dotted within that shocking band of colour were the stark white buildings of Zahedan. It was the oasis of our finish – the point we would reach tomorrow with the final short leg of our run.

We started our descent, picking our way through boulders, slipping, stumbling, but grateful for the assistance of gravity even if it pummelled our sore knees. It was when we reached the easy terrain of the plain that many of us made a serious mistake. We could see the finish, a distant camp, pitched in the middle of an enormous shimmering lake of mirage. Knowing our marathon times from London, Boston, Paris or elsewhere, and paying too much attention to watches on our wrists, we began to run faster. The sun was up with all its power and it could not forgive this enthusiasm. As we ran across the last ten miles, even the further litre and a half of water provided at the last checkpoint could not prevent our problems. Bodies were churning out waste heat so fast that we simply could not shed it all. Simultaneously the ground, the air and the sun poured more heat into us. Hyperthermia became inevitable.

For me, the sensation was strange but familiar. As I made ever slower progress towards the finish, I found my thinking becoming detached once more from the discomforts of my body. Yet with this detachment there came a deep sense of unease. By the time I crossed the line I was dazed, nauseous, and so very restless that I was unable to sit down in the shade. I ended up pacing around like a man with an insoluble problem. I felt as if I were melting from both within and without and might soon be reduced to a pot of tallow. All the same, feeling hot was a good sign that my brain was still working. It understood the danger that I had put myself in and it was doing its best to let me know how stupid I had been. When seriously overheated, this is not always the case, and in others things were going seriously wrong.

My colleague Richard, the man with no taste for urine, was a younger and better athlete than I, but he turned out to be particularly vulnerable to the heat. He had suffered terribly on many of the earlier days and now lay well behind me in the rankings. This hurt his pride and so he had been determined to run a good time in this marathon. He did, and entered the camp way ahead of me. Yet as he crossed that final unforgiving plain, he ignored the messages from his own body even more than the rest of us. Rather than slow down, he maintained a very fast pace as his core temperature soared. Eventually it reached the point at which the control systems began to fail, and here he became exposed to an extraordinary design fault in our evolution – a fault that has caused many deaths in the past.

★

When Richard entered the camp he was close to collapse. Recognising that something was wrong, officials took him straight to the medical tent. There he joined others suffering from a wide variety of problems; some were hot and dehydrated, while many suffered from painful feet, muscles or tendons. Many of the heat victims appeared to be in a worse state than Richard – sweating, flushed all over, and complaining vigorously while calling for water. Richard said little. If anything he felt and looked cool, and it was easy for the young French medics to ignore him. Although they had all attended a special course to teach them about the sort of problems they might meet in this unusual race, the course had been brief and held two months back. Now they had forgotten a small but important point. When people overheat they usually suffer from a condition known as heat exhaustion, which is largely due to salt and fluid depletion. They feel sick and weak, but continue to complain of the heat, and when examined they are flushed and sweaty with the fast pulse and low blood pressure of dehydration. But, if left untreated and they become hotter still, things can change. They go on to get heat stroke and behave quite differently.

With heat stroke the body temperature rises so far that the control systems become wayward and inform the brain that the body is too cold instead of too hot. As they do so, not only do attempts to keep the body cool cease but the mechanisms designed to maintain body heat are activated. Victims suddenly feel cold, shiver and their skin becomes pale and dry. These are changes that can spell disaster for a patient in an intensive care unit, let alone for a man who has just

1, 2 At 72, Helen Klein undertook the first Eco-Challenge – a 300-mile multi-sport adventure race designed to test accepted physical limits with wading and swimming through miles of flooded canyons as well as horse-riding, running, white-water rafting, rock climbing and canoeing.

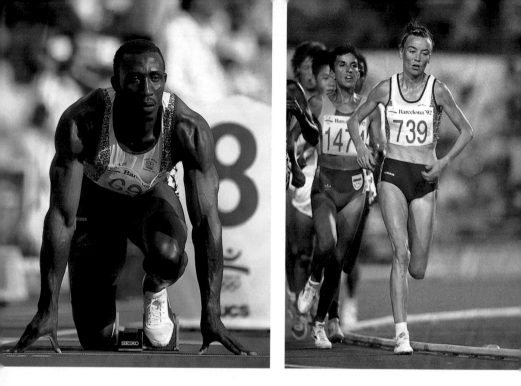

3, 4 Differences in muscularity – Linford Christie, a typical anaerobic power
athlete, and Liz McColgan, a lean oxygen burning endurance runner.

5. Thousands of ordinary people participating in the London Marathon
testify to the endurance capacity within us all.

6. Sir Ranulph Fiennes during the crossing of Antarctica – a 95-day expedition on foot which tested the limits of human endurance.

7. The journey came close to a premature end when Mike Stroud and his sledge went into a crevasse on Day 3.

8. Approaching camp during the 130-mile Marathon of the Sands.
The ability to run in the Sahara heat illustrates our
extraordinary physiological capacity to cope.

9. Chaos in the medical tent at the end of a day running in the Sahara.

10. We have far less physiological capability to cope with the cold, but can use clothing and heat generation from hard work to protect ourselves.

11. Mike Stroud with Emperor penguins in the Antarctic winter darkness.

12, 13. We share 98.4 per cent of our genes with the chimpanzee and have bodies designed for regular exercise and a largely vegetarian diet, but modern lifestyles can lead to obesity and heart disease.

14. The British Columbia Eco-Challenge team – Ranulph Fiennes, David Smith, Rebecca Stephens, Vic Stroud, Mike Stroud.

15. Crossing a river using a Tyrolean traverse.

16. Mike Stroud with his 70-year-old father
at the end of their race. It proved that age need not be a bar
to physically demanding pursuits.

17. Crossing a high ice-field in the British Columbia race.

run a marathon across the Sahara and is lying in the corner of a tent.

If the victim has progressed to heat stroke through the more usual heat exhaustion, there will be strong clues to their condition, but sometimes it is possible to get heat stroke without going through the heat exhaustion stage. If you heat up quickly enough, or drink adequately while under great heat stress, the controller fails while you are still quite reasonably hydrated, and this is exactly what had happened to Richard. His core temperature had probably been quite reasonable until he reached the plain and he had felt fine, maintaining a fair fluid intake. Then, when he stepped up the pace, his core temperature rose swiftly. Entering the camp less than an hour later, his temperature control had failed, before very much fluid had been lost. By the time a doctor examined him, he had also been encouraged to drink and was marginally rehydrated. He was not thirsty, did not complain of being too hot, did not appear to be dehydrated or sweating and had a relatively normal pulse and blood pressure. Certainly he looked unwell and appeared pale and vague, but nevertheless he seemed much better than many around him. The doctor told him to lie down and said she would attend to his sore feet later.

Richard was critically ill. Over the next couple of hours his temperature control continued to malfunction and slowly he became worse. He began to feel increasingly cold and his automatic anti-cooling systems became more active. They cut the blood supply to his skin still further, and the command went out for his muscles to start heat production. Noticing that he was lying covered in goose pimples and shivering, the doctor attending him became more concerned. She thought he was getting hypothermic and sent for his sleeping bag. A nurse arrived at our shelter to look for it, but this was the last thing he needed.

After two hours in the camp I was beginning to feel better and had just cooked some supper. Hearing of Richard's predicament, I left the meal and went along to see if I could help – more with a mind to make encouraging noises than to instigate any medical remedy. When I saw him, I was horrified. From several years' research on survival I knew the dangers of heat all too well. A shivering man in a hot environment rang all alarm bells, and instead of wrapping Richard up in a warm sleeping bag, I quickly took his body temperature. Although only an under-arm reading, the thermometer rose above 40°C, and it was nearly three hours since he had finished the race.

I called for water urgently. It would have been better to immerse him completely – not in a cold bath, which would have shut down skin blood flow further, but in cool water that would extract a lot of heat yet allow blood to continue coming to the surface. Here, however, a bath was out of the question. All we had were the bottles of mineral water dispensed for drinking. They would have to do. While Richard complained bitterly – mumbling about being cold from chattering lips – we soaked him and fanned him as best we could. It would be touch or go. He really needed hospital care, for with shivering putting heat production up to many times the normal, and sweating and other skin losses shut down, breaking the vicious cycle of heat overload can prove impossible. Sometimes it is necessary to paralyse and ventilate such a patient to stop them making more heat, and even then the outcome can be disastrous – an overheated brain may never recover from the insult. Here, as we tried our meagre best, I feared we would be unsuccessful, but then things changed. Richard began to complain of the heat and started to sweat. I sighed with heartfelt relief.

★

We all wanted to complete the last day's run as fast as possible. It meant additional pain for feet, legs and mind, but through it all the green band and the white houses drew closer, and we knew that there food and rest awaited us. We had worked hard and were about to be rewarded.

To make the finish more dramatic, the organisers set the race off in three groups at fifteen minute intervals, with the slowest first. I was in the middle group, and so caught up with others and was caught up myself. Here I saw runners from previous years handing on a fine tradition. Whether passing or being passed, it was convention to reach out and touch hands. It re-emphasised the feeling of unity that had been so much a part of the race.

About two miles out from the town, we reached a green field and then a river. The river was broad and fast, carrying water down from the Atlas snows, the source of irrigation for the swathe that cut through the desert colours. There was a bridge, but none of us seemed to find it. Instead the line of runners plunged into the water, wading across waist deep, encouraged by the whoops of local children who had come out to wave us in. The water was cold as ice but oh so lovely. We had run without washing for a week and were filthy and sore. It was a great sensation.

Beyond the river we entered narrow unmade lanes that wound through more fields and passed scattered white buildings. Then into the town proper, and we began to climb one last lung-bursting, leg-crunching hill. At the top, from a roundabout filled with flowers, the main street ran a mile or so towards a blue-tiled, domed mosque. Both sides of the street were lined with people – men, women and children – who clapped and cheered us on. I doubt that I or many of the others have ever run the final mile of a long distance race so fast. Psychology is what it is all about, and it would be difficult ever to feel better than on that final few hundred yards. As we reached the finishing arch we were met by a band playing strange pipes and drums. We cried with joy.

★

The race was over and there was a great feeling of satisfaction, but it was not long before the joy faded and was replaced by depression. The death of Mauro cast a long shadow over us all, and I think we all felt that it was time for the Marathon of the Sands to be laid to rest. To lose someone while climbing or motor racing seems acceptable. The participants in these sports make their choices while aware of the dangers. Running races should be safe, and nothing was worth the loss. As we left Morocco and headed for our respective nations, the thought of Mauro's body buried in the sand destroyed most of what we had gained from our participation. It was with sadness rather than satisfaction that we finally reached home.

Then, after being back in England for ten days, I heard some extraordinary news. Mauro was alive, unwell and in hospital, but expected to survive. The news was from the press and I simply could not understand it. His survival after so long did not seem possible. It transpired that after wandering for nine days he was found in Algeria, more than sixty miles from where he had disappeared. For me, the story remains incomprehensible. How did he get to Algeria, south of where he was lost, without noticing either the main tarmac highway or the border fence and its associated tracks? Stranger still, how do you move for nine days in the Sahara without water?

Mauro claimed to have lived on plants and drunk his own urine, but the former would provide little hydration and the latter could only make things worse. While in no way wishing to cast doubt on his story, I can state with confidence that when the body gets dehydrated, it becomes phenomenally good at conserving water. Within hours, the urine is so concentrated that to drink it can only do harm. The salts and metabolic products within it would immediately demand fluid to flush them out, and every litre drunk would need more than a litre of water to get rid of it again. It cannot work. If you do ever have the urge to drink your own urine – don't. For that matter, just as Richard resisted the idea in training, don't drink anybody else's either!

SEVEN

The Heart of Darkness

ANTARCTICA . . . Not those summer months of my crossing with Ran Fiennes but mid-winter on the earlier 'Footsteps of Scott' expedition. Roger Mear, Gareth Wood and I were ten days out and struggling through frozen darkness, relieved only by the writhing serpents of luminescent green aurora. The sledges we towed behind us weighed just 50 kilos (110 pounds), nothing like as much as I would pull on the later trans-Antarctic trip, but they felt just as heavy. It was so cold that the snow underfoot was more like non-slip sand and every step we took drained all of our effort. In eight hours of pulling we could not make eight miles, and we were utterly exhausted.

Only seventy-five miles separated our winter hut at Cape Evans from our goal at Cape Crozier, but that was still far away. So slowly were we moving, we became anxious about running out of food for the return journey, yet none of us felt like turning back. The Cape was not just a point on the map; it was the home of the Emperor penguins – a place steeped in Polar history.

Although the pulling was hard, it was at the end of each day's efforts that our real difficulties began. The work of hauling kept us warm, and with warmth we were safe. Slow down or stop and the most bitter weather on earth was out to get us. The Antarctic night can freeze your skin and chill your body in just minutes if you have no shelter. Ours was the tent, but it took time to erect it and tie it down securely enough to withstand a big storm of gale force winds, sweeping across the ice-shelf and carrying all before them.

By the time we had finished pitching camp and dug snow to pile on the tent valance, all warmth we had generated while pulling the sledges had drained from our bodies and was lost to the Polar night. Speed was the key to survival. Inside the tent it was not warm but at least it was tolerable. If the stove was on and we kept the pressure lamp going, we could cook, eat and rest, and while we did so our clothing would begin to defrost. We had to brush it down before the ice melted. Our garments would be rigid in the morning, and it would be a struggle to get them on.

Morning is hardly the right word. Morning denotes sunshine or at least some light to lift one to a new day's work. Here, two months after the sun last set, only our watch alarms told us that it was time to rise. It was pitch dark when we stirred and showers of hoar frost fell from what had become overnight a crystal dome. The vapour from our breath had frozen, precipitating on the tent fabric in myriads of growing geometric forms. As they fell, they went down the back of our garments and added to the chill of awakening. It was so cold that our hands would get frostbitten while just lighting the lamp. Crawling from our sleeping bags and into more clothing was another race. But then the tent would warm up a little and breakfast could be taken at leisure. It often was. None of us were keen to take the next step. The time between exiting the tent and getting away with the sledges would be ten minutes. It was ten minutes too long.

Exposed to temperatures of below minus 60°C, even our thick down duvet jackets were useless. Only after we started the heavy work, would the warmth generated by our muscles offset the heat sucked from us. The tent had to be cleared, swept out and then collapsed, and our sledges packed up and made weather-tight. Perhaps worst of all, we needed to answer calls of nature – an awful experience in the bitter darkness. Gloves were a hindrance, but if we removed even one of our mitts to seek dexterity for a knot or a zip, our fingers would begin to freeze in seconds. I could feel the ice crystals form within them, and even if I was careful, by the time we were ready to depart, both my hands and feet would be screaming with pain. Sometimes beyond pain, an anxious thirty minutes of numbness would follow before we knew if they were actually damaged. Each of us sported several blackened and blistered fingers and toes.

When we set off, it was not entirely dark – our 'day' was timed to have the moon as our companion. It shone with glory, and its pale grey light added to the auroral flickerings above, and both were reflected from every ice surface. They allowed us to see something of the way ahead, although there was not much to see. Out in the 'Windless Bight' – the huge bay that we were crossing – the shores of Ross Island were just a dark shape rising ten miles to the north. We travelled on a flat and featureless ice-shelf that stretched hundreds of miles from here to the Trans-Antarctic mountains which were all that stood between us and the South Pole. From that direction, the vicious storms might come – the winds that had so nearly killed our only forbears.

★

Humans can withstand the coldest of environments as long as they continue to move and are well clothed, but to go out in the Polar mid-winter is to expose yourself to the severest conditions our planet has to offer. The fact that we were there at all provides an example of mankind's enormous flexibility. How did we cope with such conditions?

Africa 100,000 years ago was a place that was mainly warm and was certainly never cold for prolonged periods. The ancestral populations that moved from there to areas such as Europe must therefore have been poorly adapted to the cold, and survived by using behavioural means such as clothing and lighting fires. Today we use essentially the same techniques. Even though humans reached the far north and cold climes in pre-history, it was only in the last ten thousand years or so. Just as there has been no time for any race to lose their evolutionary adaptations to heat, there has been even less time to develop new physiological defences against the cold. That is not to say we have none, but if we go to extremes we need our intellect to extend them.

★

Apsley Cherry-Garrard's classic book, *The Worst Journey in the World,* is one of the greatest Polar epics ever written. A week after mid-winter in 1911, three men from Scott's expedition set off to seek the eggs of the Emperor penguin – a bird that was then a mystery to science. The Emperors had been discovered by members of Scott's first expedition in the spring of 1903, when they had travelled along the southern side of Ross Island in twenty-four-hour sunshine. At Cape Crozier they found penguins of a size that they could scarcely believe, but although they collected a few adult specimens to take back to their base, they could find no eggs or newborn chicks. All the young were already many months old, and Scott's men were baffled by this timing. The other more familiar and smaller species of penguins were only just hatching their eggs, yet the chicks of these enormous birds were already fully fledged. It seemed impossible, but they must have hatched out in winter. The colony must stay in the Antarctic through the months of darkness when, with the sea thickly frozen for hundreds of miles, there would be no food available. How could they survive?

The three men of the 1911 winter expedition aimed to find out, but the journey became a living nightmare. The cold of the Antarctic winter is scarcely imaginable, and as they walked through the darkness, they were wearing little more than English tweeds with an extra woolly. Their suffering began quickly. Their clothing rapidly accumulated water from their breath and this then froze. Even their reindeer sleeping bags turned into rigid frozen boards, and their nights became as bad as their dark-filled days. But it was not just the inadequacy of their garments that was to bring them close to death. The Antarctic is the windiest place on Earth, and it was the teeth of the wind that so nearly closed around their hearts.

The outward journey went fairly well, although they too were surprised by how hard it was to pull the sledges, and it took longer than they had anticipated. Over a couple of weeks they made only slow progress across the 'Windless Bight', taking the shortest route eastwards rather than following the line of the coast. Then they picked up land at a low rocky point and knew that it would not be far to Cape Crozier. To escape from the confines of their tiny tent and create a shelter that they felt would be stronger if the winds came, they built an igloo – not made of ice but formed from a low circular wall of rocks with snow pushed into cracks as mortar. A tarpaulin was used for the roof, spread across the stones and anchored down with heavy rocks. Inside it seemed both strong and weather-proof, and they made their plans for the trip to gain the eggs. Outside they pitched their tent to give them extra room.

The next day they left the igloo and their sledges and travelled straight across the base of the Cape to reach the place where the penguins had been found back in 1903. In complete darkness they found the Emperors once more, but not what they had expected. Instead of a mix of sexes, only the males stood in groups on the ice, backs to the weather and huddling together for warmth. The outer birds were obviously at great risk, but as the explorers watched, they saw the strength of altruism. After a few minutes exposed to the worst of the elements, the outer animals moved back into the middle of the huddles to recover, while other birds accepted their turn in the breach. And all the time that this went on, each bird carried an egg upon its feet, holding the precious future of their species up and away from contact with the ice, insulating it with a warm flap of fat and feathers.

The men made their unique observations and returned to their shelter, taking with them a few of the precious eggs. During their journey back, the weather began to break and visibility became so poor that they were lucky to find their igloo before the full force of the storm hit them. The hurricane-force wind attacked with violent fury from the mountains across the ice-shelf. As the hours went by, the roof tarpaulin of the igloo began to pull away. Outside, their tent was also ripped from its guy lines and disappeared into the maelstrom. The men could do nothing. They huddled deeper into their frozen bags, trapped by cold that they knew could not be faced. If they got out and tried to repair the shelter or find their tent, it would kill them in no time. They were still lying helplessly when, with a final wrench, their roof departed. They now had no shelter at all for the long journey home and, as they lay shivering in the continuing furore, each man must have thought that he would now perish.

Yet instead of death, the outcome was a miracle. The storm abated the following day and the men decided that, rather than die where they lay, they would go out and search for their shelter. Far from their search becoming a passage to oblivion, it was only a short time before they literally stumbled across the tent in the darkness. Instead of being blown far out to sea, it had been caught on an icy outcrop less than a mile from where it had been torn away. The worst the wind could offer had not freed it again, and once more they had a shelter.

It was not to be the end of their troubles, for they had little food and fuel remaining, but they survived and brought the eggs back to their hut. They had probably gone through more cold than anyone who had ever lived to tell the tale, although two of the three were then to die when they accompanied Scott on his fateful journey to the South Pole during the following summer.

Environments as hostile as Antarctica cannot be faced without danger, yet the winter journey proved that man can travel under the worst conditions our planet has to offer.

★

As with the heat, temperature regulation in the cold is under the control of the hypothalamus, the thermostat of the body, acting when cooling occurs to diminish heat losses and increase heat production. Many of the cold sensors lie in the skin, adjacent to the heat sensors, and they too are much more sensitive to change in temperature than to the temperature itself. A rapid change sends the most intense signals to the brain, explaining why the toe-dip in the swimming pool gives the strong message: 'This is really cold'. It is exactly the same as the toe in the bath warning, but this time leaving you wandering around the pool-side, trying to look casual rather than gingerly advancing into the tub.

Perhaps embarrassed, you then pluck up the courage to jump into the pool, where once again you are given a dramatic demonstration of the signal strength resulting from rapid temperature change. As you splash into the water, heat is conducted so fast that skin sensors all fire wildly. The hypothalamus interprets this as water so near to fatally cold that a second message is passed to consciousness along the lines of 'This is bloody FREEZING!' You have to fight the desire to get straight back out and run for the changing rooms, but bear with it for a minute or two and all changes. The temperatures of the skin and pool tend to equilibrate and, even though the absolute temperature of the sensors is now lower than ever, their frenzied firing dies away. The water feels comfortably warm and the fuss is over. With almost no hint of malice, you call to your friends to 'Come on in, the water's lovely.'

The hypothalamus is not solely reliant upon skin signals. It also receives impulses from deeper sensors elsewhere and some of its own cells register the temperature of the blood flowing around them. It is the integration of these inputs that informs consciousness whether we are cold and triggers both behavioural and physiological responses.

One of the first things we do as we begin to grow cold is to start moving around. We either get on with the job at hand or add pointless vigorous movements to our actions, such as jumping up and down. Alternatively, we alter our clothing insulation, although the interactions between environment, clothing and activity are remarkably complex.

The effectiveness of your clothing is usually at its best when you are stationary, for in addition to the protection provided by layers of air trapped within the garments, insulation also comes from the layer of still air over the outer surface. Nevertheless, keeping still is rarely the answer in cold conditions and even at a modest plus 5°C with no wind, the insulation needed to stay warm when completely still is equivalent to about five times the average set of indoor working clothes – impractical to wear in most circumstances.

When you start moving, cold air is forced both over and through your clothing and much of the insulation is destroyed. However, this problem is usually more than offset by the increases in your heat production from working muscles, especially as cold weather clothing restricts movement, and walking over snow is also especially taxing. Well wrapped up, you often find yourself too hot, and this has even been demonstrated by men living and working in research bases in Antarctica. Overall, in contrast to the enormous amount of clothing needed to keep you warm while stationary at plus 5°C, the equivalent insulation of normal office dress will do the job at minus 18°C if you are working moderately hard.

Of course, there is more to the weather than the temperature alone and more to clothing than just insulation. During a stay in Antarctica in the 1950s, Admiral Byrd invented the concept of windchill, with which we are now all too familiar. His experiments were simplistic in the extreme, most of them involving observations of the time it took for empty baked-bean tins filled with water to freeze under different wind and temperature conditions. Still, the results were striking. For example, a temperature of plus 5°C with a stiff wind of 25 m.p.h. is equal in cooling power to a still air temperature of below minus 50°C.

The use of such windchill figures can make cold windy days sound highly dramatic, and they are widely quoted in modern television weather broadcasts. In reality this is ridiculous. We behave differently to baked-bean cans, not least because we usually wear clothes on cold windy days. These protect us from some of the additional cooling, and so to take windchill as the effective temperature is as foolish as ignoring the effects of wind altogether. During our Antarctic winter journey, the windchill temperature sometimes fell to less than minus 100°C on days when winds were combined with still air temperatures in the minus forties. Yet it did not feel half as cold on those days as it did in black, fog-filled stillness when temperatures dropped as low as minus 67°C.

At the same time, wind effects cannot be ignored. Exposed hands have fingers that behave more like bean cans than the rest of the body, and a windy day in Antarctica will freeze bare skin in a matter of seconds. The reality is that the cooling power and dangers of the environment are somewhere in between the still air temperature and the more dramatic windchill. You must bear this in mind if you are to operate safely.

★

Through the seventy-five years that followed Cherry-Garrard and his two companions' *Worst Journey,* nobody repeated any significant Antarctic winter venture. Then, in the autumn of 1985, we arrived at Cape Evans with the 'Footsteps of Scott' expedition. As with Scott's 1911/12 venture, the overall aim of the 'Footsteps' team was the journey to the South Pole in the long Polar summer of the following year, though the freezing of the seas around the continent had forced us, like Scott, to come down to Antarctica early and spend the winter waiting. Living in a hut through the winter on the same beach as Scott's team had done, it was inevitable that we should attempt to repeat their mid-winter journey. It would be both an historical re-enactment and an ultimate test of the preparations for the coming South Pole trip. If you can live, work and travel in the Antarctic winter, you can do so in any circumstances.

The three of us had therefore set off from our winter hut at Cape Evans on the same day of the year as the original trio. After travelling south and crossing Discovery point, we had gained the ice-shelf to strike eastwards out and across the Windless Bight. Now, in the moonlight ahead, we could just discern a low line in the darkness. It was the rocky promontory of history, and we knew that there we would find the remnants of the circle of stones in which our predecessors had faced their mortality. We approached in awed silence.

★

Modern thinking about clothing has gone down the road of using multiple layers to achieve protection, with flexibility to cope with a variety of activity and weather conditions. The addition of water to the cooling equation can be as dramatic as the addition of wind, since water evaporation from the surface will extract enormous amounts of heat, and any water penetrating the garments will much reduce their insulating capacity. In order to survive in bad weather, we usually need windproofing, water proofing and insulation, and a huge range of garments has been developed to meet these requirements. The choice can be bewildering when deciding what to wear, especially if you plan to walk through the Polar night.

In the Antarctic, as in other very cold regions, rain or melting snow is not usually an issue; it is actually drier there than in the central Sahara. Clothing need not therefore be waterproof in the way it would be for more normal winter conditions. Even so, while pulling sledges, our bodies would give off water in both sweat and breath. There was therefore a risk of condensation and wetting of the clothing from the inside. This had been a source of terrible problems during the original journey when their clothing became rigid. To avoid the same difficulties, we had our garments specially made to be waterproofed on the inside. Our sleeping bags were similarly protected, and even one of our layers of socks was waterproof, protecting our boot insulation from the sweating of our feet. The proofed clothing also allowed us to choose duck-down for insulation which, despite modern innovation, remains the warmest material for weight as long as it stays dry.

★

It was colder than ever and our thermometer had gone off the scale. After our return to base, we would hear that temperatures had dropped to minus 67°C, but now we only knew that it was dire. It was difficult to make things out as we drew nearer to Igloo Spur, for along with the darkness we were enveloped in thin mist. We guessed that it must have been emanating from an open tide-crack on the other side of the

Crozier peninsula, for nowhere else would there be a water surface to generate it. In such cold conditions, mist must be a rarity indeed.

Our plan was not to stay at the igloo, but to move on to Cape Crozier, leaving two out of our three sledges behind. The way ahead was largely over the rough rock of the Cape and it would be difficult to pull the sledges without damaging them for the forthcoming South Pole journey. The third sledge was a rough and hardy pulk made of indestructible moulded plastic and equipped with metal runners. The three of us would go on pulling it together with just a few days' supply of food and fuel. But before going on to the Cape, ten miles away, we wished to visit the shrine of Polar exploration. And so, as we reached Igloo Spur, we set up camp in order to warm up both ourselves and our cameras.

The way to the igloo was down a slight slope and we kept to the shallow ridge line as we descended. The low stone circle was only a couple of hundred yards away and we found it easily, spending an hour there photographing and filming. It was a moving experience but it left us chilled to the bone. By the time we set off back to our tent, we were badly in need of shelter, the stove and a hot drink.

As we walked back up the ridge it became quite difficult to see anything. A slight breeze had risen and more mist had been blown across the Cape, but the rock spur was only a few yards wide and we felt confident that we could not miss the tent. We were wrong. We found nothing, just an empty snow slope above the spot where the rocky ridge petered out. Puzzled, we turned back and made another sweep downwards, but once more to no avail. We did not even return to the igloo but went right down to the end of the rock and then on to the flat ice-shelf without coming across any familiar features. Obviously the layout of the land was more complicated than we had thought. We paused and considered our strategy.

With our head torches we could see each other when we were quite far apart. The obvious thing to do was to form a broad line and then sweep back up the hillside so that we could not miss either tent or igloo. We spaced ourselves about fifty yards apart and climbed once more – with no success. The mist was getting thicker and it became hard to discern rock from snow and up from down. It seemed impossible that we should not find our camp. Using a compass to move at right angles to our first line by about a hundred yards, we spread out again to work back down the slope. Still no tent, no igloo, and this time no rock. We were truly becoming disorientated, and we each began to feel rising panic, but self control was the only answer. If we were logical, we had to be able to find our things even if it took some time. We had to quarter the whole area, making the search square bigger and bigger until we met success.

It was now obvious that we had made a fundamental mistake when we first left the igloo. Perhaps it had something to do with the reading of the slope, which can be notoriously difficult in whiteout conditions. If we had taken a spur off the promontory as we headed back towards our camp, that might have led us in any direction. The best thing to do now was to ignore any pre-conceptions about the terrain and to simply search in an organised fashion. The tent could not be far away.

In the end, we quartered the ground for the best part of another hour of rising fear before there was a sudden cry from Gareth.

'The igloo – I'm at the igloo.' Roger and I hurried towards his bobbing light.

He was right, we were back at the stone circle and the tent was close if we took the right ridge line. We then found our camp quite easily. Our mistake must have been made through lack of concentration. It is often such mistakes that kill.

★

The choice of clothing for our winter Polar journey would not be effective in less extreme conditions where a good outer waterproof layer is needed to stop rain or melting snow from getting in. This can be either a totally impermeable fabric, keeping out both water and water vapour, or a vapour permeable material such as *Goretex*.

The vapour permeable waterproofs have microscopic pores, large enough to let water vapour pass, but too small to allow liquid through. Their aim is to limit condensation inside the garment building up from vapour coming off the body. When working hard, inner garments may become as damp from condensation as they would if it were raining and you had no waterproof. However, vapour permeable garments are expensive and even the best will not let much vapour through when it is cold. At low temperatures, water vapour condenses or even freezes when it first meets the inner surface of a jacket and, once condensed, it will no longer be able to penetrate. In winter cold, the specialist vapour permeable jackets are really little better than their cheaper non-breathable alternatives.

Less extreme cold also influences the best choice of insulation. Down is useless when wet, and so is best avoided unless in conditions where all water is guaranteed to be frozen. If not, duvet clothing and sleeping bags are better filled with modern synthetic materials which, although heavier than down, are fairly warm when wet and dry out quickly. Artificial fleece jackets also absorb less water than woollen jumpers, and polypropylene underwear does not become damp like its older cotton counterparts.

Another important consideration is what to wear on your head. When you are well wrapped up in cold conditions, as much as ninety per cent of your remaining heat losses can come from your scalp, and so putting on or taking off a hat can change thermal equilibrium substantially. Without the problems of fiddling with buttons or zips, you can regulate you overall comfort and may even influence thermal balance in distant parts of the body. If you have cold hands, you should put on your hat.

★

After finding our tent, we thawed out and ate a meal, but soon it was time to move on. Despite the fact that we had already been up and out for more than twelve hours, the Igloo Spur was far too exposed for safe camping – as our predecessors had found to their cost. We packed up, anchored down two sledges that we would pick up on our return, and set off pulling the rough pulk between the three of us.

On the broken ground we kept stumbling into each other and, cold and fatigued, none of us were in a mood for tolerance. Tempers frayed as we argued about the best route and we spent some time quite lost. The first sign that we had reached the seaward side of Ross Island was ground littered with the corpses of frozen penguin chicks. These were not Emperors but the smaller Adelies which dwelt in the summer months on the hills immediately behind the Cape. It was difficult to believe that the Emperors could really be here in the heart of winter's darkness until we heard through the darkness eerie, distant, echoing calls.

We stayed the night on the Cape before looking for the penguins in earnest. Armed with head torches, we descended from the land and searched along the cliff where the ice-shelf met the frozen sea. It was a difficult task, for the shelf was riven with deep cracked inlets, in any of which the birds might be sheltering. After hours of searching we found, not a host of birds, but just a couple of small groups standing in the darkness. Although few in number, I was not disappointed when I saw them. Disturbed by our approach, they turned and dark backs which had faced the deep cold were replaced by white breasts that shone in the light from our torches. From their curved stiletto beaks rose haunting melancholy calls which echoed from the walls of the ice-cliffs behind them. We were in the court of the Emperors of Antarctica and the trumpets had sounded. It was an extraordinary moment and a great privilege to be there.

★

Our physiological responses to cold are limited to two mechanisms: the first involves cutting down heat losses and the second increasing our heat production. Heat losses are reduced by shutting down warm blood flow to the skin which limits convection, conduction, and radiation. Furthermore, a simultaneous decline in the blood flow to deeper layers of fat, and even to superficial muscle, helps to prevent heat being lost from deeper, warmer regions.

The insulation we gain from the fat under our skin is one of the most unusual aspects of our evolution. All our primate cousins – near and far – rely on warmth from fur, and there is considerable debate as to when and why we went down a different road. Conventional wisdom suggests that the change improved our heat tolerance, since bare skin permits efficient heat loss. On the other hand, fur grants considerable protection from the rays of the sun in hot climates and if it had no advantages, other tropical animals would also be bare skinned.

A striking alternative suggestion is that we developed sub-cutaneous fat as an adaptation to life in the water – an idea fundamental to the 'marine theory' of human evolution. I have mentioned previously the possible drives that brought our ancestors to their feet – the advantages of reaching for fruit, protection from the midday heat, or the need to move fast and efficiently over open plains. At first sight, all these suggestions seem credible but the development of upright posture must have been a slow process, and any change in anatomy or behaviour can only evolve if it confers significant survival advantages at every step on the way. This calls into question the conventional explanations of our uprightness. If becoming fully erect took many generations, there must have been a prolonged period when our ancestors were only partly erect – a posture which would grant few of the advantages that being upright would eventually confer and many disadvantages. Moving half bent while losing your ability to move on all fours would seem like a recipe for disaster.

The 'marine theory' of evolution attempts to get round these problems with a radical re-interpretation. It suggests that instead of a move on to the savannah dividing the common ancestors of chimps and man, the separation was much more definitive – a change in sea level around five to seven million years ago that led to our own ancestors becoming trapped on an island located roughly where the southern uplands of Ethiopia are today. As a result, our ancestral group spent much of the time living in and around the water and, when danger threatened, they simply walked out as far as they could go. They would also have spent long periods wading in search of seafoods, and so had good cause to come to their feet. The theory is attractive because it can cite progressive advantage to the change of posture, especially as the buoyancy of the water would have helped us not to fall over when we were still less than good at walking upright.

An integral part of the marine theory also suggests that, if we spent long periods in the water, we would need protection from cooling, and that the development of sub-cutaneous fat insulation was effective for this. Certainly the only other mammals that have a skin structure similar to our own – whales, manatees, hippos, and elephants – are either fully aquatic or spend a great deal of time immersed. Yet, however tempting, the ideas remain highly contentious, not least because there is no real fossil evidence to support it.

Whatever the origins of the fat layer beneath our skin, the insulation it provides is very dependant upon its thickness. Due to hormonal influences, females tend to have more than males and this may reflect a different evolutionary need, although not necessarily for its insulating qualities. It may be that women may have bigger sub-cutaneous fat stores to help see them through the demands of pregnancy. Nevertheless, it raises an interesting question: would women be better suited for Polar exploration than men?

This proposal was once made by a female correspondent writing in response to a newspaper article I had published on the science of Polar travels. She pointed out that, despite all my interesting observations concerning hardships, women might have done better. Not only were their body fat stores larger, but they had a greater resistance to discomfort and cold. She qualified the point, however, with one penetrating observation: '. . . since all Polar journeys are prime examples of the heroically daft,' she wrote, 'what makes me think that where men will see the heroism, women will only see the daftness?'

Increased heat production in the cold occurs chiefly through higher levels of activity but, if stationary, we can turn to shivering. When we shiver, the brain switches muscle fibres on and off at very high rates so that there are a lot of brief uncoordinated contractions. These produce no useful work but do burn fuel and produce heat. The trigger is usually a cold skin, especially in rapidly changing temperature situations. This is quite different to heat regulatory changes which respond mainly to a change in core rather than skin temperature. Whereas sweating does not start until the body has overheated to some extent, shivering can take place immediately, even in quite modest conditions.

Sitting out on a summer's evening and feeling quite warm, you may even shiver if the sun is suddenly hidden by cloud, although there has clearly been no fall in your core temperature or any significant heat loss from your body. The response is made solely to the rapid change in skin sensor temperature, although your core warmth is not entirely irrelevant. It exerts a strong inhibitory action on the shivering process, so that if your core temperature is normal, you can never shiver vigorously. The summer's evening experience is mere passing *brhhhhh* . . .

The heat generating capacity of shivering is fairly limited. Even with teeth chattering and body shaking like a leaf, you only increase your resting heat production by about five-fold, taking you up to around 500 watts – rarely enough to keep your whole body warm. Perhaps the evolutionary purpose of shivering was, as some experts suggest, less for the production of heat than to prevent resting muscles from becoming too stiff with cold. This would allow our ancestors to sleep through the African darkness, with core temperature dropping but still with muscles plastic enough to face a nocturnal threat. Although this is an attractive hypothesis, I am not sure that it stands up to closer scrutiny. As we begin to cool, shivering always starts in the muscles of our trunk – the ones that give us the *brhhh* in the upper chest and shoulders. Only later, when core temperature really drops, do the muscles of our arms and legs become involved, and by that time the shivering is so violent that one could never rest anyway, let alone sleep. To me, this suggests that the main teleological purpose must be to limit falls in core temperature rather than to maintain limb suppleness. Its ineffectiveness can be attributed to the fact we evolved in climes where it did not need to be efficient.

★

Early in 1979, Major Mike Kealy D.S.O., based at the Special Air Service headquarters in Hereford, was appointed to head up a squadron of elite armed forces performing counter-terrorist operations in Northern Ireland. He was an experienced and well respected officer, used to being out on winter hills and mountains, but his duties had just taken him through a period of easy desk-based work. Having lost some of his previous fitness, he needed get back into good physical shape and decided to go on part of the course that the SAS uses to select new recruits.

Only the toughest and most experienced members of British Army units can apply to join the SAS, and the selection course is run over a period of five weeks in order to whittle down 150 applicants to perhaps a dozen or two. These then face a final gruelling week, and although all the men who remain are the hardiest of the hard, many fail on the final hurdle. The last exercise of the course is known simply as 'Endurance', and it entails carrying backpacks of more than 25 kilos across nearly fifty miles of rugged and difficult mountainous terrain at a time of year notorious for bad weather. The potential recruits must move as fast as they can, and sometimes nobody is deemed to have completed the test satisfactorily. If that happens, nobody joins the SAS that year.

Satisfied with his performance on some of the earlier test exercises, Mike Kealy thought that he too would take on 'Endurance'. If he did well against recruits that were almost half his age, it would confirm that he had come far and was truly back to strength. So, at about one in the morning, fully dressed for the hill, he joined the others for breakfast – a massive, traditional English fry-up of eggs, sausages, tomatoes and bacon. The meal completed, they all got into the back of a truck and were driven to the foot of the Brecon Beacons.

The Brecons are not big mountains but they are bleak and very exposed. That morning, high winds were hurling snow across cloud-covered tops. The men set off individually, with ten minutes between them, beginning soon after 2.30 a.m. First they climbed through the darkness towards higher ground and, as they gained in altitude, the wind rose and the temperature fell. Soon it was blowing at more than 60 knots, and the thermometer was down to minus 9°C. The combination produced a wind-chill well in excess of minus 50°C.

Everyone found the first couple of hours incredibly hard. The route took them off all paths and up through dense tussocky grass, which takes a heavy toll on strength even when bare of snow. The work, however, did have compensations – nobody felt particularly cold. At 1,500 feet above their start point, the going became easier as the ground levelled out. It was here that the conditions began to take effect. For men working less hard, the environment became a vicious and dangerous enemy, and everybody began to suffer terribly.

Nobody knows exactly what happened to Mike Kealy. At around 5.30 a.m. he met and spoke to a number of other recruits who were briefly taking shelter beside some rocks. They were contemplating dropping down from the high ground to take longer more sheltered routes, but Mike exhorted them to go on. He obviously felt that crossing the high Brecon was still possible, and certainly he should know for he had done it many times before, and even in similar weather conditions. Every foot of the way was familiar to him, and furthermore he had been an instructor on the dangers of bad weather and hypothermia. If anyone would be safe, it would be him.

As the night turned into an equally savage day, Mike Kealy's body was found just a few miles further from the point at which he was last seen. It was 9 a.m., and he was resting against a rock with some of his clothing and equipment discarded nearby. It looked for all the world as if he had simply sat down and waited for his fate. The formal enquiry discovered nothing more. The only lesson to be learned was one that had been heard many times before – even the fittest and most expert can be overcome by cold.

★

If you choose to face hardship, your clothing, behaviour and physiology cannot always be relied upon to protect you. Throughout history men and women have become victims of excessive cooling, often due to a dangerous combination of cold, wind and water rather than especially low temperatures.

If the body fails to maintain core temperature, a number of things happen both consciously and subconsciously, depending upon how cold you are. When core temperature first begins to drop, you consciously feel chilled, and this is a useful signal to prevent further deterioration. It makes you actively try to improve your protection, to move around more to generate warmth, or to get out of the situation, and while you do so, your body cuts down on surface blood flow and sets you shivering. These reactions occur at core temperatures of between the normal 37°C and a low 34° or 35°C, and are usually enough to stop things from getting worse. Often you will over compensate and end up feeling hot.

Sometimes, however, these defences will not be enough and cooling will continue. Then some rather strange responses take place. First, your thought processes begin to fail, with some individuals becoming withdrawn and uncommunicative, while others exhibit wild disinhibition. The variation seems to be analogous to getting drunk, when one might end up either drowsy or overactive. The difference is down to whether the essentially sedative action of either cooling or alcohol knocks off the centres that inhibit innate boisterousness before they knock off the centres that keep you alert and active. Unfortunately, neither being dopey or irresponsible is helpful in a cold survival situation. Both serve only to increase the level of danger.

If things get worse, further changes make it even harder to recover. At core temperatures of around 32° to 33°C, hypothermia victims begin to stumble or stagger and the mild personality changes exhibited earlier become aggression, confusion, or pre-coma. Any thought of saving themselves disappears at this point, and sufferers often exhibit particularly strange behaviour where they try to undress. If asked what they are doing, they will often say that they feel too hot, and this reflects a change of great seriousness. The hypothalamus is packing up and misinterpreting the situation in a manner analogous but opposite to that seen when the body is overheated. It begins to tell consciousness that heat is the problem and, simultaneously, it turns off both shivering and the insulating blood flow responses of the skin. The victim is left near defenceless and rapidly heading for the point of no return.

Once a casualty has lost hypothalamic control, he or she has entered a vicious downward cycle. At around 30°C, drift into coma begins, and while there is still some way to go to the temperature at which death will take place, unconscious, inert and with protective mechanisms switched off, it will not take long to get there. At around a blood temperature of 24°C the heart stops.

Every year sees fatalities of this nature, even in Britain where it is never exceedingly cold. People ask me for advice about what action they should take if they are with someone who is becoming hypothermic. It is a difficult question to answer. The correct, perhaps lifesaving action in one situation may be entirely and fatally wrong in another. When someone in a group starts getting cold, you must always act quickly to increase their clothing insulation. After that you are faced with a more fundamental decision. Do you stop or do you go on? The problem is the balance between the victim's heat production and heat losses. You need to try to keep up the first while minimising the second, but the two aims are incompatible.

Most hypothermia victims out walking or climbing in Britain will not be so far from civilisation that you cannot guide, encourage or push them to proper shelter and warmth. To my mind, you are therefore asking for trouble if, at the first sign of hypothermia in a member of a party, you decide to lie up in temporary shelter and perhaps spend a night or more waiting for rescue. In most cases, it would be better to keep on moving. This advice runs contrary to that often given for, generally, it is recommended that if one person is too cold, the rest should halt, keep the victim as warm as possible, and send for help. I believe that groups sometimes follow this recommendation rather too literally. Inexperienced parties without full equipment have stopped when one of them has been a bit cold and then tried to shelter in awful conditions. The hypothermia victim, and perhaps some of the others, have then died when, had they pushed on, they might have got everybody off the hill quite safely.

Of course there are numerous situations in which stopping and lying up is the correct choice. If the victim is more than just cold and is actually showing signs of staggering or confusion, the situation has become so dangerous that it is probably better on balance to stop. Continuing when you don't know where you are going also tends to be unwise, and hypothermia problems often arise because bad weather has led to a party getting lost. Civilisation can be infinitely far when walking round in circles.

If forced to lie up, you must attempt to reduce everybody's heat losses to less than the meagre warmth they will be producing. This is impossible if you are ill-equipped. Ideally, one should have enough protection to cope with getting stuck out in the worst possible conditions that could be encountered but, for practical reasons, this is rarely the case. Nevertheless, all groups should carry something to make a windproof and waterproof bivouac if they are out in winter and must never rely on those lightweight, silvered survival bags or blankets. Although temptingly light and small for the backpack, they have been shown to be almost useless. They are certainly less effective than a decent plastic bag.

Hot drinks are also part of standard teaching, and these can be of great value. Once again, however, do not treat them as too important, for it is asking for trouble to sit around preparing them instead of keeping moving. Usually they are not worth the effort unless ready made in a vacuum flask. Food, on the other hand, is always of value. Even if not heated, it will rev up your metabolism and, more important, will help to sustain the supply of fuel to muscles so that you can keep on working. It will also make you less likely to develop a low blood sugar, a factor which I believe may have been important in some of the apparently inexplicable exposure accidents.

The death of an expert like Mike Kealy can be difficult to explain and makes one wonder if he experienced some unusual physiological problem. To my mind, his breakfast before departure may be relevant. Traditional English breakfast fare is sorely lacking in carbohydrate, and after many days of arduous activity, his muscles were probably low in carbohydrate when he came into the dining room and little better when he stood at the foot of the hill. By the time he had climbed the steep slope, more would have been burned through his strenuous efforts, and with depleted glycogen stores and continuing work at a level demanding carbohydrate, it seems quite possible that blood glucose levels fell. With this, he would have slowed down and so produced less heat. At the same time the low sugar levels may have affected his hypothalamus and turned off his physiological protection. Unfortunately, we can only speculate, and if we do so, we can think of other possible ways in which defences may be overcome.

Large fluctuations in heat production occur when working in the cold, particularly when walking in the mountains. Going uphill requires more effort than going down or walking on the flat, and as you climb you produce so much heat that you will often need to activate your heat loss mechanisms to get rid of it. Approaching the top may therefore see you flushed, sweating hard, and with much of your clothing in your pack. Then, as you reach a crest or summit, the slope changes, work and heat production are suddenly cut, and you can enter a period when heat losses are enormous. With your skin blood flow set at rates suitable for the climb, clothing damp with sweat, and no inclination to stop and dress up again, core temperature can plummet. Some experimental work with which I was involved suggests that the combination of a falling core with still high skin temperatures can cause some confusion in our temperature regulation.

These experiments simulated just this sort of situation in a climatic chamber. Using an artificial wet, windy and cold environment and military volunteers as our guinea pigs, we discovered that when our subjects moved from hard exercise to easy exercise, some found it very difficult to keep going, and several very fit men had to give up because they felt so unwell, despite having skin that was still warm and core temperatures, although falling fast, above normal. On the other hand, if they managed to continue, they felt better again once their core temperature had dropped to subnormal values and their skin was cold and shut down. It seems as if the abnormal combination of warm skin and hot but falling body temperature was confusing. It is easy to imagine that, if they been out on the hill, some of them might have slowed down so much while feeling awful that got themselves into trouble. This provides an alternative explanation for Mike Kealy's downfall as he finished his brisk, warming climb and switched to a tramp across the freezing tops.

★

The accident occurred after we had ditched the sledges and were struggling along with crushing backpacks. My bindings had broken some weeks before, and I was following on foot where Ran had just passed on skis. The Arctic pack ice was most untrustworthy and even here, close to the North Pole, the thickness varied from a few inches to tens of feet. The problem was judging the difference as you walked. Without my weight distributed along the length of a ski, I was much more likely to break through, and so it was with no surprise that I suddenly found the snow crust giving way as I followed on Ran's track. The shock came as I continued down. With horror I realised I had broken through into an open water crack.

There had been no sign on the surface of this fissure despite it being about five feet across. Wind blown snow had built a delicate carapace to disguise it, but now that had gone and just thin ice lay upon the dark water surface below. I went straight through and under, swallowing cold brine as involuntarily I gasped. Immediately, I popped up, buoyant with the air that was trapped within my clothing. I spluttered and wiped the water from my face. For a moment, I wondered why I didn't feel cold, but that soon changed. Within seconds, icy water poured through my clothing and a tight band grasped my chest. It was so cold that it hurt, and I knew that it could squeeze me to death. I had to get out quickly.

I shouted desperately to Ran while looking around and searching the sides of the crack for an escape route. The sides were vertical and smooth, and although I was only four feet or so below the top, it might as well have been four hundred. There was no way to climb out, and I shouted again more urgently. There was no reply. Ran had been close when I fell – stopped only twenty yards ahead – but he had been facing the other way. With his back to the wind, he might have set off walking without looking back, and I began to feel the fear and panic welling up inside me. Then clearer thoughts of self-preservation punched through the pointless whirl of fear.

I couldn't climb out with a backpack on. It was heavy enough already, and rapidly becoming sodden. I had to get it off before it dragged me down, but every time I tried, I sank as I attempted to free my arms from the straps. The cold water over my face was too much to bear as each repeated attempt became more difficult. My fingers were turning numb and they would soon lose their grip altogether. Time was running out and terror again took hold. It was then that I heard Ran calling.

'Mike, Mike! Are you okay?'

Thank God! He hadn't disappeared after all, and the urgency in his every syllable pushed back the fear once more. Then he was there, above me.

'Here, grab the stick.' He lay down and lowered a ski-pole full length. I seized it but as he pulled I realised the futility.

'I can't! I've got to get this sack off.' My voice was querulous with fright but the hope he offered gave me strength.

There was only one way to get rid of my load. I had to let go of the ski-pole and go under to fight with the straps that contained me. I took a deep breath and went for my only chance. Fumbling with the straps, I began to sink, and the cold water on my face was almost impossible to bear. I felt faint; I felt like letting go. My mind began to darken. Then, just as my lungs were bursting, I broke free. The sack was off and I pushed back to the surface, still thinking clearly enough to hold on to my load as I did so. The sack contained vital food, equipment and the radio.

'Quick! Take the sack or it'll sink.'

Ran hooked his ski stick into the hoisting strap and up it went. Then it was my turn. I grasped the stick and hauled once more, but I came only slightly out of the water. The weight of my saturated clothing was now too much, and there was no way that Ran could heave me out. I fell back exhausted within moments. My whole body was rapidly chilling and my muscles were losing their function.

'Come on, Mike. Come on!' Ran shouted.

It was all very well for him to shout. How could I come on? I could feel my strength ebbing away with every moment that passed, and within seconds there would be none left at all. Desperately I looked round again. This time I saw a chance. To my right, perhaps twenty yards away, the crack narrowed. Perhaps there I could get some purchase. If I bridged with my legs at the same time as Ran pulled, perhaps there was some hope.

'Quick, Ran. Where it's narrower.' I pointed.

We both moved along, me half swimming and half pushing off the walls. Then Ran lowered the ski-pole again. I held it with both hands and heaved once more, this time thrusting one foot on to the ice wall in front of me and one on the wall behind. I moved up and jammed my feet hard to rest for a moment. Then I repeated the action – up a few inches, move my feet, up a few inches and rest. Slowly it paid off and I gained height. Then with a heave from Ran I flopped over the lip and lay gratefully on the ice-floe – my sopping clothes instantly transformed into rigid frozen boards. My muscles also seized up from the effort and the cold and I realised that I needed the tent and warmth quickly. I had been saved from an untimely burial at sea.

★

Full immersion in cold water, rather than just wet clothing, increases the danger of hypothermia immeasurably. The waters of the Arctic ocean are close to zero, but even much warmer water will conduct heat away so efficiently that you will become cold rapidly. Keep still long enough, and you will be close to death in the local leisure centre; fall in the sea away from land and even in the summer you will succumb. Each year more than 140,000 people throughout the world die in the water. Few of these fatalities are due to drowning alone; most take place in water that is cold. People die of either cold shock when they first fall in or more slowly from the inexorable onset of hypothermia.

Cold shock is ill understood. Pain and reflex responses triggered by sudden cold exposure lead to rapid increases in blood pressure accompanied by a radical slowing of the heart. At the same time, the acute exposure to cold triggers an involuntary gasp which may make the victim inhale water, followed by further uncontrollable breathing which soon washes out carbon dioxide, and so changes the acidity of the blood. The result is a combination of metabolic upset, confusion and impending heart failure, and many cold water deaths occur with victims just a few yards from safety. Even good swimmers can be totally immobilised.

I have experienced something of this myself when I was one of several volunteers in a series of tests on helicopter escape equipment. It has long been recognised that as soon as a helicopter ditches on water, the rotating blades tend to flip it straight over, the cabin floods, and many potential survivors drown before they are able to escape. The tests were evaluating the use of tiny compressed air cylinders, designed to give about two minutes of air and so increase the chances of people getting out. They took place in a 'dunking' facility in which, strapped into a typical small helicopter seat, with cabin wall on one side and aisle on the other, I was suddenly inverted into the water and had to undo my belt, get into the aisle and exit through one of the escape doors.

It was surprisingly difficult, even though I knew exactly what I had to do, and the water was warm and light. It was very disorientating to turn upside down like that, but with the short-term breathing device, escape was possible, for it contained plenty of air for the time needed. I and all the other volunteers were successful. We then repeated the test in water at 5°C, in semi-darkness. It was quite different. As the rig inverted and water flooded in and over us, the urge to gasp was near to uncontrollable and needed to be fought as we tried to put in the mouthpieces of the breathing cylinders and trigger the flow of air. Then, once breathing, and in considerable pain from the cold on our faces, we struggled to work out what to do. At or before this stage, many of the volunteers failed and needed to be dragged out by safety divers, while the rest of us managed to make sensible decisions and to head for the lit up exit. As we did so, however, the cold water drove our respiration wildly, and no doubt our hearts were also slowed by the diving reflex. The result was a gradual loss of concentration. The air supply lasted for less than thirty seconds when breathing so quickly. Only a few of us actually managed to get out without help. I, for one, had no idea what was happening when I first broke surface.

Beyond cold shock, and again even when individuals can swim, survival can be short unless the victim is in the warmest of the world's waters. The fact is well known, and the likely time to fatal hypothermia often dictates the duration of active searching for the survivors of marine accidents. Research to identify likely survival times was pursued in earnest during the Second World War when dog-fights over the English Channel and the North Sea led to large numbers of both British and German aircraft crashing in the water. Although the planes were lost, many of the airmen baled out and survived, and because these men were valuable assets to their respective nations, great efforts were made to save them from a watery grave. Nevertheless, no one wished to waste precious resources searching for men who had already perished. It was necessary to know exactly how long they might live and swim in such seas at different times of the year. The research in Britain was strictly limited by ethics but, in Germany, some of the Nazi scientists had no such qualms. The experiments performed became the yardstick of depravity as inmates of concentration camps were dropped into cold water tanks and observed as they died. The results from this inhuman work are still used, although not without ethical dilemmas.

Some insist that data from Dachau and elsewhere should be ignored for all time, but personally I could not disagree more. If anything useful can be salvaged from the horrors of that era, I believe that it should be. At least then the deaths may not have been in vain. There is still a need today to know how long a man or woman might survive in the ocean, for planes still crash, ships still sink, and yacht crew are still lost overboard. To provide answers, we must use all available sources and, to my mind, ignoring evidence from the concentration camps is to perpetuate the crime against humanity.

The interactions between water temperature, clothing, individual body fat and physiological responses are very complex and, even using all available data sources, final cooling rates are still very difficult to predict. Modern experiments, rightly constrained by the need to avoid risks to volunteer subjects, can only observe the early phase of cooling in water. To avoid these limitations, scientists have tried to develop computer simulations of cooling to extend the available information. Some of these simulations are among the most complex 'mathematical models' in existence, utilising some of the world's most powerful super-computers. They have produced predictive tables used by the authorities to guide them when searching for victims of accidents at sea. Yet, although they give some indication of when it is time to stop hoping, even the most sophisticated predictions can be proved wrong. Some individuals survive whatever the odds against them.

A recent case, of which the whole world heard, was that of a competitor in a single-handed round the world yacht race. Crossing the Southern Ocean, bound for Australia, his yacht was capsized by a freak wave and he was trapped beneath it. His vessel was spotted by a searching aircraft several days later, but it was to be many more before a ship could reach him. With the water temperature at only a little above freezing, survival had to be impossible. Listening to news reports as the days went by, one had to wonder why the authorities were bothering.

Underneath the hull, Tony Bullimore felt otherwise. Wearing a waterproof survival suit over considerable clothing, he was also lucky enough to be overweight. The resulting total insulation both natural and manufactured, along with the fact that he could spend much of the time wedged out of the water under the boat, meant that he was not robbed of his warmth. Certainly he needed sheer willpower if he was to live, but nobody who sails around the world alone is lacking in that type of determination. More than a week after his boat was knocked down, a Royal Australian Navy vessel pulled alongside to retrieve the body. Tony swam out to meet them. It was an extraordinary case, but at least it could be explained logically. Some other water survival stories are less amenable to analysis.

★

In the early hours of a March night in the late 1980s, a fishing vessel, the *Hellisey*, was upturned without warning off southern Iceland by a combination of snagged nets and a large swell. Of the five men aboard when it happened, two were below decks as the boat turned turtle and were presumably drowned, or died of cold shock, immediately. The other three found themselves together in freezing darkness, clinging to the upturned keel. They knew that staying with the boat provided the best chance of rescue but the option was rapidly withdrawn when, after only a few minutes, the hull sank. Stranded more than three miles from the shore, with the air temperature below freezing and the water at 5°C, the three men's situation was impossible.

All predictive models would give them no chance of living, and likely survival times were of the order of ten to twenty minutes. Oddly, the models would also predict that if they swam for the lights they could see on the distant shore, their time to death would be cut further still. It is a strange fact, observed in both computer models and experimental data, that people immersed in cold water will cool faster if they move in an attempt to try and keep warm – their increased heat production is more than offset by greater heat losses from the cold water pushing through and around their clothing with the swimming action. It is a consequence of the conductivity of water being more than twenty-five times that of air.

Yet, even if computers pronounced them dead, the men themselves did not agree. Perhaps not knowing the theoretical risk, perhaps recognising that rescue wasn't coming, they struck out for the shore, swimming as strongly as they could for their chance of life. Only one, the 23-year-old Gudlaugur Fridthorssen, survived, the others lasting for about as long the computer models would have predicted. According to Gudlauger's account, it was less than ten minutes before he found himself alone in silent darkness. He then swam on, and although he admits that the pain in his legs and arms made swimming difficult, he obviously managed to work hard enough to prevent his core and muscle temperatures falling. In the end, he swam for more than six hours until, as the sky lightened and the sun rose, he found himself approaching a beach. Half crawling, half washed up by the incoming tide, he landed not far from a farmhouse and stumbled there to tell his tale.

★

Gudlaugur Fridthorssen's survival raises several questions, only some of which can be answered. The two most important factors in his favour were his size and his physical fitness. The Icelander was as big and strong as they come, and along with his very considerable muscle he also carried a great deal of fat. There is no doubt that if you wish to survive in water, the fatter you are the better. It is just about the only positive thing that can ever be said of being overweight. Unlike clothing, the insulation under your skin never becomes sodden, and it seems that he had enough to limit the rate of heat loss even when he started to move and the cold water swirled around him. His warm woollen fishing kit would have also helped somewhat, as would his huge muscle bulk and fitness, which not only produced a lot of heat but allowed him to keep going through the night. Above all, he too had sheer determination.

Ironically, many miraculous escapes from cold water occur for precisely opposite reasons. Young children, small and thin, and unlikely to struggle for long in the water, cool so rapidly that hypothermia can protect them. Every year, there are dozens of reports of children successfully resuscitated after being dragged from pools, ponds, rivers and seas after tens of minutes, or even an hour, of complete immersion without breathing. The exact mechanism of such miracles is unclear, but it seems to be down to a combination of the diving reflex with the fact that a cold brain has such a slow metabolic turnover, oxygen demands are virtually nil.

Seals and dolphins, while diving for very prolonged periods, slow their heart rate right down, and we humans seem to have a similar, perhaps vestigial, response to immersion. In our case, the reflex is triggered by water on the face, and if that water is very cold, our heart beat may slow profoundly. A small child falling into cold water may drop his or her pulse rate to just a few beats a minute, and this will limit the extraction of oxygen from the blood. At the same time, the cold will seep through the small body very quickly and within only minutes all the tissues, including the brain, will have cooled markedly. Their oxygen demands are then so tiny that the children enter a state of suspended animation, and the smaller you are, and the colder the water, the more likely you are to be revivable.

This message should never be forgotten. If anyone, particularly a child, is recovered from cold water, resuscitation should not be abandoned even if it appears to be hopeless. No reliance can be placed upon feeling a pulse, hearing the heart, or recognising breathing, for it is impossible to pronounce death with confidence while still out in the field. Recovery may yet occur if the victim can be rewarmed. An old adage among the cold experts has been proved time and again – nobody is dead until they are warm and dead.

★

Living on islands off the coast of Japan are the Ama people who have harvested kelp and pearls from deep cold waters for centuries. It is the young women who do the diving – demonstrating not only a remarkable capacity to swim vigorously under water while holding their breath for more than two minutes, but a capacity to resist hypothermia while they do so. The waters are often at just 7°C, and the women dive repeatedly through long working days, traditionally wearing nothing but thin cotton garments. Their apparent ability to withstand such cold stress has given rise to a number of studies of their metabolism, aiming to answer one question. Do they show an ability to adapt to the cold in the same sense that we all show adaptive changes to the heat?

Theoretically, humans might acclimatise to cold at the short-term, individual level or over evolutionary timescales at a population level. In either case, the adaptation would need to diminish heat losses or increase heat production so that a smaller fall in body temperature occurs with cold exposure. Studies seeking evidence for such cold adaptation, however, have generally found the opposite to be true. Men and women repeatedly exposed to cold, or originating from cold regions, tend to cool more easily than normal individuals. They are tolerant rather than resistant to cold.

Such cold tolerance is known as habituation rather than acclimatisation, and it can be very striking. The Australian Aborigines, Kalahari Bushmen and the Arctic Indians can all sleep outside with little or no clothing, even if their core temperature drops to less than 35°C. At such temperatures, you or I would be shaking too violently to rest and at the same time would be burning valuable food. A capacity to let your body cool can therefore be seen as a useful survival strategy, for not only do you get some sleep but you lose less weight while doing so. Nevertheless, such habituation does not appear to be a true evolutionary adaptation. The aboriginals' capacity to tolerate cold waxes and wanes with the seasons, and in the summer they shiver as easily as centrally heated Caucasians. Furthermore, Caucasians, if deliberately exposed to repeated cold, develop the same degree of tolerance.

The Innuit Eskimos of northern Canada also show cold habituation with no convincing evidence of true metabolic changes when compared to men or women from warmer climes. Nevertheless, some authorities suggest that they show a different type of adaptive change in that they are short and rotund, which reduces the ratio of their surface area to their body mass and heat-generation. This would make evolutionary sense, but not everyone is convinced. Growing up in a cold climate can also influence body development directly. Young pigs, for instance, develop short legs when raised in cold rooms, due to less blood circulating to their limb growing points. A similar effect may underlie the stature of the Innuit while, conversely, warm rooms following the spread of central heating may partly explain the increasing height of men and women from developed countries. In both cases, however, nutrition must play a part.

Returning to the Ama, we find the only convincing piece of evidence that genuine cold adaptation does occur. During the winter, the women divers not only become habituated to cold, but also demonstrate improved resistance to cooling. In part, this follows their becoming fatter at that time and so improving their insulation, but this seasonal change in body fat is seen all over the world, even in people who have central heating in their homes, and probably reflects lower activity levels in winter rather than climatic adaptation. In the case of the Ama, however, there is also a more marked shut down of blood vessels to the surface so that the flow to the skin and even the muscles immediately beneath is limited. This further reduces heat loss, and appears to be an example of genuine, short-term, physiological acclimatisation to cold.

★

I cannot leave the subject of cold without some discussion of hands and feet which, in many ways, are more of a problem than general hypothermia. During our Polar expeditions, Ranulph Fiennes and I have both suffered from frostbite in fingers and toes which have large surface areas for volume, low heat production, and are, quite literally, out on a limb. The blood they receive has therefore been cooled by travel down the arm or leg, but physiological defence mechanisms make matters even worse.

Shutting down blood flow to the skin and periphery, such as fingers and toes, is a good thing for core temperature protection in a true survival situation. Yet it is quite the reverse when it comes to coping with life and work in cold regions. The level at which evolution has set our hypothermic defence mechanisms is quite wrong. It is fine to sacrifice function of the hands and feet when a matter of life or death, but close down blood flow at the slightest hint of cold and the chances of getting into danger are increased.

The problem is that the temperature levels for cutting peripheral blood flow were set on the plains of Africa. Out on the savannah, the worst exposure to cold would have been at night, and then never extreme. Our sleeping ancestors would have benefited from reduced peripheral blood flow which retained heat in the centre of their bodies so that they could rest easily and burn less food. Cold hands and feet in the morning would have been a small price to pay. But following our late move into temperate and cold regions, the price escalated. Painful, sometimes frozen hands and feet are now a reality in extreme climates for we are stuck with a climatic protection system suitable for Africa, and entirely unsuited for the Poles or even northern Europe or the USA. Our hands and feet get cold in conditions which are otherwise quite comfortable and certainly not threatening, and this can be a major liability. Long before they actually freeze, they become too cold to be useful. With no dexterity, it is difficult to perform any tasks, and I have been caught out several times in situations where I have been unable to put on warmer clothing quickly because my fingers have already stopped working. Our cold defences literally threaten our lives.

There is, however, one brighter point. Unlike our very limited ability to adapt our cold defences in general, repeated exposure to peripheral cold does improve the situation with our hands and feet. The change involves a mechanism known as cold-induced vaso-dilatation, which operates as blood vessels in a finger or a toe drop to a stiffeningly low temperature of 12°C. Then they tend to dilate intermittently so that the limb is flushed with warmth. It is the explanation for that sudden and sometimes painful heating-up of fingers after twenty minutes or so of throwing snowballs with poor gloves on.

When first exposed to cold, the flushes of blood to fingers and toes are too infrequent to make much real difference, but after exposure for a few weeks marked improvements occur. Flushes come more often and for longer and overall fingers are warmer and work better. The adaptation is seen in all cold-dwelling races such as the Innuit and Lapps, but equally good responses are seen in the hands of fishermen, fish filleters and even postmen who work in cold climates. Its occurrence does not support the idea that any races have spent enough time in cold climates to have improved their physiological defences to the cold.

Compared with our fantastic abilities to adapt to the heat, exposure to cold can only induce meagre increases in heat production from shivering and a limited improvement in our insulation and the blood supply to our hands and feet. This, coupled with the hopelessly incorrect setting at which the blood flow to our periphery becomes dangerous, fits the theory that our ancestors left Africa just 100,000 years ago, that their progress to true northerly climes was later still and that evolutionary change runs slowly. The reality is that we remain adapted for sun worship, and it is only our clever behaviour that sees us through the heart of darkness.

EIGHT

★

Cries of the Heart

THE CLASSIC heart attack is unmistakeable. Out of the blue comes crushing chest pain, often followed by death. If the victim does survive, further attacks are likely in the next year or two and, even with no other episodes, there is a fair chance that the heart can no longer act as an efficient pump. Yet, despite the distinctiveness of such symptoms – recognised instantly today by anyone whatever his or her medical background – there were very few descriptions of such events in the medical literature before this century. Furthermore, the early physicians failed to write much about angina, the typical band-like chest pains that come on with exertion and frequently serve as a warning of a full blown collapse. The term was coined by the famous eighteenth-century physician William Heberden, but references to it are scarce. This is extraordinary. In centuries past, doctors may have been limited to cupping, bleeding and leeches as their therapeutic options, but they were generally assiduous and accurate recorders of disease. As late as 1892, Sir William Osler still described coronary artery disease as being 'relatively rare' yet less than two decades later, one in eight of all deaths in advanced nations could be attributed to it. By the 1980s, this figure had risen to a third of all deaths, and for every two people dying from a heart attack, three others suffered one. The increase in incidence has been terrifyingly swift. Coronary heart disease, unusual before 1900, has today become the primary peril, the biggest single killer in the western world.

What changed to make heart disease such a serious problem? The underlying cause of a heart attack is that the blood vessels supplying the muscles of the heart become clogged and narrowed. The blockages are due to inflamed fatty deposits forming both on the inside and within the walls of the arteries through a process known as atherosclerosis. Once narrowed, they cannot maintain a sufficient supply of blood to the heart, especially if it needs to work harder to meet the pumping demands of exercise. Even modest walking may cause cramping anginal pain, but this is just a prelude. When the vessels become narrow enough, a clot can form to block the artery completely. A whole area of muscle wall will then lose its blood supply, causing both intense pain and great danger. If the area of damage is extensive enough to stop the pump working properly, or it interferes with the electrical control of the heartbeat, death may follow swiftly.

If we examine what underlies the clogging process, a prime cause would seem to be smoking. The recurring theme of this book is that we no longer live in harmony with our evolutionary design and there can be no better example of this than drawing smoke, laden with chemicals and particulate irritants, into our delicate lung linings. It is clearly so unnatural and likely to cause harm that it is hard to understand how smoking was ever accepted as a reasonable thing to do. Yet, while nobody believes the contorted arguments of the tobacco industry that try to dismiss the link between smoking and damage to the lungs in the form of bronchitis and cancer, many people do not realise that the danger from cigarette smoke does not stop there. By mechanisms not entirely clear, smoking markedly contributes to atherosclerosis in the coronary arteries, as well as the furring up of blood vessels elsewhere. Many smokers will therefore have heart attacks and strokes, and may even lose their limbs, well before their time. As a practising hospital doctor, I can state without prejudice that I almost never see a patient under the age of 40 who has had a heart attack but who is not an avid proponent of the weed. For that matter, most of those I see in their 50s and 60s are also keenly addicted.

Despite the association between smoking and heart disease being well documented, tobacco has been around for much longer than the last hundred years. It is true that there has been much greater availability of cigarettes in the twentieth century, and no doubt this, along with the continued immoral advertising, has encouraged more people to smoke enthusiastically. In many Third World countries smoking is now more prevalent than ever it was in Europe or the United States, but in the less developed nations it tends to kill more people via the lung route, and so it is difficult to blame tobacco for all today's heart ills. There must be other factors that either contribute to the damage or else protect the hearts of Third World smokers from the ravages of cigarettes. An obvious place to look is in the diet.

Some decades ago, it was recognised that high circulating levels of cholesterol make it much more likely that you will suffer from coronary heart disease. Indeed, the relationship between your blood cholesterol and your vulnerability to a heart attack is almost linear. This led to the unpopular yet powerful advice that cholesterol should be avoided in the diet, and through the 70s and 80s health-conscious individuals were busy avoiding butter, eggs and cheese. Nowadays this approach is seen as rather too simplistic. Your body itself manufactures most of the cholesterol in your circulation, and so high cholesterol levels are more often a reflection of abnormalities in your metabolic handling of many types of fat rather than the direct result of eating too much food containing cholesterol. It is still a good idea to go easy on the butter, eggs and cheese, but the restriction is less rigid and the reasons underlying the recommendation have changed. It is not specifically cholesterol you should seek to reduce but the intake of fats in general.

Cholesterol is carried around the body linked to specialised proteins. Essentially there are two forms – high density lipoproteins (HDL) and low density lipoproteins (LDL). Taking a rather simplified view, cholesterol in HDL is being carried to areas where it is needed, while that in LDL is the excess which is to be put into storage. People with high levels of HDL usually keep their blood vessels clear, while those with high levels of LDL seem inadvertently to deposit fats in their arteries. The difference is important. It is not the total level of cholesterol in your blood that matters but the ratio of HDL to LDL: high and you are relatively safe, low and you are in danger.

The ratio of HDL to LDL is little influenced by the intake of high cholesterol foods alone, being more related to excessive intake of other types of fat. Our western diet contains many different fats and oils, and it has only recently been appreciated that, far from being merely a ready energy store, their metabolism is exceedingly complex. Too much fat is definitely harmful, but the balance of different kinds in our diet seems more important than the overall quantities consumed. In recent years research has focussed on three areas: the potential dangers of saturated fats, coming mainly from animal sources; the possible protective effects of consuming specialised unsaturated fats, such as those found in fish and some vegetable oils; and the body's handling of the hydrogenated fats, which come in a form that has been chemically altered by man. Each of these can be examined from an evolutionary perspective, bearing in mind the epidemic of chest pain that started early in this century.

It is not in doubt that a high intake of saturated fat can increase LDL cholesterol levels in the blood. Your consumption of saturates is largely dependant on the amount of animal products in your diet in the form of both meat and dairy produce. The fact that such products are potentially harmful has encouraged some enthusiasts to claim that humans must have been designed to be vegetarian. It is clear from the mechanisms of our teeth and gut, as well as from our metabolic need for certain vitamins, that this cannot be true. Our ancestors must have been omnivorous for several million years, although the balance of their diet was probably quite different from our own.

In contrast to real carnivores such as the leopard, our predecessors were certainly no great hunters. Their ability to find and kill prey would have been, to say the least, inefficient. We must therefore have eaten a diet which, by modern standards, would have been towards the vegetarian end of the spectrum, and this would have changed little until the development of agriculture brought greater access to meat and dairy products. The latter, in particular, are interesting. Today the eating of foods derived from milk is often thought of as a return to the natural diets of our past, but before the domestication of cattle and goats our chances of getting much milk would have been remote. The butter and cheese that now contribute a lot to many people's diet is even less in line with our evolutionary design than a high meat intake, and a substantial proportion of the world's population lacks the enzyme in the gut to allow proper digestion of cow's milk – a fact overlooked by vegetarians advocating their practice for 'return to nature' health reasons rather than for those of animal cruelty.

On the other hand, as with smoking, if a high intake of animal fats were the prime cause of atherosclerosis, why have heart attacks become so much more common only in this century? The consumption of meat and dairy products has been widespread for many thousands of years, and although better farming methods and increasing affluence have probably led to greater intakes by more of the population, those who would have eaten a lot in centuries past would also have been those seen by the doctors. Physicians have always been the prerogative of the rich. So, while animal fats may contribute to heart disease, it is difficult to point the finger at these alone.

In recent years medical science has confirmed that consuming fats from oily fish grants some degree of protection. This was first suggested by observations on the Innuit Eskimos of northern Canada, who traditionally have eaten a diet that is extremely high in fat but who exhibit a remarkably low incidence of heart disease. Their consumption of fish and marine mammals in place of conventional meat and dairy products appears to be the likely explanation, although the Innuit might also have a genetically determined resistance to atherosclerosis.

To examine these possibilities, studies have been made of other fish eating communities. The Japanese are great consumers of sea food and have a low incidence of heart problems when dwelling in their native land. Those who migrated to the United States in the 50s and 60s and enthusiastically adopted American eating habits are not so fortunate. In fact, they face the same risk as their fellow Americans of suffering from heart complaints, and that is many times the risk of their cousins back home. Since they have refrained from much intermarriage with other U.S. citizens, the genetic make-up of the migrant groups has changed very little, and so their consumption of meat in place of fish appears to be the key. Indeed, it has been shown in Caucasian populations that those who eat a portion of fish a couple of times a week suffer from heart disease less frequently.

Of course, the health benefits of higher fish intake could just be a consequence of eating less animal fat rather than a positive effect of consuming fats of marine origin, but this is unlikely. Taking fish oil in the form of capsules will reduce your chances of suffering a further heart attack after surviving a first, even if you make no other changes to your diet. So some of the benefits must be direct, and there are a number of ways in which these may occur.

The membranes of all our cells are constructed from specialised fats built from components that are taken in through our diet. Marine oils alter the properties of these membranes, including those of small cell fragments in the blood called platelets. These are designed to adhere to breaks or damage in blood vessels before they trigger the system that builds a fibrous clot, and the incorporation of marine oils in their membranes significantly reduces the stickiness of the platelet surfaces. Since heart attacks are caused by clots forming at narrowings in the coronary arteries, less sticky platelets make it less likely that a clot will block things off.

Substances derived from fats are also involved in the body's in-flammatory system which responds to trauma, infection, cancer and so on. The system is in careful balance, with some elements encouraging inflammation in order to ensure that protective white cells get to the areas of damage to make repairs, while others are designed to control the extent of the process and to damp it down again when it is no longer needed. The equilibrium seems to be influenced by dietary fat, with animal fats tending to increase inflammation, while marine fats help the anti-inflammatory side. Although inflammation is probably not the primary cause of the deposition of fat in blood vessel walls, once formed the deposits themselves become inflamed, and this can make them grow larger. The effects of marine fats on lowering the incidence of heart disease may therefore be dependant on their reducing inflammation.

A teleological view would suggest that if marine oils in the diet are beneficial to health, evolution must have set us up to work with them, which in turn indicates that we must have spent some time in our evolutionary development exposed to a diet high in sea foods. I have mentioned previously that some anthropologists believe our divergence from the chimps was triggered by our isolation on an island where we lived as semi-marine apes. At first sight, this would seem to fit the bill, explaining why fish oil became so good for us, but since that divergence took place just five to seven million years ago, and Australopithecines were wide-ranging by four million years ago, it is difficult to place our ancestors in a littoral setting for more than two or three million years, even if the theory is true. Although this was clearly a long enough period to effect a change in our posture, it was so because becoming erect conferred very considerable survival advantages precipitated by the move from the forests to a different environment – whether the sea-shore of the marine theorists or savannah of conventional thinking. It seems inconceivable, however, that so short an evolutionary period would alter our inflammatory processes for they are fundamental to mammalian biology. In fact, the way that inflammation works is virtually identical throughout the higher animal kingdom, and so it seems much more likely that the benefits of fish oils are a remnant of our truly distant marine origins, before the evolution of mammals. Yet, whatever the history, a change in our intake of fish could not account for the relatively recent increase in heart disease anyway. Many may have eaten fish every Friday before 1900, but very few of the population would have done so more regularly.

The introduction of hydrogenated fats with an artificial chemical structure at the beginning of this century certainly did coincide with the rise in heart disease. The hydrogenation process allows liquid vegetable oils to be made into solids or semi-solids by chemically altering some of their bonds. Margarine and other related fats began to appear in many of our foodstuffs and have since been used widely in the baking of biscuits, bread, cakes, and pies. Hydrogenated fats contain structures that have never been seen in nature, and we may be poorly adapted for them. The way in which the body handles these unnatural chemicals is only now beginning to be understood, and they certainly have an adverse effect on the ratio of HDL to LDL. Most experts believe, however, that their influence is small, especially as there is mounting evidence to suggest that we should also focus on many other areas where we have deviated from our natural behaviour and food intake.

Vitamins and trace elements in fruit and vegetables, particularly the anti-oxidants, are now thought to underlie the health benefits of the so-called Mediterranean diet, and these include a reduction in heart attacks. Vitamin C, E and Beta Carotene are just some of these anti-oxidants and all are more common in diets with a high vegetable content. Oxidation damage is caused by the metabolic production of free radicals – reactive atoms or ions – which rapidly combine with the nearest proteins or membranes. It is nature's equivalent of rusting. Because free radicals are so harmful, evolution has developed a complex system of defences against them, largely dependant upon substances which can combine with them before they have a chance to damage body tissues.

In dietary research on 700 people over the age of 65 carried out in Britain in 1973, it was found that the chances of their dying from a stroke in the next ten years were highly correlated with their vitamin C intake from citrus fruits and vegetables. Those who ate a lot were less than half as likely to have a stroke as those that ate a little. It is difficult to be sure that vitamin C was solely responsible since diets high in one anti-oxidant often contain high levels of other vitamins as well. An alternative explanation, not recognised at the time, is that the benefits of diets rich in micronutrients may be linked to the metabolism of a substance called homocysteine.

Homocysteine is an amino acid, both consumed in our diet and made by the liver. Reasonable amounts are needed to build our proteins, but it is now known that excess quantities contribute in some way to the furring up of our arteries. Recently it has been shown that many people have levels high enough to be a danger, not due to eating more than is healthy but because the liver makes too much. Excessive homocysteine production in the liver results from a group of chemical pathways not working properly due to an inadequate supply of one or more of the vitamins B6, B12 or folate. In the past, a lack of these vitamins was thought to be significant only if sufferers developed sore mouths, bad skin or anaemia. Now it is clear that long before this level of vitamin deficit is reached, liver malfunction can disturb the balance of amino acids in the circulation. Today's evidence suggests that one third of the 30-year-old population already has a high homocysteine level, while among the elderly the proportion rises to two thirds. It is encouraging to know, however, that even the elderly can reverse the abnormality if given the vitamins as supplements, although this does not mean that we should all start popping pills. Eating a healthy and mixed diet is a sounder approach.

Could then a change in vegetable intake in this century have been the predominant cause of the increase in heart disease? I think not. Vegetable consumption has certainly diminished in the developed world, and this is cause for concern, but the decline began after the Second World War rather than earlier, and so it would be a mistake to ascribe the majority of our problems to that change.

Another dietary factor that has received much attention recently is our intake of salt. Back on the African plains, salt losses in the heat would have been significant despite acclimatisation limiting the concentration of the salt lost in sweat and urine. In those conditions we did well to eat as much as we could get, and so evolution gave us a taste for salty things. Now in the era of snacks and canning, salt has become a significant problem. Men and women in modern societies eat up to one hundred times more salt than our early ancestors, and for some decades it has been thought that this excessive intake might contribute to high blood pressure. Recently this has been confirmed by the 'Intersalt' study which compared blood pressures in more than ten thousand people in 32 different countries with their salt intake. The results were irrefutable; the more salt people ate, the more likely they were to suffer high blood pressure, and the effect was more marked with age.

One of the difficulties in treating hypertension is that, unless it is very severe, when it may cause headaches or illness from kidney damage, it usually displays no symptoms. There may therefore be long delays in diagnosis and it is sometimes hard to persuade individuals to accept treatment when they feel nothing wrong. Medication can also have side effects. Nevertheless, it has to be taken seriously. A high pressure in the arteries leads to thickened walls which are more prone to internal damage and the fatty deposition of atherosclerosis. Men and women with high blood pressure are at much greater risk of heart attacks and strokes. To make matters worse, the blood vessels that supply the kidneys are particularly prone to atherosclerotic damage and normal blood pressure is largely under renal control. When the kidneys sense that blood pressure is low, they release hormones in order to raise it so that more blood will be received for filtering and cleaning. This is fine while the pressure they are sensing is that of the whole body and not a lower pressure caused by local narrowing of the vessels that feed the kidneys themselves. But often the kidneys are misled into releasing hormones to raise pressure when it is already high, and then a vicious cycle ensues. In the end, levels can become so high that strokes, heart attacks and kidney failure are almost inevitable.

Unfortunately, throwing away the salt cellar would make little difference. Our taste for salt has not gone unnoticed by those employed in the food production industry who know that the more salt they put into their products the more their customers will like them. Salt content is often high where least expected. There may be more salt in a bowl of cereal than in a packet of crisps. Furthermore, salt in foods such as sausages is associated with a higher water content, making them heavier and effectively cheaper to produce. The net result is that in Britain, one in four adults over the age of 45 and one in three over the age of 60 are hypertensive; the cost to the National Health Service in drug treatments is some £300 million pounds a year. Industry knows this but has never been keen to put health before profits. The only solution is to use more fresh produce and to cook and eat without adding salt. If you do this, you can at least take encouragement from the fact that the foods will not taste bland for long, and soon it will be the salt-rich foods of industry that taste nasty.

Overall, when it comes to our diet, there are clearly many aspects which are out of step with our biology, and these contribute to our propensity to develop heart disease. Nevertheless, taking account of the changes in our diet between this century and the last, it is difficult to identify a single reason for the increase in heart attacks. We are therefore left with an incomplete explanation for the problem and must look elsewhere to find the cause of our ills.

★

I discussed earlier in this book why, in the distant times of the hunter-gatherers, all men, women and children would have been active for most of their lives and, although less marked, active lifestyles must have continued for the majority right up to this last century. Even if occupation did not entail much action, most people would have walked from place to place. Since the Industrial Revolution, however, the development of urban conurbations, with housing close to places of work, and the invention of motorised public and then private transport has seen this progressively change. Are low activity levels at the root of our heart problems?

In the Allied Dunbar health survey of Britain carried out in 1994, adults were rated on a scale that assessed the number of episodes of moderate to vigorous exercise undertaken in a typical week. The results showed that for middle-aged groups between 45 and 64 years old, those who did no exercise at all had heart problems in 16 per cent of males and 11 per cent of females, while the figures dropped to just 4 per cent and 2 per cent of the same age who exercised at least five times a week. Even exercising just once or twice made a considerable difference. Such observations are striking, and it is reasonable to assume that, before this century, most adults would have been getting enough exercise to qualify for the five times per week groups. A change in activity levels can therefore provide an explanation for about a four-fold increase in the incidence of heart disease.

Exercise can influence the heart in many ways, primarily making it larger and stronger in order to meet demands for it to pump blood faster. With the increased thickness and strength of heart muscle walls, the blood supply vessels also enlarge so that any furring within them has less of a narrowing effect. At the same time, the expansion prompts the development of new blood vessels which make any one region of heart wall less dependant upon the supply from a single artery. Further benefits are gained from a change in platelet stickiness. Just as the consumption of fish oils reduces the likelihood of platelets adhering to areas of damaged artery wall and generating clots, regular exercise has the same effect. An experiment with which I was involved while working for the Ministry of Defence illustrated this most strikingly.

At that time, I was advising the Army on the types of food that would best sustain soldiers in the field. There is a difficult balance to be struck here, for troops must often carry their food on their backs, and so eating more to feel better and stronger has to be offset against the increased weight of their loads. Using high fat rations would allow more energy to be carried for the same weight, and over the likely period of active operations should not be detrimental to long-term health. But sometimes soldiers must work at high intensities, the sort of levels at which muscles prefer carbohydrate for fuel. My research examined whether using high fat rations to save weight might reduce their working capacity.

To answer the question, we took two comparable groups of soldiers and made them compete against each other in performing hard military exercises over bogs, moors and fells for many hours every day. The training continued for a whole week, and when they arrived back at camp each evening, we took them into a temporary laboratory and made them work on exercise bikes until they were so utterly exhausted that they could do no more. The time that they could keep going on the bikes in the evenings was used as a measure of their endurance reserve at the end of each day's efforts.

During the biking, we took blood samples in order to monitor the soldiers' metabolism. Little plastic cannulae (the type used to give drips in hospital) were inserted into the veins in their arms so that small samples could be taken every fifteen minutes without stopping their work on the bikes. On the first evening of the study, I found in ten of the sixteen subjects the cannula blocked off by a clot which had formed between samples. By the evening of the third day, the number blocked was down to three, and for the last couple of days there were no blockages at all. Quite apart from the main purpose of the study, the subjects had shown that a vigorous exercise programme completely changes the tendency for blood to clot and block off narrow tubes. It was clear evidence as to why an active individual may not have a heart attack even if he or she has diseased and narrowed coronaries. A more important benefit, however, is that regular exercise makes atherosclerotic narrowings far less likely anyway.

Muscles burn fat at all but the highest intensities of work, and regular activity modifies the fat profile of your blood and eliminates a good part of any excess intake. Furthermore, when endurance trained, the muscles have enlarged fat stores within them, ready for use as an immediate source of fuel. As part of this adaptation, the HDL transport system becomes more efficient in order to carry fat to the muscles where it will be needed. Regular exercise will therefore produce a healthy ratio of HDL to LDL and so reduce the risk of fat being deposited in the wrong place. Such benefits are evident even on lazy days. Studies of volunteers who ate a fatty meal on the day following a bout of vigorous work showed that fat levels in the blood an hour after the meal were 30 per cent lower than when eating the same food on the day following no such activity.

A similar but more extreme example of the effects of exercise on fat metabolism were to be seen in the blood samples taken from Ranulph Fiennes and myself during our Antarctic crossing. As explained earlier, we ate a very high fat diet during our walk in order to minimise the weight of the sledges we dragged, and the fat source used was mainly butter. Our food was not only high in total fat but very high in saturates. The World Health Organization recommends that you eat a diet containing no more than 30 to 35 per cent fat, of which the larger proportion should be unsaturated. Our Antarctic rations contained 57 per cent fat, and that was as a proportion of a diet containing more than twice the normal adult consumption. The result was an intake of more than four times the saturated fat normally considered wise, yet our total cholesterol stayed at healthy levels and the good HDL type went up while the bad LDL went down. The message was clear. If you exercise enough, the fat gets used for the purpose that nature intended, and you can eat the most awful of diets and get away with it. And there is another reason why this is so. Since exercise burns calories, people who undertake it tend to eat more food. This gives them more micro-nutrients, which in turn can protect their health.

It is a serious matter for the medical profession as well as the public to find two thirds of the elderly deficient in vitamins to the extent that they develop unhealthy homocysteine levels. In some cases, the shortages can probably be ascribed to factors such as illness, poverty and social isolation, which lead them to eat poor quality food and a diet lacking fresh vitamin-rich produce. Yet, on the whole, the elderly adhere to a diet which would contain plenty of micronutrients if they ate anywhere near enough of it. The problem is that their micro-nutrient requirements are very similar to younger more active adults but their sedentary lifestyles demand little food to meet their energy needs. Evolution simply did not expect anyone, at whatever age, to stay still for so much of the time. It set us up to extract vitamins from a high throughput of food. My fear is that the problem will get much worse. Today the older generation tends to eat a healthier diet than the young, with more fish, vegetables and fruits. If they cannot extract enough vitamins from that type of food, how will the younger generations fare as they grow older, consuming packaged foods of low micro-nutrient content and exhibiting a tendency to exercise even less?

In conclusion, optimal health is to be gained only through attention to both food and exercise. If we want to avoid the cries of the heart, there is no doubt that we should never smoke, and we should reduce our intake of fats from animal sources and eat more fish, fruit and vegetables. But along with these changes, an increase in activity is needed. This is an uncomfortable fact for those who currently do little, and many people will be unable to alter their lifestyle to counter what they see as a theoretical risk of that chest pain somewhere in the distant future. However, for such people there is perhaps a more powerful motivating factor close at hand. You are unlikely to become obese or even unfashionably overweight if you exercise regularly.

NINE

★

Survival of the Fattest

EVEN MORE than the epidemic of heart disease, our propensity to get fat reflects a biology out of step with our lifestyle. Wild animals, including all of our close relatives among the primates, are virtually never plump, and the same must have been true of our ancestors through the whole of human evolution. Now more than half the adults in Britain are overweight and one in seven are obese, not just too heavy according to some whim of fashion but fat enough seriously to endanger their health.

Recently, some groups have been trying to promote the idea that being overweight is really nothing to worry about. Particularly vociferous in the United States – where one in four adults are clinically obese – has been the punchily named National Association for the Advancement of Fat Acceptance. The members of this organisation proudly claim that fat is both good and beautiful, suggesting that the only problem is an unfair dislike of fat people. There is some truth in their highlighting this prejudice. The social slur of being too fat is easy to dismiss as exaggeration, but the consequences can be cruel. It is particularly harsh for overweight young women, who complete fewer years at school, marry less frequently, end up with fewer friends, and have more episodes of significant depression than their lighter equivalents. However, while rating fattism as bad as racism, I also believe that we should avoid promoting the message that becoming overweight is all right. In reality, this is very far from the truth.

It has been estimated that diseases caused by obesity cost the National Health Service more than two billion pounds a year, and that a further similar sum is spent on the largely ineffective slimming industry. It is also reckoned that Britain loses up to fifty million working days per annum due to ailments related to obesity. The cost to industry amounts to a further £2.5 billion. These are very significant sums, yet it is not the economic cost that is the real problem. Being too heavy produces all sorts of medical problems and, ultimately, is life-threatening.

The most striking additional risk to health for fat people is their greater tendency to fur up the arteries to their hearts, brain and other organs. Being over ideal weight by just 3 kilogrammes (6.5 pounds) makes a measurable impact on the chances of your having a heart attack. If you have reached the point where you are classified as clinically obese, your vulnerability is more than doubled. Many of the factors damaging your blood vessels will be the same as those causing atherosclerosis in normal weight men and women, but occurring more frequently and severely. The overweight frequently have high blood pressure and adverse ratios of HDL to LDL accompanied by abnormally high levels of other unwanted lipids in the blood. Fat abnormalities in the blood are partly explained by the simple fact that the obese eat far more than their thin counterparts. The idea that they have a low metabolic rate and became overweight despite eating little is a myth. Those who are heavy tend to be less active than those who are light and so gain few of the benefits of exercise. They also develop abnormal metabolism of nutrients and this may effect the body's handling of sugars even more than disturbing levels of fat. Indeed, many of the overweight become diabetic as they grow older.

The diabetes of obesity is not usually the form that occurs in children or young adults. That type follows the destruction of cells that produce insulin in the pancreas by an inflammatory process with an unknown trigger. It is usually very severe as soon as it starts and almost inevitably requires long-term insulin for treatment. The overweight, on the other hand, develop a much less dramatic form, coming on slowly in middle or older age. It is due to the pancreas being unable to cope with a lifetime of high demands from an oversized body, combined to some extent with a propensity to eat too much sugary food. Unlike complex carbohydrates, such as starch, which would have been important components of our diet for millions of years, simple sugars are in relatively short supply in natural habitats. Our sweet tooth – an evolutionary development to help us identify ripe fruits – did not lead us into trouble until refined sugar became widely available. Analogous to salt, the situation was then worsened when food manufacturers started using it indiscriminantly without considering whether it might be unhelpful to health. The amount of sugar in a can of baked beans is a prime example of exploiting an evolutionary weakness for profit.

To be fair, not all 'age-onset' diabetics are overweight, but it has been estimated that more than 70 per cent of older diabetics would not have developed their illness if they had not been too heavy. Fortunately, many cases can be treated by diet and weight loss alone, although it does require considerable determination. For those who do not have it, control of sugar levels by tablets or insulin injections is usually highly successful, although excessive sugars are only one element of the metabolic problem.

In most diabetics, the abnormalities in the body's handling of fat are almost as great as those of blood sugar control. Frequently they also have too much circulating cholesterol, with markedly adverse HDL to LDL ratios, and the difficulties these create are not easy to treat. Even on medication, diabetic patients remain prone to developing athero-sclerosis, and few manage to reduce the risk by controlling their diet and weight or by taking up exercise. Years of being overweight, complicated by the diabetes it has caused, then leads to angina, heart attacks, strokes, aneurysms and kidney damage – all dangerous medical conditions. Furthermore, the sheer load imposed by a large body, with or without diabetic complications, can simply overwhelm the heart. With a circulation too sluggish to clear fluid from lungs or limbs, breathlessness and swollen ankles may herald an early death.

Obesity may threaten health in more ways than just metabolic derangement. Snoring is usually thought of as harmless. It occurs when the muscles of the throat relax in sleep, obstructing the flow of air. A heavy neck makes it much more likely, and fat people snore more often and more loudly. This is not only inconvenient for partners; heavy snoring carries health risks beyond the dangers of a sleepless, irritated spouse losing self-control. In severe cases, almost always in the fat, the airway can obstruct completely and prevent air flow. Often this occurs in cycles through the night, when louder and louder snoring culminates in a complete cessation of breathing that can last up to half a minute. The snorer then wakes, fighting for breath and, without becoming fully alert, their sleep is disrupted. Then they lapse back into somnolence and repeat the cycle once again. It is more than just annoying. Nights become so disturbed for snorers that they are sleepy during the day and prone to accidents. Furthermore, periods of not breathing at all cause low oxygen levels in the blood which slowly damage heart and lungs.

Apart from this, what of the impact of carrying all that extra weight around? Walking with only a light backpack considerably influences any task and, for most of us, going up hills or stairs is considerably more taxing when carrying just 5 or 10 kilogrammes (10 to 20 pounds). Yet the obese may carry an extra 30 kilos around with them all of the time. To make matters worse, they also tend to overheat due to the insulation of the fat beneath their skin. The result is like working hard in winter clothing. Hot days – a delight for most of us – are no fun at all for the overweight and summers can become a nightmare of discomfort. Apart from the survival advantage when escaping from a sinking ship in a cold ocean, being too fat has nothing but downsides. At best, it interferes with most days of your life, at worst it will rob you of many years.

★

If being too fat is so harmful, we should have evolved regulatory systems to prevent it. Why then is it such a common problem, and what factors dictate why some people stay thin while others grow so large? The answers must be matters of balance. If the energy contained in your food exceeds what you need, you will gain weight; if it is less, you will lose it. This is an absolute truth, yet it has proved remarkably difficult to pin down all of the factors on either side of this energy balance equation. When I first developed an interest in weight regulation, I found this astonishing. Now I realise the difficulties inherent in accurately measuring all the factors that contribute to energy intake and output over long periods, and have come to appreciate that tiny discrepancies between them can lead to huge changes in weight over the long term.

In 1995, newspapers in the United States reported the death of the fattest man in the world. He had come to weigh 465 kilogrammes, or around 1,000 pounds (in excess of 70 stone). On a visit to hospital he had to be transported by fork lift truck, and after his death the wall of his bedroom had to be demolished in order to remove the body. Obviously he was an extreme example of the obesity problem, yet it would have required a weight gain of only 37 grammes per day to take him from a normal 70 kilogrammes (155 pounds) at the age of sixteen up to the grotesque proportions of his death at just 45 years of age. This is the equivalent of eating an extra 250 calories per day – less than one small bar of chocolate.

To complicate energy balance assessments, factors on both sides of the equation are liable to change. The alterations occur because evolution has developed weight regulatory systems and these constantly attempt to match intake and output to prevent either excessive weight loss or gain. Unfortunately, when it comes to the latter, some aspects of modern life seem to deceive our controls. Many find weight gain all too easy and then face considerable difficulties when it comes to reversing the trend. It is a sad fact, but once you are too fat, your body does its level best to keep you that way.

Weight control mechanisms can act on either intake or output, and in animals they influence both. On the intake side, animals are rarely exposed to prolonged periods of excess food because of adjustments in population. When food is plentiful, they become more fertile and are more successful at raising their young. The result is a rapid increase in the number of mouths to feed, although usually there is simultaneous growth in the numbers of competitors and predators. Among people living in the under-developed world, such population factors may still have some effect. A successful harvest can influence birth rates, since human fertility is also improved when we are well nourished. Infant survival is also better when food is plentiful, and a good supply permits earlier weaning, whereas times of shortage may see breast feeding, with its contraceptive effects, extended for several years. In the developed world, however, no such factors operate. People's breeding habits, the success of their pregnancies and the survival of their young are entirely independent of the number of ready meals in the fridge.

Intake control in animals is not just a matter of population adjustments. Even when there is a glut of food, individuals tend to eat only enough to match their needs. They must therefore have mechanisms to match appetite to requirements, but it is not altogether clear how these work. Day to day, the amount of food in their stomachs and the level of sugar in their blood are known to trigger signals which make the animal feel appropriately hungry or satiated. Beyond this short-term regulation, a background control also seems to operate, keeping body weight stable over weeks or months. Recent evidence suggests that it is seated in their fat stores.

Genetically obese mice have been shown to lack the normal form of a hormone produced by fat cells which suppresses appetite. It has been called 'leptin' – after the Greek *leptos*, meaning slender – and is released by fat stores when they are full. After the abnormality was identified, it was found that injecting normal leptin into these fat mice made them lose weight quickly. The discovery led to a frenzy among bio-technology companies and a down-payment of $20 million was made for the rights to exploit the gene. The hope is that an artificial human leptin will be God's gift to dieters, and the dream has been boosted by the recent discovery that human fat cells do carry a leptin gene. Development of the discovery will take a lot more investment, but countering human obesity is very big business.

From an evolutionary point of view, leptin must be part of a system that ensures fat stores are fully stocked whenever possible in order that either animals or humans can face the next period of famine. It makes survival sense, and although, as yet, there is little evidence that human leptin is effective in suppressing our appetite, most dieters recognise that as soon as they relax their self-imposed discipline, they are driven to eat more until they have regained all the weight they have lost. Leptin could well underlie this drive, and my own experience certainly lends credence to its power.

When Ran Fiennes and I returned from our Antarctic journey, body composition studies revealed that our fat stores were completely empty. The very hard work had led us to use far more energy than we had eaten and our weights were down by nearly 25 kilogrammes (55 pounds). With this weight loss, we were absolutely ravenous which, to a non-expert, might not seem surprising. Nevertheless, it was actually unexpected – most people who starve to such low weights are profoundly anorexic. Starvation probably causes deficiencies in essential micro-nutrients as well as deficits in energy, and a stack of fat, protein and carbohydrate is the last thing a weakened body needs if it lacks the nutrients required to process the food and build new tissues. Anorexia after starvation must therefore have evolved to protect us from refeeding dangerously quickly. Some micro-nutrient deficiencies, such as a low zinc level, even cause complete loss of taste in order to dissuade us further from over enthusiasm. Before these issues were understood, medical refeeding of famine victims sometimes did as much harm as good. Even today, some doctors thoughtfully provide artificial feeding for patients who have starved through illness, without attending to the micro-nutrient deficiencies first. They are then surprised when the patients unexpectedly die.

Ran and I were not like famine victims. We remained hungry in the face of great weight loss because, despite eating too little to meet our needs, we had maintained a very large throughput and so had no marked vitamin or trace element deficiencies. Indeed, rather than anorexia, our drive to consume was total and we ate both day and night for several weeks after coming home. This was well illustrated by a situation arising about a month after our return. Waiting one lunchtime to give an interview for BBC radio at a London studio, I found myself in one of the hospitality suites faced with a large plate of sandwiches. I proceeded to eat them all, and ended up both uncomfortable and embarrassed. The sandwiches were meant for the five other guests on the show as well as myself.

Following my interview, I was to speak on a different programme, to be recorded a short distance away. With plenty of time to get there, I decided to walk along Oxford Street in central London. Although I am no hamburger fan when it comes to health, I am none the less hypocritical enough to eat one occasionally. During the crossing of Antarctica I had dreamed of such an opportunity, and now I had my chance. I went straight into the first McDonalds I came to, despite already feeling full from lunch at the first studio, and ended up eating six hamburgers before I finally reached my destination – not all at the first McDonalds, but three at other branches of the same chain and two at Burger Kings. I simply could not walk by these hamburger joints without going in, although discomfort turning to pain did limit me slightly. Only the first four burgers came with fries.

What drove this feeding frenzy? Like animals, humans have short-term triggers for hunger, including those that report how full our stomachs feel and our levels of blood sugar. But often these natural triggers are overridden by cultural considerations along the lines of 'it's twelve-thirty and so it must be lunchtime'. In the case of my hamburger blow-out, none of these applied. My stomach was full before the first burger, my blood sugar levels must have been normal, and my recent sandwich meal should have allayed the 'time for lunch' excuse. Something else then was driving me to eat, and empty fat stores with low levels of leptin must be a likely cause. The notion is also supported by the fact that such dedicated piggery lasted for six weeks after the expedition before it switched off, almost overnight. That was the time it took my fat stores to return to normal, and Ranulph Fiennes behaved very similarly.

If leptin, or something like it, is responsible for controlling human appetite, it should suppress our eating when fat stores are full and so prevent them from becoming overloaded. What then goes wrong in so many men and women?

One suggestion is that fat people have defective visual cues for eating. In an experiment some years ago, students were invited to participate in a study which involved completing some questionnaires, but when they arrived at the study centre and were asked to wait their turn, they were each offered hospitality in the form of drinks and sandwiches placed on a table in front of them. They were also told that there were more sandwiches in a nearby fridge if they wished. Unknown to the students, the study had nothing to do with the questionnaires but instead examined the number of sandwiches they ate while waiting. They were brought back several times to the study centre at various intervals from their last meal, and there they faced different numbers of sandwiches on the table – sometimes too many, sometimes too few, but always with plenty in the fridge. The results were startling. The number of sandwiches that students of normal weight consumed varied according to when they had last eaten, and so corresponded to how hungry they felt. Fat individuals, on the other hand, showed no such regulation. They did not necessarily eat more, for they appeared reluctant to raid the fridge, but they did eat whatever food was on the table, regardless of how much it was or how hungry they felt. Their eyes appeared to rule their stomachs.

Although this phenomenon may contribute to men and women becoming overweight, I doubt that it lies at the root of the problem. There is strong evidence that the type of food we eat can confuse our intake regulation. Indeed, it can even override the intake regulation of animals. Laboratory rats, fed on standard chow, control their appetite very accurately and never get fat however much food is put in their cage. But if offered what is known as a 'cafeteria diet' – meals consisting of snacks such as hamburgers, sweets and chips – they completely lose this self control and can become positively enormous. Several reasons probably underlie this: the diet has an attractive mix of tastes and so encourages feeding; it has a high fat content so that more calories will fit into the same degree of stomach filling, and the fat, unlike protein and carbohydrate, can slip into the body fat stores with virtually no metabolic cost. Is the human propensity to become overweight analogous to that in 'cafeteria' fed rats?

Many studies have shown that if we eat a meal with a high fat content but the same number of calories as a more balanced dish, we perceive the high fat meal as less filling. This is not surprising. The high fat foods contain more calories for weight, and so the meal is physically smaller. They also contain less sugar, and as a result cause a smaller rise in blood glucose. Our ability to assess the calorie content of the higher fat meal is poor, and when our diets are rich in such foods, we will tend to eat more than we need. Nevertheless, this is only a part of the problem.

The fact that we derive particular pleasure from high fat foods is probably not just chance. Cream cakes or curries bathed in oil are far more popular than any insipid healthy equivalent. Fatty foods taste good. There is probably an evolutionary basis to this. The body can assimilate additional fat at almost no metabolic cost, for it is simply transported directly into stores for future use. By eating as much as possible when supplies were good, our ancestors would have been set up for the bad times that might lie ahead. Fat gains are also a good way to prepare for winter, when the cold can be countered by improved insulation. Evolution has therefore designed us to overeat whenever possible, although this does not mean that our forbears were fat. Overeating arose as a problem only when the availability of food became almost unlimited and the famines no longer arrived. Today we build our survival stores but never put them to use. Civilisation has made our lives as unnatural as those of the caged laboratory animals.

Additional factors may also operate. Contained in our foods are those micro-nutrients essential for our well-being and little is known about how their consumption is regulated. The processes of natural selection, however, should have provided us with means whereby our bodies could identify shortages and take measures to correct them. These responses might be non-specific, driving us to eat more of any food when a vitamin is lacking, or targeted, making us want to eat more of the types of food rich in the missing nutrient. Either way, such responses could contribute to obesity. Most diets of today are less nutrient rich than those of our evolutionary past, and by cooking them we further diminish their vitamin content. Hamburgers and chips may make us overweight through two very different routes. On the one hand, the fat content appeals to our tastes and slips past our regulation system, while on the other, the food lacks so many of the things we need that we are driven to eat more in a bid to sustain our health. Perhaps this is best described as a double whammy.

*

As well as by intake controls, animals alter the energy they use to help maintain balance in the face of food excess or shortage. Some adjustments are simple: to gain more food in the natural world will take more effort, and if the animal does become heavier, it will need more energy to get around. Other adjustments are more complex. Animals can change their metabolic efficiency.

The amount of energy used by an animal or a person depends on three elements: their basal metabolic rate – the energy needed to maintain life when resting; thermogenic responses – increases in metabolism triggered by factors such as cold, food intake or fever; and the energy cost of activity. All three elements can be altered to maintain stable body weight.

When food is in short supply, animals lower their basal metabolism, partly because they lose body weight and metabolically active tissues, and partly because cell activity is damped down to conserve energy. When it comes to over-eating, similar but opposite changes take place. An increased body size uses more energy and metabolic activity is uprated. Other interesting alterations are to be seen in the animals' thermogenic responses, which include the metabolic reaction to cold and the increase in metabolism following a meal as it is absorbed, digested, used and stored. In many small mammals, some of these thermogenic responses occur in tissue known as brown fat which – unlike the normal white fat that is a fairly inert storage tissue – can have a very high metabolic activity. It seems that this activity can be turned on and off as needed, and that animals can use brown fat either to produce extra heat when they are cold or to burn off extra energy when they are over-eating. The fact that brown fat evokes both these responses leads to an interesting interaction. If a rat is repeatedly chilled, it adapts by revving up brown fat activity to produce additional heat and, simultaneously, it becomes very resistant to weight gain. Conversely, if a rat is overfed on a cafeteria diet, it will not easily give way to cold.

The presence of energy use regulation in humans is less definite. In modern societies the simple compensations hardly operate. It remains true that, if you are overweight, you need more energy to run basal metabolism and get around, but much of the extra effort can be avoided by the studious use of chairs, cars and elevators. Similarly, finding plenty to eat is no longer matched by an increased energy cost, for loading more into the supermarket trolley has few metabolic sequelae. Some alterations in metabolic efficiency do occur in humans, although they are not as marked as in animals, probably because we have less brown fat.

Changes in energy efficiency are a familiar experience to many. If you try to lose weight through dieting, the body compensates by lowering your resting metabolism. For a couple of weeks you tend to get thinner quite quickly but after that the diet becomes ineffective. The reason for this is the suppression of both basal metabolism and your thermogenic responsiveness. It makes evolutionary sense. Your body detects that you are eating less than you are using but fails to understand that this is exactly what you had in mind. It interprets the situation as a time of relative famine and does all it can to help you get through it. Despite your wish to go on burning extra calories, it acts to block your plan by cutting down your needs. Soon you are so efficient that your diet becomes useless. It is a response designed to meet times of hardship

Although a less obvious experience, the opposite can occur if you consume more than you need. Basal metabolic rates and thermogenic responses then increase by a few per cent, making a big difference in an average sedentary person's overall daily expenditure, equating to the burning up of many calories. This change should dispel the myth that the overweight are cursed with a slow metabolism. In reality, they have the opposite, with considerably higher demands for energy than normal. However much it is denied, the overweight have to eat more than others simply to maintain their size.

Of course, everybody knows some fortunate people who can eat and eat without ever getting fat, and formal studies confirm their existence. Different individuals of the same age, sex, and activity levels may have calorie intakes that vary by as much as 40 per cent and yet are of similar weight and fatness. It is not clear why this should be so, but it could be due to the thin individuals having more brown fat than the rest of us, although the importance of brown fat in human weight control is often debated.

Some experts claim that the amount of human brown fat is so minimal that it cannot possibly influence our body weight regulation; others point out that you need to burn only a few more calories each hour, day in day out, to keep trim in the face of a considerable food excess. This seems logical to me, and leads to an interesting point. If brown fat is important in man, the overweight might benefit from being cooler in the same way as is seen in small mammals. Experiments have shown that when room temperatures are maintained at about 4°C lower than usual, subjects increase their energy requirements by a few per cent yet do not actually feel cold. However, the observation may have been due to the cool conditions making them move around more, rather than any change in brown fat activity. Nevertheless, it suggests that wearing less clothing and turning down the heating can help if you want to be closer to the lucky eat-all skinny types.

★

Although the level of the heating thermostat in our homes may slightly influence our daily energy expenditure, far greater changes are effected by our activity levels. Animals vary how much they do depending upon food supply, becoming less active when food is scarce and roaming more widely when there is a glut. Although effective in the wild, it is a feedback control that fails when they are domesticated. Just as the cafeteria diet can lead to rotund rats, spherical dogs can be created by owners who provide plenty of food but unnaturally few opportunities for exercise. It raises an important question. Which of these two processes – freely available dietary intake or low exercise levels – predominate when it comes to spherical humans?

After my long explanation of our failed intake controls, one might expect that people eat more now than they did in the past. This is not the case. Over the last twenty years, the average adult in Britain has decreased his or her daily intake by about 700 calories, and the figure would probably be much higher if we could make comparisons with the last century. Modern diets may have confused the controls set up to match our intake to needs, but those needs have declined so strikingly that, to my mind, today's large number of overweight men, women and children can be explained only by reduced activity levels. It has to be said that not all experts agree with this, and many claim that exercise has little influence on body weight.

It is true that even strenuous efforts can be maintained by surprisingly little extra food. A single small chocolate bar can fuel a run of many miles, and one would have to put in many hours a week at running or other sports to make much difference if the effects of exercise were limited to burning the little extra food required to undertake it. But this is far too simplistic a view. Following strenuous work, metabolism stays elevated for several hours, allowing considerably more calories to be burned off, and beyond that there are other effects on metabolism which lead to even greater benefit.

One of these is an increase in resting metabolic rate, which in most people normally accounts for around three quarters of their energy expenditure, and so a small percentage rise can be very useful for weight control. Performing regular activity stimulates muscles to become larger and stronger and, unlike fat, this increase in tissue will use energy even when doing nothing. After a few weeks of exercise your body is more metabolically active, burning more fuel throughout the day and night. The suppression of metabolic rate that usually accompanies dieting is also largely abolished, and your body is not so prone to enter the self-preservation famine mode even if you do go on restricting your food intake.

The influence of exercise is not confined to the output side of the equation, and this too can be useful if you wish to lose weight. When you first take up regular physical activity it will often suppress appetite, contributing to weight loss early in an exercise programme. I have frequently experienced this myself. Being of the type with a propensity to grow fat, I have found that exercising at meal times is a good way to regain control of my weight after a period when I have let things lapse. If I go for a run at lunchtime, my appetite disappears until well into the afternoon, by which time it does not take too much willpower to abstain from eating until the evening meal. If I don't run, I am ravenous and have to eat lunch. Unfortunately, this pattern does not last for long. Within a month of restarting exercise, I come back from the run very hungry, and this seems to be a common experience. It probably reflects increasing fitness. When you are unfit, your whole body's efforts are

concentrated on keeping you going, and blood flow to the gut is minimised during your exertions. This turns appetite off. Become a little
fitter and the gut receives a better share of the blood while you work.
Then, instead of returning faintly nauseated, you come back ready to eat.
Overall, people who regularly exercise eat more than those who do not.

Losing weight through exertion may seem disappointingly slow. If
you go on a rigid diet, it is possible to shed pounds quite rapidly at first.
If your body's needs are around 2,500 calories a day and you restrict
yourself to 1,500, the deficit adds up to 7,000 calories a week, which is
equivalent to more than a kilogramme (or two pounds) of weight loss,
since both fat and muscle will be reduced. When you try to slim down
through exercise alone, however, a number of factors operate against
you. It requires a couple of vigorous exercise sessions every week to use
up an extra 1,000 calories, and so it can take up to a couple of months
to match the sort of deficit generated in one week of dieting.
Furthermore, during that time, working out twice weekly will lead to
a gain in muscle weight and an increase of up to half a litre in circulating
blood. The latter will weigh 500 grammes in itself and so, overall, you
can often find yourself losing no weight at all. This does not mean
that you won't lose fat. In the last few months my wife, who has never
been overweight but who was not as slim as she was in her youth,
decided that she was taking too little exercise. She started going to the
gym two or three times a week and has on occasions come running
with me. Greater fitness has definitely resulted. The bathroom scales
have not changed one jot, a fact she finds frustrating, but at the same
time she has regained much of her youthful figure and is met by
questions as to how she managed to lose so much weight.

Are diets, then, of any use at all? The answer must be yes, if rapid
weight loss is your aim, but they should be used in addition to and not
in place of exercise. Of course magazines, newspapers and bookstores
are full of claims for diets that work wonders. Grapefruit diets, 'F-plan'
diets, diets stopping you from mixing protein and carbohydrate – can
they help to control weight? Here too I must also answer yes, but not
for the bizarre reasons claimed. There is nothing extraordinary about
consuming either grapefruits or exclusively high fibre foods, and the
idea that you will benefit from keeping protein foods separate from
carbohydrates is a myth. Most of the benefits are simply due to the
imposition of restrictions which lead to lower intakes through both
limiting choice and increasing consciousness of every item that you eat.
They offer no magic, but a diet plan gives you strength.

★

So how – with job, family and social life – can a reasonable amount of exercise be fitted in to a modern lifestyle? Clearly many of us have too little time to be active for much of the day. We certainly cannot reach the levels for which evolution designed us. The question is: how little exertion can we get away with and yet retain our health and waistlines? Do we need to do vigorous training, or can we simply incorporate lower levels of activity into everyday life?

Young adults who do not have a manual job use about a quarter of their daily energy sleeping, with most of the rest divided about equally between sitting or engaging in light activities such as walking, housework, or shopping. Even if fairly enthusiastic, they spend less than five per cent of the time indulging in any sport or vigorous exercise. At that rate, even a doubling of sports activities will make little difference to the overall energy use. The best strategy for using up more calories would therefore seem to be changing some of the sitting time into light work. Two hours of easy gardening twice a week, for instance, is much the same in energy terms as running a half marathon. Most people could easily accommodate a little more movement into life without disrupting any routines.

For some years I worked in a laboratory complex that investigated all aspects of human performance. Many of the employees who were conducting research programmes for the British Armed Forces ran or played sports in their spare time, occupying an hour or two each week. Our headquarters had just ground, middle and upper floors and was equipped with one small and very slow elevator, situated at the far end of the building from the main entrance. It was also at the wrong end for collecting mail from the central registry, which was situated up two flights of stairs above the main door. It was with constant amazement that I would watch my sporting colleagues' choice of route. Every day I would witness one or another enter the building in front of me and turn to their right to walk the fifty or sixty yards to the elevator where, presumably, they pressed the button, waited for some time, entered the lift and waited again for the doors to close before the thing slid slowly upwards (perhaps stopping at the middle floor for some of the even more spectacularly infirm staff members). Finally they would reach the top floor to emerge and walk back along the top corridor to the registry itself – all to avoid climbing two floors. Even stranger, they would repeat the process to get back down again.

The same happens everywhere. At present I work in a large hospital with seven floors, and to walk up from the ground to the top is a demanding task – although still one I would encourage. But it is not those journeys that worry me. Just as in the research laboratories, all day I will see a variety of health professionals waiting by the overused lifts to travel just one or two floors up or down. It is very sad. Walking for just twenty minutes each day should yield weight reductions of two to four kilogrammes (4 to 9 pounds) over three or four months, and many of my hospital colleagues could achieve that within their daily work – and while saving time. The same, I am sure, is also true of many office workers. Most people could add some exercise to their daily routine, getting to and from their work place either on foot or by bike, or walking (instead of driving) to the local shop for that small forgotten item. Others could walk on escalators or those moving belts at airports, two places where standing seems even more mindless than waiting for the lifts.

Yet while this incorporation of movement into everyday life is important, I am not convinced that alone it is really adequate. From the point of view of fat control, it will certainly help to burn up more food, but there will be little effect on muscle size and advantageous changes in metabolism will be negligible. Exercise in life is not only about weight regulation. A little extra walking a day is very worthwhile but it is not going to grant a much stronger heart, lower blood pressure, improved HDL to LDL ratios and less sticky platelets. Our ancestors would have done more than merely walk, and so we too need some vigorous exercise if we want *all* the benefits. This, inevitably, takes commitment, but if you make the effort, in addition to staying thin, you will also see a healthier old age.

TEN

★

Fit for Life

HELEN KLEIN, the great-grandmother team mate of my first Eco-Challenge, may be exceptional by any standards but she is none the less an example of what can be achieved if you remain active into later years. After the race, Helen appeared in dozens of magazines and her remarkable abilities were broadcast throughout the United States, with reports on our team's progress screened on three consecutive nights on television's *NBC Dateline*. Inspired by her achievement, thousands of people across America realised that it was not too late to do something about their fitness. It was an awakening witnessed by the sheer scale of Helen's postbag.

Back in Britain, I told her story repeatedly, and that prompted an idea. My father, an active man approaching 70, had recently retired from full-time work. What if he were to undertake the next year's race? Perhaps he would provide some inspiration for people in Britain, and even if not, he would get a lot out of it. Several times while watching Helen hang from that rock wall I had wondered how my father might fare in her position. I did not know the answer, but was sure he would love trying. In many ways the idea was ridiculous. Whereas Helen was a world-class ultra-distance athlete, my father was a weekend walker, and had done no serious training for decades. Yet, unable to dispel the idea, I began to take it seriously.

We had been warned that the next course – to be held in the Rockies of Western Canada – would be much harder than it had been in Utah. Instead of ten days crossing deserts and canyons, the equally long route would cross glaciated mountains and would be completed in eight. When I explained this tentatively to my father, he was not deterred. A decision was made there and then to give it a go. It left me with a quandary – I needed to find three other team mates.

There were many factors to consider in making the choice. Each had to be capable of completing the race, and ideally some would have special skills. In particular, we needed someone who could read white-water, for this year the rapids would be run without guides. As before, each team had to include at least one female member and – most important – everyone had to be easy going. In Utah I had witnessed several teams destroyed by friction as fatigue and tetchiness set in. Furthermore, we all had to accept that we were not out to win, for with my father on the team that would be impossible. Finishing the course had to be the challenge in itself.

Ranulph Fiennes was my first approach: it would be great to do something together again. He was also well known, which would help us to gain publicity and sponsorship, and I recalled that he had been a canoeing instructor for the army and so should know about white water. Next, I spoke to Rebecca Stephens. I did not know her personally but she had been the first English woman to climb Mount Everest, and after that she had gone on to ascend the highest peak of every other continent and to complete the 'Seven Summits'. She had since become a presenter on BBC television's popular science series, *Tomorrow's World,* and so she too came with a celebrity badge. Both Ran and Rebecca immediately agreed to join us. Now we were four, looking for a fifth. It was then that something struck me. I wrote down the ages of my team members so far – Rebecca in her 30s, me in my 40s, Ran in his 50s and my father in his 70s. If we could find someone in their 60s, we would have a real pattern, and consecutive decades would give an enormous boost to the promotion of fitness for all ages. What we needed now was a strong, determined 60-year-old. But who would fit the bill?

A few months earlier I made a train journey back to London from Leeds, dining with Chris Brasher, the chairman of the cold weather conference we had both attended. Chris was an Olympic gold medallist from many years back and had founded the London Marathon. I knew he remained heavily involved in athletic and outdoor activities and that he still participated in long fell races. He had to be in his 60s, and he not only confirmed my guess but agreed to join the project. Team 'Fit for Life' was born.

★

Just downstream from the town of Lillooet, the river has carved a huge straight valley through the Rockies, its floor covered in a patchwork of fields and homesteads. Above them, the towering valley sides are cloaked in dense coniferous forest, leading up to dramatic glacial peaks. Here the second Eco-Challenge was to begin – a twenty-five-mile stage along the river banks. As in Utah, we were to start with both riders and runners, but this time we had only two horses per team.

Waiting for our mounts in half darkness and early morning mist, I was nervous about the ride ahead. My team was as strong as it could be, although not quite as I had envisaged. My hope for a balanced representation of age had had to be foregone. Chris Brasher had not appreciated how much preparation would be needed, and when he realised exactly what this entailed, he decided that participation was unrealistic. When he finally dropped out, with just a few months to go, it was too late to find another 60-year-old who could train up in time. I turned to a doctor friend, David Smith, who like myself was in his 40s. It rather spoilt the pattern, but except for Rebecca, we all remained far older than the majority of other entrants. My father, of course, was the oldest man in the race, but he was not the oldest participant. With another team, Helen, now 73, had come back for more.

The obvious pair to ride the horses were my father and Rebecca, for Ran, David and I were used to running long distances. If we did the opening marathon on foot, the team as a whole would be in better shape for Stage 2. All the same, after Utah I appreciated that riding such a distance is not effortless. My father had never been on a horse in his life before he entered for this race. He had hurriedly taken a few riding lessons but I had no idea how he had got on.

Fearing another stampede at the start, I begged my father to rein back his horse across the first half-mile field. It was a pointless suggestion. Although the horses were better schooled than those in Utah, they were still frisky with the excitement, the dawn, the noise, and the numbers. As the countdown finished they were away and I could do nothing but watch with dismay as my father galloped across the meadow, looking pretty unstable on an unfamiliar western saddle. I was sure he would fall and be badly hurt, even if he was not trampled. But he did not come to grief. At the far end of the field, where the route went through a gate and up on to the top of one of the river levees, a bottle-neck brought everyone back to a trot or a walk. It gave us a chance to re-form, and we moved up the bank as an organised team.

The next hour or two went well. The horses maintained a steady trot, as long as the riders kept a tight grip on the reins, and we found an easy loping pace. At the halfway watering station our animals were passed without comment, though the vet paused to look at my father, a question written across his face. Was this man, already hobbling, fit for the tasks ahead?

I too was anxious. Although confident enough to set out in the saddle, my father had not become an accomplished horseman. It was painful to watch his bouncing progress on a high-trotting mount, and when he got off, he was obviously in great discomfort. He ruefully admitted that his backside was already badly beaten and blistered. Clearly he needed a break, but if he had one, we would slow down considerably. Shamelessly I asked if he could continue in the saddle, and although his face dropped, he did not complain. Slowly, stiffly, he remounted, grimacing as he returned to torture.

His steadfastness lasted for another hour, by which time we had reached the last quarter of the run. Then, suffering from cramps, he simply had to dismount. David took the horse while father, obviously tired, began walking and then, after loosening up, broke into a jog. Over the next few miles we maintained a steady pace, Rebecca on one horse and David, my father or I taking turns on the other. Ran stayed on foot, sometimes disappearing ahead in his enthusiasm. A lifetime of competitiveness made it hard for him to hold back.

At the end of this ride/run stage, our route had to cross the river, even though after heavy rains, it was in full flood. This made it far too dangerous for the horses to cross, but that did not mean we would be spared the challenge. We handed our steeds to some wranglers and dropped down to the water through a small wood, emerging from the trees close to the bank. Nothing would have prepared me for what lay before us. Instead of the peaceful river of a few miles back, the waters had left the constraints of the levees and had become a violent cauldron of grey-white power. We saw from the efforts of a team ahead – all young strong tri-athletes – that crossing it was to be a ferocious battle. With the water just a little above freezing, hypothermia was a great danger and my father would be particularly vulnerable. Unlike most older people, he had never gained fat with age.

★

It is accepted as inevitable by many that the older you become, the more fat you will carry, even if you do not put on any weight. I am not convinced. Although I accept that it is the norm, I do not believe it to be a natural consequence of ageing. It is more a reflection of declining activity.

The tendency to increase internal fat with age was first demonstrated in studies comparing measurements of the thickness of fat layers beneath people's skin, with estimates of their whole body fat made from underwater weighing assessments of body density. The results showed that although both skin and whole body fat tended to increase with age, the latter changes were more marked. Indeed, there was even an increase in whole body fat in subjects who showed no change in skin fat thickness at all. Some of the elderly subjects looked slim, but they had more internal fat than the young. It seemed that muscles and other organs developed higher fat contents with increasing years, even in those who did not put on much weight.

As a result of many such measurements, scientists were able to write equations to predict whole body fat from skin measures alone, with varying co-efficients to account for age. Equal measurements of skin thickness in a young as opposed to an older subject would therefore predict that the older had a higher total body fat. Such equations have been used widely in research, but when I first saw them, it struck me that the lower activity levels of older men and women might explain at least a part of this apparent age-related change, and these suspicions were supported by measurements of skin and body fat made on Ran Fiennes and me before our Antarctic crossing.

At the time of our crossing, Ran was in his late 40s and so, according to conventional dogma, should have been internalising fat at a steady rate. When I measured his skinfolds, they corresponded to the expectations one had when looking at the man – small pinches of a lean, fit individual. But the skinfold to body fat conversion equations suggested that, at his age, this limited skin fat thickness still reflected a total body fat of more than 24 per cent. I doubted this, and we had underwater weighing measurements which were very different. In reality, Ran was still a lean 16 per cent fat, with a relationship between his skin fat thickness and internal body fat closer to that expected in a 20-year-old. To my way of thinking, the reason was self-evident. He had refused to become sedentary. Instead of depositing fat in his muscles and other organs, he was still using it for the purpose nature intended – refuelling his regular exertions.

These results prompted me to examine others who might have delayed their 'age-related' changes. I looked at skin and total body fat measurements in a number of active men in their 40s and compared the results with men of the same age, weight and height who were essentially sedentary. The results were as I anticipated. The fitter men had similar or less fat under their skin than the inactive, but much less total body fat. Clearly exercise can alter an accepted part of ageing.

★

We had been warned that if there were any weak swimmers in the team, it would be advisable to have a safety line for fording the river, and I was carrying a full length of climbing rope for the purpose. My intention had been to rig it right over the river but now I realised this was impossible. Even a full length climbing rope was far too short, and it was clear from the team before us that the whole width was to be hard swimming. So fast was the water running that as competitors jumped in they were picked up like flotsam. Nobody made it to the far bank within three hundred yards, let alone within a fifty yard rope span. There was, however, one hope. Trapped in midstream, about a hundred yards down river, was a large log. It would be impossible to reach it from this bank, but it looked as if one could get to it from upstream on the far side. I decided that if I swam the river first with my kit and the rope, I could work my way back up the far bank and then swim and drift down to the log. From there I should be able to throw a line to my father as he went past.

My worst fears of the water were nought compared to the reality. As I leapt from the bank I was engulfed in the raging torrent, and it was so cold that it crushed all breath away. It took enormous will to concentrate and strike out for the far bank, a difficult task when washed sideways in a liquid roller coaster. Progress was also hampered by my pulling the kit and rope in a large waterproof sailing bag which dragged terribly. Eventually, however, my feet touched bottom near the far side. I was then reminded why we had been advised to keep on swimming until in real shallows for, with the water running thigh deep, I was simply knocked back down. By the time I emerged on the far beach, I was an exhausted cold man, and I am neither thin nor weak when it comes to swimming.

After resting for a couple of minutes, I took out the rope and set off up the broad, shingled bank until I was much higher upstream than the log. Gritting my teeth once more, I then re-entered the freezing torrent and allowed myself to be carried down, making my way towards it. The plan worked and I got there, but each time I tried to climb on to the dead tree, the power of the river's flow and the slipperiness of the surface defeated me. With my legs and chest screaming from the cold, I simply had to get out of the water soon. So I released the log and swam back to the beach. Shattered and shaking, I put down the now sodden rope and waved to my father to come on. He would have seen my failed attempts and must realise that I could do nothing. Yes . . . he and David were heading down for the water.

They jumped simultaneously. I saw David strike out strongly for my bank while my father appeared to do nothing. There was little sign of movement and certainly no progress – just a bobbing head racing down the river a few yards out from the far bank. With both his own and my father's kit, David was already too far away to help. Another few seconds and my father would be passing where I stood but still the whole river's breadth away. I had seen around the next corner that, less than half a mile on, the river narrowed and steepened into a boiling boulder run. I had to do something. A little downstream, there was a small spit that stuck out into the flow. I sprinted down and out along it, diving from the end into the swirling rapid that it generated. The rapid pushed me out towards midstream, where I pitched into hard front-crawl, trying to head across the flow. For thirty seconds I did not look up. I knew it was a race to make contact quickly and that it would take at least that long to reach my father's stretch of water. If I were too slow, we would both be in trouble. The water was too cold to sustain activity. If we were still in it when we reached the rapid, we would be unable to defend ourselves at all.

Powering on as best I could, I literally counted the seconds, face screaming from swimming with my head down. When I reached thirty, I changed to breast stroke and tried to look around. I was perhaps two thirds of the way across the deep segment of water, in big waves and running down the inside bend. With the water in such turmoil, it was difficult to see any distance. For a few seconds I thought that I must have completely mis-timed things and that my father had already passed. Then I caught sight of him just a few yards away and roughly level. He was staring straight at me, but he made little sign of recognition or movement. He looked for all the world like a man day-dreaming in the bath rather than one about to go down the plug.

I reached him in just a few strokes, grabbing at his collar and immediately setting off back towards the far beach. The nearer bank was too steep and inaccessible. It had been years since I did my lifesaving course at school, but I remembered how to pull a man to safety while maintaining his airway. My father kept getting a faceful of water, but it made him cough, so he was obviously all right. Progress was desperately slow and became slower still as fatigue and chill poisoned my leg muscles. It was so cold around my chest that I could scarcely breathe. Then, just as I thought I could go on no more, my foot touched bottom. Simultaneously, I also noticed another figure in the river below us. Roped to the shore and wearing a black rubber suit, the man was clearly a safety backstop. It was welcome reassurance, but I did not want his help. Rescue would mean disqualification.

Stupidly, I tried to stand as I had done before but was again reminded of the futility. I went under and in my tiredness lost grip on my father.

He began to drift away but, coming to my senses and revitalised by the shallower water, I grabbed him again and made another final effort. It brought us into shallows where I could wade, and my father also made an effort to stand. We could not help the river from pushing us on downstream, but we were now out of trouble. In the end, we were helped over the final few yards by the man in black, but I had done my bit and had no qualms about his assistance.

Of course I should have realised that help would be at hand. The race organisers knew there might be problems with such violent water and had strung a cable across the water above the rapids. In the heat (or perhaps cold) of the moment, the presence of such a safety net had not occurred to me when I saw my father heading for serious trouble and needing help. In the event, I need not have bothered. So many competitors had to be rescued that the rules about disqualification were abandoned.

★

At the time we decided to enter the race, my father had about eight months left for training. He started immediately and, a few weeks later, when I was over at his home for the weekend, I suggested a run together. We went out early one chilly and wet Sunday morning, with the trees bare and paths sodden through field and wood. The mud made the going hard, but I was still dismayed when, after only a mile or so, my father had to stop on a slight incline and walk. He was carrying nothing, we had not gone far, yet he was out of breath. Had I made a dreadful mistake?

It was too late to change my mind and steadily the weeks and months went by. I tried to organise dates for a few weekends when we could all go to South Wales and have access to canoes, rock walls, horses, mountain bikes and open country. Practice for glaciers and white-water rafting was not going to be possible, but I felt we could rely on my own skills for the former and Ran's for the latter. Unfortunately, when I spoke to Ran about this, he denied all knowledge of rapids and suggested that, as I had run the Colorado the year before, raft captain should also be my job. In any case, he had no time to spare for the weekend sessions and in the end made none of them. While Rebecca, David, my father and I spent at least some time hanging from cliffs or paddling up rivers, Ran's refresher course in abseiling and climbing was held on a rope slung from the chimney of my house. It was hardly ideal training for big rock walls, but it had to do.

If concern remained over aspects of our specialised preparation, I was encouraged by my father's progress. He committed himself in earnest, and it was great to see him paddling a canoe up the river Wye all morning, battling on ropes up a huge cliff all afternoon, and then tramping over the hills as the sun went down. The transformation in his fitness was also astonishing.

★

After staggering from the Lillooet, it was some time before my father and I were in a fit state to continue. Besides the physical exhaustion and the cold, we were both quite shocked by what had happened. Indeed, the events in the river were to replay themselves in my mind with vivid intensity several times over the next few days. It was very disturbing. Had I been grossly irresponsible?

From the river bank to the first stage checkpoint was only a few hundred yards. There our support crew – Morag, who had been the radio operator for Ran and me in Antarctica, and her daughter Moyra – gave us a late lunch while we studied the next part of the course. The huge climbs over glaciated peaks and passes, separated by lower segments through wooded slopes and valleys, would obviously be taxing but the terrain did not look too bad. We reckoned we could complete the stage in a couple of days and nights.

It was still hot when we left the checkpoint at around 4 p.m. and made our way up and into a side valley. The lower slopes fell precipitously into a narrow river gorge, and so we had to climb more than three thousand feet in order to contour in above the cliffs. There was a logging track to help us make the climb, which would have been easy had we been fresh but, after a day which had seen us run a marathon and then battle with the river, the hairpinned route was hard work, and we baked in the heat. Darkness was falling before we had finished the climb and were ready to enter the forests.

Here, orange flags were meant to mark the way, but under the dark canopy we soon lost sight of them. Traversing across the steep hillside, there were the prints of boots and trainers all over the place under the trees where teams had cast up and down in their attempts to pick up the trail. We decided to maintain height by sticking to our altimeter, but the slope was often cliffed and we were forced to climb or descend to find ways onward. At around midnight, with the slope so steep that we had to hold roots and branches in order to make headway, we decided to stop and rest. Progress would be safer and faster in daylight, and our water was in short supply. Below us, we could hear the river raging. It sounded quite close, and so we dropped down to find a place to stop. We had not gone far when we reached a broad ledge of rock and mossy earth large enough to accommodate us. The spot was somewhat spoiled by a dead tree trunk lying across the middle. Beyond the ledge was steeper rock – perhaps a small cliff down to the river itself.

We sat down on the tree trunk and nibbled some bits of food without cooking as our rations needed rehydration. We then made more room by heaving the tree from the ledge. It slipped easily into the darkness but, to our amazement, several seconds passed before we heard its distant crash. Clearly the cliff was higher than we thought, and we all moved a little further from the edge. As we lay down to sleep, Rebecca, who was close by Ran, took out her ice axe and planted it firmly between them. She saw him eye its vicious blade nervously.

'Don't worry, I do trust you,' she said with a smile. 'It's for the bears, not you.'

★

After the age of twenty-five or so, it is accepted that your aerobic capacity will decline by about one per cent per year. It means that by the time you are fifty, you will be very much less fit than when you were in your prime. Although this is the accepted norm, many studies have shown it need not be so. The rate of decline in the aerobic capacity of older endurance athletes who continue to compete is only half as fast as normal, and their decrease does not begin until well into their thirties. Indeed, a couple of studies on elite veteran athletes, who have gone on training into middle age at the same sort of intensity as when they were younger, show no decline at all until they approach their sixties. As they aged, they became no more breathless as they ran, and there is no reason at all why we too cannot limit our declines in this way.

Even if you have not maintained activity throughout your life, but come to it when older, the benefits from training can be quite extraordinary. My father's improvements over the eight months leading up to the Eco-Challenge provide a good example of the benefits to endurance performance, but even more marked improvements can accrue when it comes to strength. Even at the age of eighty, people can benefit as much as young adults if they take on a strength-building programme. This is not to say that they become equally strong but that they improve by a similar proportion. A twenty-year-old who has done little training can double his or her strength in just a few weeks. So can the elderly, and in their case it makes a greater difference. Instead of a change that simply improves sporting prowess, increased strength in older men and women can work miracles. Much of the unsteadiness of age is simply down to weaker musculature, and training can literally put people back on their feet. Overall, it can take the person who needs help when getting out of the bath to a life of more independence and dignity, the person who can only get around the house down to the shops, and the person who can make it to the shops up into the hills. There is no doubt, exercise can be a fountain of youth.

★

My watch alarm woke me at 5 a.m. – less than four hours after I had dozed off. It was still dark but the moon illuminated the scene. I lay there for a minute or two, savouring the moment. It was great to be out in the mountains, surrounded by awesome beauty, and good to be there with my father. We had to move with daylight, and so I woke the others and made ready to depart. While we dressed, my father asked me to have a look at the sores he had acquired from the riding on the day before. I thought he was being fussy until I saw them. Then I was horrified. His backside had been rubbed by the saddle until areas the size of saucers were left bloody, raw and devoid of skin. It was obvious that they would not heal quickly and would give him hell, but he would not hear of stopping. I covered them as best I could with adherent pads, but I knew the wounds would weep so much that the dressings would soon become useless.

By the time we were ready, there was enough light to see down the drop over which we had sent the tree and, for the first time, our precarious position was brought home to us. We were camped on an overhanging ledge, high up on a gigantic cliff, hundreds of feet above the river. What was more, much of the ledge consisted not of rock or solid ground but had been created by some rotting tree trunks falling and jamming across rock pinnacles. Indeed, you could look right through gaps in the floor of this giant nest, and from the hollow which Ran had chosen for comfort, you could drop pebbles straight to the water far below. We abandoned our plans for losing height and finding water, leaving our spectacular eyrie by the way we had come.

A few hundred feet higher, we turned the top of the cliff and moved along the slope, much as we had done in darkness the night before. After an hour or more, the whole valley widened and the gradient became easier. The vegetation, however, became increasingly difficult to negotiate. I had always thought of coniferous forests as bare straight trunks rising to a canopy with little undergrowth. The early part of the woods had been just that, with trees of all ages – some young, thin and tender; others ancient and of enormous girth and height. As well as those that lived, there were many that had died and dead branches and trunks were strewn on the forest floor. Some were so rotten that they shredded beneath a booted foot or were nothing but a line of decomposing dust. The whole of life and growth, death and decay was written clear on the woodland floor and the place had an air of primaeval mystery.

As the slope eased, the nature of the forest changed with the addition of vibrant lush green undergrowth. Two plants were particularly common, both designed to make life difficult. One had numerous thin branches that grew like a horizontal lattice to bar progress. Our feet constantly caught upon them, whereupon their bark stripped instantly, turning them from straightforward obstacles into slippery wands that lay across the slope. They were aptly named slide alder, and encounters often brought a crashing fall and a short roll downhill to meet the Devils Claw – a plant with barbs so sharp that they went straight through our clothing. The Devils Claw grew avidly beside the slide alder and was always the plant that one grabbed to try to stop a fall. We took wide detours to try to avoid these evil plantations, often walking along giant fallen trunks well clear of the ground. Even so, before long we all had cuts, scratches and hands filled with thorns, and to add to the torment, vicious hornets stung our faces. It did make me wonder why we were there.

We struggled through the vegetation all morning, stopping once as soon as we could drop down to the river to have some rehydrated breakfast, supplemented by welcome wild berries that tasted like redcurrants. Then, at around midday, and with increasing altitude, the vegetation began to thin out and we came across the orange flagged trail once more. It made the going easier, but within a few hundred yards we reached a point at which the valley branched. The orange markers went straight on, while the map suggested that we should turn right. Reluctant to leave a proper trail, we decided to follow the marks, believing that they must swing round some yet invisible obstacles. This meant fording two large rivers, wading thigh deep across the fast flowing torrents. It was a bitter experience, made even more humiliating when we discovered that the flagged trail finished just a hundred yards further on at a hunter's cabin. The path was nothing to do with the Eco-Challenge course, and miserably we had to accept our mistake and recross the icy waters.

We spent the next few hours slowly gaining height, either fighting with plants or climbing up screes, finally to exit all vegetation in a high snow-filled valley. It had taken the best part of twenty-four hours to get there, despite being little more than twelve miles on the map, a rate of progress that did not bode well for the remainder of the stage.

At the next checkpoint, we learned that more than two thirds of the field were still behind us. The news gave us considerable encouragement for clearly others were also having problems. Later we heard that some teams had become so lost in the forests it took them a week to make that point.

Here David confided that he was feeling dreadful. Perched on a rock beneath the checkpoint, he was clearly in a bad way – shivering, despite strong late afternoon sunshine, and looking extremely pale. A lack of food and perhaps hypoglycaemia seemed the likely cause, and so we halted for long enough to make a brew and cook a meal. The diagnosis proved correct and David recovered rapidly.

The route continued upwards to cross a col before dropping steeply down a glacier. Our second night fell as we descended, and at around 9 p.m. we left the ice and skirted a glacial lake. There I called a halt. It was useless to become exhausted so early in the race, and we were about to re-enter the forests. Tackling the plants in darkness would be a nightmare. Instead, we lay down on the moraine and went to sleep.

The following day, the third of the race, began with another six hours of forest-bashing before we finally approached the bottom of a valley, where we needed to cross the river. It was clear from the map that this was likely to equal the size of the Lillooet, and my heart sank at the thought. If it was to be anything like the Lillooet crossing, I was not prepared to try it. One close call with my father might be considered bad luck, another would definitely seem careless. Fortunately, I need not have worried. The river had been rigged with a Tyrolean rope traverse, and we all had the excitement of working our way across, dangling inverted a few feet above roaring rapids.

Beyond the crossing was another enormous climb, four thousand feet on a single unbroken slope. In the hot afternoon sunshine, it was very hard work, and soon Ran, David, my father and I were sweating wrecks. Rebecca, on the other hand, remained cool and composed. Throughout the race she was to demonstrate extraordinary physical resilience, coupled with humour that was never bowed by fatigue. She had an amazing ability to look more like a model on a fashion shoot than a competitor in one of the world's toughest endurance races. Beyond doubt, she was a tremendous asset to our team.

The climb took us from forest to pasture, to scree, to ice, with the vista behind becoming ever larger. Finally we arrived on a high alpine summit, surrounded by glorious views. We stopped and savoured the mixture of effort, achievement and beauty – what being in mountains is all about. While sitting on the peak, we reviewed the rest of the hiking section. It made us troubled. We had greatly underestimated the difficulty of the forests, and now the two nights and days that we had thought would take us to the next checkpoint had passed. There was still a long way to go with barely any food.

During the evening we descended across large snowfields, but we were still high as darkness approached. Turning the end of an ice-bound ridge, we came to a steep drop. It looked a dicey venture in the dark, especially after fourteen hours of travelling. I glanced at my father and saw that he was limping. I did not know if this was due merely to fatigue or whether he had pain from his hip – a considerable problem in the past. Either way, to continue seemed unwise and I called another halt. After we had settled, I asked him about it. His reply reassured me. Although tired, his hip was fine. Indeed, it was giving him less trouble than it had done for years.

★

A common criticism of too much exercise, especially as you get older, is that it leads to damage from long term wear and tear. This is largely a fallacy. In itself, exercise strengthens bones, joints and cartilage as much as muscles. Naturally, if you are untrained and overdo things, it is easy to wreak terrible damage to joint surfaces that are not hardened and ready for it. The time taken for bones and joints to adapt to exertion is longer than that taken by the muscles that move them. Over enthusiasm can therefore do harm, and I have been guilty of this many times. Even when not in training, I know that I can run twenty miles or more on determination alone, but while determination can drive muscles, it cannot take care of joints and tendons. Suddenly returning to exercise with another expedition to prepare for, I have paid the price, and there is nothing more frustrating than being forced to rest a sore ankle or knee when wishing to get fit.

Although injuries are unavoidable in sports, some later infirmities are unfairly attributed to them. There are many ex-athletes with painful ankles, knees and hips, for the same reason as there are many in the sedentary population with the same complaints. Training does not remove the possibility that someone might suffer from arthritis. The trouble is that the former activities of athletes invariably take the blame, and so the myth that exercise causes joint damage is perpetuated. The reality is probably the opposite. Exercise makes it less likely that you will end with problems, and individuals may well suffer worse invalidity if they never participate in strengthening activities.

Broken hips, wrists and ankles are common in the elderly because their bones become so very thin. Although continued exercise cannot protect the skeleton completely, it can largely eliminate the chances of fractures. People tend to think of bone as being a static, inanimate substance but the reality is far from this. Bones consist of a living matrix of connective tissues with a blood supply and nerves, and contain millions of living cells that produce the matrix and the hard calcified mineral deposits which are superimposed upon it. These cells constantly work to fashion the bones in order to meet the demands put upon them. Some build strength to make the bones more dense when required, while others do precisely the opposite, weakening the structure if demands are few so that the body's resources are not wasted and the skeleton remains as light as possible. The system worked well for millions of years but, unfortunately, it no longer does so. As with so much of our physiology, it was set up to expect daily activity and so the bone building system needs almost continuous stimulation in order to do its job properly. In many modern lifestyles, this simply does not happen.

The man or woman who drives to work, sits all day, drives home, and then watches television puts little stress on their skeleton and, inexorably, bones that were strong at the end of teenage growth begin to waste. It takes decades to become serious, and so it is in later life that disaster strikes. A trip, a twist and a fracture results. We all know people who have done it, but few realise that most cases could have been avoided. The problem is particularly bad for women. The balance of bone deposition and destruction is also influenced by hormones. Where males continue to get some stimulation to bone growth through middle and older age, females lose that stimulus with the menopause. It is therefore vital that all women approach the change of life with as dense a bone structure as possible. Although a healthy diet can help, there is only one way to ensure this. Women, even more than men, must continue exercise beyond their schooling years.

★

In many ways the fourth day of the race was similar to the third. We spent much of it at altitude, crossing a series of glaciers. We were roped at times, but there were no great difficulties, and it was not until early evening that the routine plodding was broken by our arrival at a high alpine ridge where another checkpoint nestled in front of two enormous upright stones. We signed in and passed on, having been warned that the descent beyond should be taken with extreme caution. They were not joking. Passing through the giant gateway – a natural equivalent to a cromlech from Stonehenge – we looked down a slope of broken chaos. A huge scree-field, bereft of snow and littered with gigantic rocks, stretched away below us – a mountainside laid waste, with huge angular boulders piled upon one another. It was a nightmare to descend for the rocks were loose and very sharp to hold. Several times they would shift beneath our feet, threatening us with a cascade of mammoth blades of granite that would crush a limb with ease. We were grateful when we reached the bottom unharmed.

From the foot of the scree we descended once more on a glacier. The further down we went, and the warmer it became, the more the surface was covered in meltwater streams which ran along before tumbling into deep cracks in the ice. Soon the whole surface of the glacier was more like limestone caste, and our travel was broken by the need to jump over the ever enlarging cracks, or edge gingerly round the more daunting caverns. The worst difficulty, however, was saved for the last. Without thinking, we cut to the right hand side of the glacier when not much further down the valley we had to climb out via the left. Bursting from beneath the snout was a large river – grey, cold and filled with glacial fragments. Once more, we were on the wrong side, and the day ended with another deeply chilling fight with iced waters.

After a third uncomfortable night, we passed down through a mixture of empty floored conifer forest and sunny grass-filled glades where redcurrant berries grew prolifically. Though we picked them freely to stave off our hunger, we were always wary, for the berries are the staple diet of the brown bear. Fears of bear attack had been with us from the start, despite the fact that we carried a noxious spray and were told that a noisy passage would keep the bears away. To this end, we wore around our ankles little golden bells that jangled as we moved along, but they hardly gave us confidence. For that matter, neither did the advice that we should lie still and play dead if attacked by one of these ten-foot teddies. It was in all the books, but we had no confidence that the bears had read them. Furthermore, you were meant to behave differently if your close encounter was with a 'grizzly'. Quite how, in the heat of an attack, you were supposed to tell the difference between a brown and grizzly bear was not clear. The best way was apparently through examining their droppings. Brown bear shit would contain obvious berries, while grizzly shit would contain bear bells!

Lower down, the forest thickened up again, and we had more fun and games with the vicious plants and hornets. Still, we made good progress until, late in the morning, we met a large and fast river coming in from a side valley. A short distance upstream, the river was spanned by a fallen tree, a log running high across the waters, starting broad and strong on our side but becoming disturbingly narrow before it reached the further bank. There was no other way to cross. One by one we climbed on to the unstable bridge.

When it came to my turn, I was surprised to find how difficult it was to walk along the log. I looked down and found the moving waters almost hypnotic. It took little imagination to know one's fate if one slipped. As the trunk became thinner, there was a marked increase in vibration, and the whole tree bounced up and down. Reaching the far side, I was fearful for my father and shouted back, asking if he wanted a rope. He seemed confident that he could do it without, and very slowly, with care but no hesitation, he started on his way, moving step by step with an expression of infinite concentration and self control, hiding the fears that must have welled up inside him. It was a savage test of nerve. Then, as the last few feet approached, his features began to relax and transformed into a smile of triumph and satisfaction, while we others burst into spontaneous applause.

How we all had changed. A few days back, none of us would have dreamed of making that crossing unroped, but now we all had confidence in our strength and in each other. We had become close, and we urged each other on where necessary, or held ourselves back to stay in a group. We supported ourselves with humour and camaraderie that forged strong bonds. To all intents and purposes, we had become the modern equivalent of a small roaming band – not hunting down some quarry but deriving pleasure from work as a team over days of physical hardship. Perhaps we were dipping into ancient race memories.

★

As we approached the second staging point in the late afternoon we were surrounded by photographers, reporters and TV crews – all hurling questions, mainly at my father. 'Say, how're you doing?' 'Are you hungry, or just tired?' 'D'ya reckon you can keep on truckin?' Dismayed by his rather understated answers – 'Fine', 'No' and 'Yes' – they turned our way, displaying the peculiar American obsession with team psychology and jargon.

'How does the interactive dynamic within a team containing inter-generational family members affect your role?' was one gem I heard aimed at Ran.

'It's great,' he responded smoothly. 'Whenever Vic tries to drown himself in a river, I know that I needn't do anything about it.' The interviewer was suitably perplexed.

At the checkpoint we contemplated the next event over a hasty meal – canoeing two long segments of ribbon lake, broken by portage around a dam. The whole thing was likely to take twenty hours, but there was no question of sleeping. We made our preparations as quickly as possible, but darkness had long since fallen when we made our way down to the lake shore.

Casting off and paddling into the night was tremendous. There was no wind, and before long the moon rose. I travelled with Ran and my father in one canoe while David and Rebecca took the other. It was very peaceful, and the regular dipping of the paddles was the only sound to be heard as we went gliding along. Progress was swift, and within a few hours we came to the dam where we had to put in and travel overland. It took some time to get the two heavy canoes – almost 90 kilos (200 pounds) each, including our kit – up the steep bank to where a narrow road led us a mile or so to the point at which the river could be rejoined. Dragging the boats was forbidden, so we had to carry them one at a time along the dark and deserted road, stopping frequently to rest our screaming muscles. They were not only heavy but very awkward.

Beyond the dam we faced an interesting choice. We could either put the boats into the water close by, and then run some rapids, or we could carry them an extra few hundred yards and miss the worst of the white water. With no misconceptions about our abilities, we opted for the latter and carried the first boat down to the second put-in. There we left it by the side of the road, not far from a small village associated with the dam and power station. My father opted to stay and watch over it when the rest of us set off to fetch the other. While we were away, he decided to create a comfortable berth from the kit in the canoe and settled down to sleep. When we arrived back at dawn, we met a woman walking her dog and a TV crew out filming the Eco-Challenge. Both were delighted by the cameo. Even in Canada, it is rare to find a grey-haired 70-year-old dossing in a canoe by the roadside while a woman walks past with her dog.

We were about to embark once more when some bright bags floating down river caught our eyes. Not far behind them came first one and then the second canoe of another team, both upturned with the crew trying to cling on to the hulls as they swept by. We waved gaily to them but, perhaps not surprisingly, elicited no response. It certainly confirmed that it had been a good choice to avoid the upper rapids. Mind you, the river was not exactly placid where we cast off. All eyes were watching for concealed rocks when suddenly Ran, my father and I realised that we were heading for a sandbar in midstream, just covered by water. Desperately, we tried to take evasive action, turning across the flow and paddling like demons, but it was not to be avoided. Here Ran's denial of canoeing experience was shown to be false. As the boat approached the shallows, sideways on and soon to run aground and turn over, he leapt out, turned the boat head on, pushed it over the bar and jumped straight back in again. It took him no longer than two or three seconds, so David and Rebecca assured me. At the front of the boat, I had seen nothing of what had been going on behind and so I was mystified by our abrupt change in direction and escape from capsizing.

Beyond the rapids the river entered a second long ribbon lake, down which we would paddle for most of the rest of the day. At about lunchtime we stopped briefly, pulling into the shore to show our race card at another checkpoint. We debated sleeping awhile, with Rebecca keen to rest for an hour or two. We had now gone for well over thirty hours without sleep, and before that for several days with only brief uncomfortable stops. Besides, other than my father who had snatched some shut-eye in the boat, we had begun to hallucinate.

Paddling along in a half reverie, we had all been seeing strange things in water, hills or sky – not true hallucinations in the sense that one believed in them but still very real. I had seen huge words written across forest hillsides and some gigantic animal heads carved in the cliffs. It struck me as odd that I couldn't easily make them disappear, even though I knew that they could not be there. Yet they were not unwelcome. Although mildly disturbing, they made the time go by.

Hallucinating or not, I was still concerned about our position in the race. We remained in the top third of the field, but we would be slow on the mountain bike section to come compared to many of the teams behind us. The end of that section would be a critical cut off with many teams prevented from continuing if it was felt that they could not complete the course before the eight-day deadline. I felt sure that we couldn't afford to stop at all, even though it meant another wakeful night to come. I insisted we should continue.

At around six that evening, having paddled for eighteen hours non-stop, we reached the end of the lake and the next staging point. The setting sun cast a glorious light across the water as we cooked and ate. With tired limbs and backs, each of us thought about remaining there – resting, relaxing, and enjoying the wonderful surroundings instead of meeting them head on. But the race officials confirmed my earlier fears. We had to get through the biking section by sundown the next day or our race would be over. Wearily, we picked up our bikes and packed to leave.

The heavy rucksacks on our backs – containing food, fuel, camping gear, safety ropes, and extra clothing – made cycling extremely difficult. Things grew worse when we met the start of the hill, a hairpin logging track that would take us up about three thousand feet and over a mountain pass. It soon became evident that pedalling uphill was not an option with the weights we carried and so, dismounting, we reluctantly began the long push just as night fell.

It took more than four hours to make that climb, and winding back and forth through essentially unchanging forest might have been boring were it not for the hallucinations that altogether eclipsed our earlier experiences. Exhausted from physical effort and lack of sleep, we found our minds entering a wild, free-running state of continuous delusion. All around us as we walked were men, animals, and cities, created from the play of moonlight on trees and rocks, but to our thinking, as solid as concrete. For some reason, I saw many giant Disney characters, or at least things quite like them, and they were not just transient. Some I saw from hundreds of yards away, and then watched as I drew nearer and they became larger. Try as I might to recognise the nature of my delusion, the reality that underlay them, I could not do so, and once or twice I actually walked close by my fictional beings, peering at them from every angle as I approached, passed and left them behind me. It occurred to me that here we might be experiencing something quite traditional. We were on the lands of native North American Indians, and many of them had used periods of fasting and wandering in the wilderness as part of the initiation into adulthood. They believed that, in this way, young men would meet their Gods, and now I wondered if I had met them too. Many of the spectral figures formed from trees were giant living totems.

My father still claimed that he was not seeing illusions – a denial we all accepted until he stopped to have a pee. As he returned to pick up his bike, we were astonished to see him drop suddenly to his knees.

'Look,' he said, his voice filled with wonder.

We all looked but could see nothing except his bent figure crawling forward as he brushed his fingers back and forth across the rocky surface of the mountain track.

'It's carpeted,' he went on. 'Beautiful . . . Why have they done that?'

Clearly he had not had enough sleep after all.

Sometimes we had identical visions. At one point my father, now hallucinating freely, and I were baffled to see the others standing in the road amid piles of office equipment. We could even point out imaginary items to one another. Yet, when we drew nearer, the imaginary world dissolved – desks, chairs and tables melting back into tree trunks, branches and shafts of moonlight.

A little after midnight we reached the top and crossed the pass. Now we could benefit from all the height we had gained, for down the other side would be miles of non-stop descent. Getting on the bikes and free wheeling through the darkness was a welcome respite, but it was not easy to see where we were going on the darkened road, especially with moonlight and our brains continuing to play tricks. We also had to be careful to avoid branches and sharp stones on the track. Unable to steer with accuracy, we ran over things that we might normally have avoided, and it was not long before we had our first puncture. I worked on the bike while the others lay down on the road. It only took a few minutes to complete the repair, but when I had finished and called for us to be off, I was startled to find that the others were fast asleep – not just dozing lightly but needing to be shaken in order to rouse them. The same thing happened repeatedly as we descended, stopping to deal with five or six more punctures.

At the foot of the hill our unmade track joined a tarmac road, and for some miles we pedalled along quite happily. The danger now was dropping off while cycling. Relief appeared in the form of a town which we entered in the early hours of the morning, passing blackened shops and offices, a silent railway station and some sleeping residential roads. Then we left it behind and began a long climb back into the mountains, re-joining a high logging road that would carry us on for the next twenty miles.

Our lack of speed uphill was our saviour for, away from the town lights, our spectral visions returned with a vengeance, and while we were cycling they were positively dangerous. Visions would appear on the track with no warning, and when half asleep one would take sudden evasive action. The road was narrow, bounded on one side by rising rock and on the other by an ever increasing drop. A nasty accident could have occurred had we been travelling any faster.

At the top of the hill, the safety factor of slowness disappeared. Having gained enough height to pass above the huge cliffs, the road now contoured along the mountainside, passing over precipices, rising and falling gently to match the terrain. Inevitably we became faster on the downhill sections, and when our efforts slackened off, our lapses into sleep or visions also became more frequent. After only five minutes of this, David at the front stopped and waited. As we approached, he spoke.

'This is ridiculous. I keep swerving to avoid things that aren't there. We have to rest.'

'It'll be okay once it's light,' I urged. 'Surely we can keep going till then.'

'But someone's going to have an accident,' David insisted. 'It's just too dangerous. What do you guys think?' He turned to the others. 'Should we vote on it?'

The others agreed with David and, reluctantly, I had to accept the mutiny. At the same time, I was adamant that we would not stop for long. Although we had been up for 45 hours, I knew from my previous work with the military that a sleep of just forty minutes can restore vigilance in situations such as tired men watching early warning radar screens. Longer naps – unless of many hours – tend to make matters worse. I set my watch alarm for just forty minutes' time and joined the others lying in the road. They were already fast asleep.

Two hours later I was woken by my father shaking me. My watch had failed to rouse me, and what had woken him was the splashing of rain. It was almost 4.30 a.m., and with no time to waste, we got the others up and restarted. The sleep had helped, and after an hour the light also improved, although it revealed a murky sky. The rain that had woken my father came to little, but the cloud over the mountains ahead looked extremely threatening.

Another three hours passed, during which we made fair progress, although the road became a bit of a roller coaster and we were not all able to cycle up the steeper hills. Two or three teams appeared in the distance behind us, but it did not seem to matter for we knew we would now make the next checkpoint. Indeed, it was fun to watch them as they laboured closer. When they caught sight of my father, you could see astonishment written on their faces. Here they were, nearly seven days into an event for which they had given their all, and only now had they caught up with a 70-year-old. If it knocked their egos, it certainly made me smile.

The storm hit us at around 9 a.m., coming quite suddenly. The drizzly rain had been and gone a few times earlier, but now a sudden wind developed and was soon followed by an intense downpour that turned to vicious hail, beating on our clothing. For about half an hour it was unrelenting, then the wind died away and left regular heavy rain which continued for the next few hours as we descended from the mountains into another town. From here it was just twenty miles along a surfaced road all the way to the stage finish. We had as good as completed the ride and began to wonder what was next.

An hour later, while pedalling flat out, a sudden movement on the other side of the road caught my eye. It was a large white dog, coming out from a farm gate, and the sight of my bike seemed to arouse its ire. Barking furiously, it ran straight for me as I swerved. Although I got past the dog, at the speed I was going I could not regain control and was soon off the edge of the tarmac and on loose shingle. This removed any remaining chance of steering and, still going at great speed, I left that as well and plunged down a ten-foot rocky ditch.

For reasons that have never been clear to me, I took the whole impact on my right shoulder, leaving everything else unharmed. The shoulder pain was so intense that I retched repeatedly. After a few seconds, the waves of nausea settled and I could think more clearly. I was sure that it was broken. Slowly, however, as the initial shock subsided, I realised that I could move it a little, and that made a fracture less likely. I had been fortunate, and as the others caught up and halted anxiously, I reassured them that we could carry on.

The reassurance was fine but pretty optimistic. My shoulder throbbed terribly. If the white-water rafting came next, how could I steer? The weather deteriorated further, rain falling from leaden skies, driven by the wind, and we were utterly sodden and cold by the time we turned from the roads on to a track leading to the finish. The track was a sea of mud, churned by the race support vehicles which had preceded us. In fact, those vehicles had been that way twice, for we had returned to the very field by the Lillooet from which we had started. More than a week of travel had seen us complete an enormous circle through the mountains, and now we would have to repack quickly and be off again.

Coming into the field, all was strangely quiet. A couple of teams that had passed us earlier in the day were crouched by their tents. They stood up to clap and whistle as my father appeared. But where was everybody else? The majority of teams were still behind us, and so most of the support crews should have been here and waiting. We caught sight of our own vehicle, close to the river and the administration tent. Mo and Moira were nearby but, although they too waved and clapped as we approached, they had an air of dejection about them and wore frowns instead of smiles.

'What's happened?' I asked.

'It's off,' said Mo gently. 'Nobody's going any further.'

'Why?' I demanded. 'There's plenty of time yet.'

'The storm,' she replied. 'It's hit the teams in the mountains and they're all lost or stuck. They won't let anybody else up there.'

Slowly, with many interruptions, the details emerged. The rain and hail which we had encountered that morning had caused havoc in the mountains ahead where the next stage was a short but technical climbing section. A foot of snow, driven by 70-mile-an-hour winds, had pinned down all but the leading three teams who had already crossed the highest ground when it hit. Tents had been destroyed, and big crevasses on the glaciers had become extremely dangerous. A rescue operation was in hand to lift the teams off the high ice before another storm struck. The weather forecast remained appalling.

At first, I was furious. I had not come so far to be told that I had to stop because it was breezy and snowing. Ran and I had crossed the largest ice field on the planet in weather that made this look like a summer's afternoon. We were not going to be dictated to about what we could or could not do here. I raced up to the administration tent and started to argue. My impression was that the storm was not their only problem. The whole race had been badly misjudged in terms of difficulty, so that teams were now scattered over hundreds of miles. The reason for the absence of many support vehicles was that some teams had not even started the biking section and several, including Helen Klein's, had still not finished the big hike. It all struck me as highly incompetent, and I'm afraid the poor junior staff at the checkpoint got an earful from me.

Slowly my wrath subsided and I began to see sense. If we had come this far, moving relatively slowly, the teams that were a long way back should never have entered the race. Ahead, the snow was wet and heavy and, drifting with the winds, would make deceptive crevasse bridges. Avalanches would also be a threat, and I had to accept that it would be wildly irresponsible to allow any more teams, some of them with little true mountain experience, into the area. If they had to stop some, they had to stop all. I tried to persuade them to let us cycle round the range by road, so that we might still do the white-water rafting the next day. The ride would be long and my shoulder would probably make the rafting impossible but, anyway, they would not hear of it. Their resources were far too stretched.

There was nothing more to be done; we simply had to look on the bright side. Of the twenty-five teams that had gone faster than we had, only fourteen had left this camp before the storm struck. Eleven of those were now being rescued, and only three of more than eighty teams that had started were to finish the whole course. We had done well.

I looked across at my father and saw an exhausted man in pain from those dreadful sores he had carried from the first day. His eyes were sunk in pits of sheer fatigue, yet from within them came tremendous sparkle. He had done what for most, whatever their age, would have seemed impossible, and now he was probably grateful that we had to give up. Yet I also knew that if circumstances had been different, his pack would have been on again in the hour and we would have departed once more. I felt enormous pride. He may have been an elderly weekend walker, but he had taken up horse-riding, canoeing, rope climbing and glacier crossing, and he had kept on going to pass through a stern test of fitness. Above all, he had shown that action was not only the preserve of the young or the super-normal. I dearly hope that I may do the same when I reach his age.

CONCLUSION

★

In Sickness or In Health

Positive health requires a knowledge of man's primary constitution and of the powers of various foods, both those natural and those resulting from human skill. But eating alone is not enough for health. There must be exercise of which the effects must likewise be known. If there is any deficiency in food or exercise, the body will fall sick.

Hippocrates – 5th century BC

ALTHOUGH at times I have strayed far from the theme, this book has tried to examine human performance and health in the context of our evolutionary design. I have argued that the whole balance of our physiology is set up to overcome the rigours of a mobile life with a variable, mainly vegetarian food supply – the life led by our ancestors, who were genetically almost identical to ourselves, until just 10,000 years ago. As a consequence, the resilience of the human frame is far greater than many people realise. This particularly applies to endurance. Although few are blessed with the muscle fibre mix, hearts and lungs to become champions, most could run a marathon if they wished. Indeed, they could go much further, and I have no doubt that many could run over the Sahara or walk across Antarctica if the opportunity arose. I have also suggested that age is less of a barrier to physical fitness than is generally accepted, and indeed it is in later years that the benefits of a continued active lifestyle are felt most. Exercise can truly help to maintain youth but, unfortunately, most people go out of their way to avoid it.

Our evolutionary heritage has also granted us amazing attributes when it comes to environmental hardship. Because the ancestors of all modern humans were still living in the heat of Africa as little as 100,000 years ago and most evolutionary adaptations move very slowly, we all have an extraordinary capacity to cope with the heat if given some days in which to acclimatise and a reasonable water supply. Simultaneously, despite the fact that it has been only ten thousand years or so since some races came to dwell in much colder climes – inadequate time for evolving better physiological defences against the cold – our intellect allows us to operate in the most frigid regions of our planet. Yet the idea that normal members of the population today should ever be exposed to extremes beyond those met while sunbathing or starting the car on a winter's morning, is anathema. We hear of these capacities in tales of people who either deliberately test themselves through sport or expeditions to the wild and remote landscapes, or who meet genuinely life-threatening situations through accident or misadventure. Everyone who hears of them is duly impressed, but they fail to realise that they too have the capacity to survive.

There is also the down side. Although natural selection honed our metabolism to deal perfectly with our ancestral diet, the development of farming profoundly changed the nature of what we ate. At the same time, the advent of civilisation allowed people to pursue less active lifestyles, a change which has accelerated greatly since the Industrial Revolution and even faster in the age of silicon. Now the majority of adults and children in the developed world are increasingly sedentary and eat foods that are unhealthy. The consequence is an epidemic of ill health. A surge in the numbers of deaths from heart disease, strokes and some cancers in the past century has almost kept pace with the decline in mortality from infectious diseases and some other illnesses that has come about through medical and social advance. Furthermore, in the modern industrial world more than half of all adults are overweight and, if current trends persist, half of those will qualify as medically obese by the year 2005. This creates a burden upon health and leads to great physical and financial cost.

So, what can be done about it? It seems to me that, for any individual, there is a clear choice: either one can hope it is not true, that it will not apply to oneself, and carry on regardless; or one can recognise the problem and try to rectify it as best one can within the constraints of modern society. With modern medicine, you probably stand a good chance of reaching old age either way, but your choice may still have considerable influence on whether you live in sickness or in health. It is all a question of risk.

Although a lazy lifestyle with no dietary constraints has wide appeal, it is a route that leads directly to the possibility of getting too fat, becoming infirm, or of experiencing the crushing chest pain of a heart attack. Nobody would be surprised if the engine of a Formula 1 car were to clog and break down after being left ticking over for most of the time with just an occasional run to the supermarket. The same applies to humans. We too are set up for high performance and will clog if we are idle for too long. Furthermore, because of that inbuilt propensity to eat as much food as we can get hold of, particularly fat, we are not just idling but have filled the tank with heavy oil instead of light high-octane fuel. It puts us on to the road to sickness and may cause far more trouble than a breakdown. In our case, blockages can literally affect our lifeblood.

Where, then, is the road to health? The most obvious answer – a return to the lifestyle of our evolutionary roots – would be both impractical and unwelcome. There is no possibility that current society will go away and very few of us have any sane wish to return to the hunter-gathering of our Cro-Magnon past. If we did so, not only would we forego most recreational pleasures and comforts but many benefits to health reaped by technology would disappear. Nobody would be fat or die from coronary heart disease, but millions would perish from starvation, infections and injuries. It would be a truly retrograde step and far more people would suffer. Instead, what is needed is compromise, changes that can be incorporated into our modern lifestyles without imposing too many dietary restrictions or activity levels that are either too unpleasant or too time consuming.

On matters of diet, we could arguably get very close to the ideal. It is entirely practical to consume meat rarely, eat fish fairly often, and otherwise consume a predominantly vegetarian diet. Evolution, however, has given us such a strong liking for meat that many, myself included, would find it hard to cope with eating it only once or twice a week. The compromise for many could perhaps be having meat on alternate days in reduced amounts, while avoiding too much fat from other sources, especially the frying of foods, which turns inherently healthy products such as potatoes into such health hazards as chips. Fortunately, these changes need not solely rely on an act of willpower. The increased proportion of carbohydrate and protein in your resulting diet will allow your control systems to work properly. Your body will recognise much more of your intake and your appetite will be better satisfied accordingly. You will also get a more plentiful supply of micro-nutrients to help with other aspects of health, as well as eliminating any question of hunger being driven by a shortage of vitamins or trace elements. Eat at least five portions of fruit or vegetables every day, include some fish in your diet twice a week if you can, and avoid too much salt and sugar – best achieved by steering clear of too many manufactured foods – and you should be well on your way to health.

Would this be enough? Unfortunately, the answer is no. As was realised by Hippocrates, we cannot maximise our chances of avoiding illness and infirmity if we remain inactive. Even the most healthy of diets may provide inadequate vitamins if you are eating very little because your demands from activity are low. Exercise also brings many other benefits unrelated to food intake, including increased strength of heart, lungs, muscles and bones, but it requires time and effort. Both men and women find it hard to make either.

As a first step, more activity needs to be incorporated into everyday tasks. Unless you are living or working in a tower block, you should avoid lifts as a matter of principle, and when going short distances, you should always walk. Every day in our larger cities, huge numbers of commuters and shoppers take subways and buses for just a stop or two when it is often easier, quicker and more interesting to stroll down the street. Don't always sit. Many jobs may tie you to a desk or screen for much of the day, but informal discussions with one or two colleagues can be made while standing, and the same goes for telephone conversations. Put your phone out of reach of a comfortable chair. Never mind if people think you odd, you can add considerable physical exertion to the day with no cut in efficiency.

All these measures will help your overall energy expenditure and calorie control, but they will still not be enough. For optimal health, you need to stress the system properly. Those of us with muscles close to a fifty-fifty fast/slow fibre mix and average hearts and lungs need surprisingly little vigorous exercise – by which I mean work at a level hard enough for the heart rate to rise and make you short of breath – to achieve 90 per cent of our genetic fitness potential in terms of aerobic capacity. Just one hour a week of aerobic work will see most people up to this level. It would see your oxygen delivery systems ready for a marathon run, even if it would not be quite enough to make muscles, bones and joints ready for such a race. If you finished, you would suffer terribly. Of course, many might feel that such a consideration is irrelevant for they have no plans to run in marathons anyway. But only by exercising frequently can you develop muscles that will clear excess fats from the circulation, produce platelets which are less sticky, bones strong enough not to break, and joints lined with tough cartilage that cannot easily be harmed. To do that will take two or three hours of hard work a week and, if you are anything like me, it can be hard to make yourself do it.

Why is exercise so hard to undertake? Boredom, discomfort, fatigue and lack of time must all be contributory, but there is an additional problem. Goals such as good health in old age are far too nebulous to provide the motivation. For that reason, it is often only those with a more definite short-term goal that succeed. People who need to lose weight, for example, are more likely to continue to put in time and effort than those who are not too fat. If you give yourself a definite aim, exercise acquires a purpose. You need to set yourself a challenge.

What sort of challenge is required? The answer must be different for every individual and its nature is probably unimportant. What does matter, however, is that you seek to undertake something physically demanding that is beyond your currently perceived limits. Rather than simply taking up jogging, you should decide to do next year's local half marathon. If you already run, or a half marathon looks too easy, decide to attempt a full length race in London, Boston or New York. Of course, it need not just be running or an organised event. Plan to spend a weekend mountain biking along the full 80 miles of the South Downs Way or decide to walk around the coastal path of Cornwall. It little matters exactly what is performed, so long as the undertaking is difficult enough that achieving it will bring pride and self respect. Take on something hard enough and achieve your goal, and you will never regret the experience. The capacity to overcome hardship is inherent in us all and, if we use it, we not only gain health but find satisfaction.

And what for me? Will I go on looking for challenges in life, or settle down and hope that what I have done up to now will protect me from the ravages of heart disease or osteoporosis? Will it be enough to allow me to live to a healthy old age? I think not. Indeed, as I write these final lines another challenge has appeared on the horizon. Just three weeks ago, another quiet Sunday afternoon was disturbed by the ringing of the telephone. When I heard the soft American drawl of Mary Gadams, I thought she must be calling to suggest another go at the Eco-Challenge. This year's race – to be held in the Atlas mountains and the Sahara – had just been announced. No, she had a quite different proposal.

'Mike,' she said. 'Are you still running?'

I assured her that I was, though the writing of this book has curtailed it of late.

'Would you be interested in joining in some marathons for charity?' she went on.

There was something in the 'some marathons' that sounded suspicious but I responded positively.

'Sure,' I said. 'How many have you in mind?'

'Oh . . . seven,' she replied, but there was teasing in her voice. 'One in each continent.'

Wow! My mind raced. It sounded like a great challenge and a good reason for getting fit. My immediate guess was that we would be entering a major city race in each, but the significance of 'seven' had not escaped me. She must be thinking of a trip down to Antarctica to run one of them. This sounded like fun.

'When are you talking about?' I asked.

'Next year . . . early November.'

That did not quite fit my assumptions. 'Early November' sounded like a short time. Too short to do the London, New York, Tokyo, Sydney, Cairo or whatever.

'What, all of them?'

'Yeah . . . That's the plan.' She laughed. 'Seven marathons, seven continents, seven days and raising a million dollars for charity.'

Now I understood. This was going to be more difficult than I had imagined, but then that is the whole point.

'Okay,' I replied. 'Count me in.'

Sources and Further Reading

THESE SUGGESTIONS are for those interested in exploring further the subjects raised in this book. It is neither a full bibliography nor a complete list of scientific references, although the list does provide useful starting points for those wishing to examine the primary literature. They are listed with comments where appropriate. Scientific references are given in standard format using recognised abbreviations for the journal titles. Publishers are in the U.K. unless otherwise indicated.

Human Evolution

Richard Dawkins' *The Selfish Gene* (Oxford University Press, 1982) and *The Blind Watchmaker* (Longman, 1986) give comprehensive accounts of the workings of natural selection and evolution, particularly emphasising how genes and the DNA dictate everything that happens in the evolutionary process. This approach has been criticised recently by Stephen Rose in *Lifelines* (Penguin Press, 1997) as being too reductionist. Rose suggests that the forces of natural selection operate at a much higher level than the single genes, selecting for the whole organism within the context of its environment.

Jared Diamond's *The Rise and Fall of the Third Chimpanzee* (Radius, 1991) sets out a thorough and very readable account of human evolution, pre-history and the development of civilisation. It also describes in much more detail than I have done the relationship between man and the chimpanzee, giving details of how the closeness of our relationship has been demonstrated. *The Naked Ape* by Desmond Morris (Jonathan Cape, 1967; illustrated edition, 1989) is now a classic, providing a summary for laymen of human development. It discusses the reasons for our becoming upright, as well as the development of human intelligence, tool use and speech. It is, however, heavily criticised by Elaine Morgan for being too male orientated in *The Descent of Woman* (Souvenir Press, 1972). Her book gives a most readable, if rather biased, presentation of the 'marine theory' of human evolution – the idea that we spent a significant period after dividing from the chimpanzees as a predominantly water-dwelling mammal. A more balanced analysis of the pros and cons of this theory is reported in the proceedings of a conference on the subject: *The Aquatic Ape: Fact or Fiction* (eds M. Roede, J. Wind, J. Patrick & V. Reynolds, Souvenir Press, 1991).

Sports Physiology and Nutrition

Two recommended books giving broad coverage of the subject of muscle function and training are: *Textbook of Work Physiology – Physiological Bases of Exercise* (P.-O. Astrand & K. Rodahl, McGraw-Hill, New York, 1977) and *Physiology of Sport and Exercise* (J. Wilmore & D. Costill, Human Kinetics, Champaign, Illinois, 1994). An excellent and reasonably up-to-date review of the interaction between nutrition and exercise is given in *Foods, Nutrition and Sports Performance* (eds C. Williams & J. Devlin, Spon, 1992) and I have recently written a review of the dietary needs for very protracted physical effort – *Nutrition for prolonged endurance exercise* (M. Stroud, Proceedings of the Nutrition Society, 1998).

Heat and Cold

By far the most complete coverage of the human physiological responses, adaptation and acclimatisation to heat and cold can be found in the book *Human Performance Physiology and Environmental Medicine at Terrestrial Extremes* (eds K. Pandolf, M. Sawka & R. Gonzalez, Benchmark, Indianopolis, 1988). A much briefer review is outlined in a chapter written by myself and entitled 'Environmental Temperature and Physiological Function' in the book *Seasonality and Human Ecology* (eds S. Ulijaszek & S. Strickland, Cambridge University Press, 1993).

Polar Expeditions

Accounts of various Polar expeditions to which I refer in Chapters 4 and 6 include: *In the Footsteps of Scott* (R. Swan & R. Mear, Jonathan Cape, London, 1987); *Scott's Last Expedition* (Capt R. Scott, Smith, Elder, London, 1913); and *The Worst Journey in the World* (A. Cherry-Garrard, Chatto and Windus, 1922).

The full story of my own expedition across Antarctica with Sir Ranulph Fiennes is given in *Shadows on the Wasteland* (M. Stroud, Jonathan Cape, 1993). Accounts of journeys across the sea-ice to the North Pole are given in Robert Swan's *Icewalk* (Jonathan Cape, 1990) and Ranulph Fiennes' *To the Ends of the Earth* (Hodder, 1983).

Scientific papers covering the physiological studies made on my own Polar journeys include: 'Nutrition and energy balance on the "Footsteps of Scott" expedition 1984–86' (M. Stroud, Human Nutr. Appl. Nutr. 41A: pp. 426–433, 1987); 'Measurements of energy expenditure using isotope-labelled water ($^2H_2^{18}O$) during an Arctic expedition' (M. Stroud, W. Coward & M. Sawyer, Eur. J. Appl. Physiol. 67: pp. 375–379, 1993); 'Nutrition across Antarctica' (M. Stroud, British Nutr. Found. Bulletin. 19: pp. 149–155, 1994); 'Protein turnover rates of two human subjects during an unassisted crossing of Antarctica' (M. Stroud, A. Jackson & J. Waterlow, Br. J. Nutr. 76: pp. 165-174, 1996); 'Thermoregulation, exercise, and nutrition in the cold' (M. Stroud) in *Physiology, Stress and Malnutrition: Functional Correlates, Nutritional Intervention* (eds: J. Kinney & H Tucker, Lipincott-Raven, New York, 1997); 'Energy expenditure using isotope-labelled water ($^2H_2^{18}O$), exercise performance, skeletal muscle enzyme activities, and plasma biochemical parameters during 95 days of endurance exercise with inadequate energy intake' (M. Stroud, P. Ritz, W. Coward, M. Sawyer, D. Constantin-Teodosiu, N. Brown, P. Greenhaff & I. Macdonald, Eur. J. Appl. Physiol. 76: pp. 243–252, 1997).

Nutrition, Exercise and Illness

Excellent sources of information on the benefits of exercise to all aspects of health including heart disease, obesity and ageing are *The Benefits of Exercise: the Evidence* (P. Fentem, J. Turnbull & J Bassey, Manchester University Press, 1993) and *Physical Activity, Fitness and Health* (C. Bouchard, R. Shepard & J. Stephens, Toronto, 1994).

The literature discussing more specific details of factors underlying heart disease is enormous. *Preventing Coronary Heart Disease in Primary Care: the Way Forward* (National Heart Forum, H.M.S.O., 1995) would be a good starting source and a review of the benefits of physical activity on heart disease is given in 'Exercise in the Prevention of Coronary Heart Disease: Today's Best Buy in Public Health' (J. Morris, Med. Sci. Sports Exerc, 26: pp. 807–814, 1994). The British Nutrition Foundation have published numerous books and articles on the interaction of diet and coronary problems which include: *Diet and Heart Disease – A Round Table of Factors* (ed M. Ashwell, BNF, 1993); *Fish and coronary heart disease – has the tide turned?* (M. Sadler, BNF Bulletin 20, 1995); *Homing in on homocysteine* (U. Arens, BNF Bulletin, 20, 1995). The *Allied Dunbar National Fitness Survey* (Sports Council and Health Education Authority, London, 1992) provides figures for the increasingly sedentary nature of the UK population, while further figures from Britain and beyond are cited in some of the works on the causes of obesity I have listed below.

Why We Get Sick (R. Nesse & G. Williams, Times Books, New York, 1994), published in the UK as *Evolution and Healing. The New Science of Darwinian Medicine* (Weidenfield & Nicholson, 1995), examines many aspects of human health and disease from an evolutionary point of view, including brief discussion of heart problems and obesity. However, the authors pay surprisingly little attention to the role of declining activity levels in any of their discussion. Thorough general histories of the changing face of disease in society can be found in Roy Porter's *The Greatest Benefit to Mankind: A Medical History of Humanity from Antiquity to the Present* (HarperCollins, 1997) and *Diseases, their Emergence and Prevention* (eds. Trowel and Burkitt, Harvard University Press, 1981.).

The Causes of Obesity

In *Energy Balance and Obesity in Man* (Elsevier and North Holland Press, 1974) John Garrow gives a complete description of the interaction of dietary intake, resting metabolism and activity levels underlying the development of obesity. Once again, articles and reports published by the British Nutrition Foundation provide many useful starting points to further reading: *What should public health policy be towards 'overweight'?* (B. Tuxworth, BNF Bulletin 19, 1994); *Body Weight and Health* (ed M. Sadler, BNF, 1996); *Fat and obesity* (J. Westrate, BNF Bulletin 21, 1996). Two excellent articles emphasising the fact that the obese do not have slow metabolism and hence abnormally small dietary needs are 'Don't Blame the Metabolism' (A. Prentice. MRC News, Autumn 1995) and 'Obesity in Britain: Gluttony or Sloth?' (A. Prentice & S. Jebb, Br. Med. J. pp. 311, 1995). A specific review of the recently postulated role of leptin in obesity can be found in *Leptin – the 'new' player in energy balance and obesity* (P. Trayhurn. BNF Bulletin 22, 1997).

Exercise and Ageing

Many of the books listed in the second and fifth section above contain much information on the benefits of exercise with age. More specific information can be found in scientific papers including: 'Endurance training in older men and women. 1. Cardiovascular responses to exercise' (D. Seals, J. Hagberg, B. Hurley, A. Ehsani & J. Holloszy, J. Appl. Physiol. 57: pp. 1024–1029, 1984); 'Effect of age and training on aerobic capacity and body composition of master athletes' (M. Pollock, C. Foster, D. Knapp, J. Rod & D. Schmidt, J. Appl. Physiol., 62(2): pp. 725–731, 1987); 'Fifteen-year changes in exercise, aerobic power, abdominal fat, and serum lipids in runners and controls' (B. Marti, M. Knobloch, W. Riesen & H. Howald, Med. Sci. Sports Exerc. 23(1): pp. 115–122, 1991).

Acknowledgements

Many people have contributed indirectly to this book – educating me in the medicine and science underlying it during my time in the Anthropology Department at University College London; the Medical School at St George's Hospital, London; the Centre for Human Sciences, Farnborough; and in my current post at the Institute of Human Nutrition, Southampton. Particularly important in this respect were Prof. Mike Stock, who first pointed me towards research in human physiology and metabolism, and Prof. Alan Jackson, who, in his teachings on nutrition, has made me think in terms of teleology. Thanks are also due to all those who made my expeditions and other outdoor activities possible: my father, who started it all off by taking me to the hills as a child and teenager; Robert Swan and Roger Mear, who set up the 'Footsteps of Scott' and so involved me in major expeditions; Ranulph Fiennes, who was both an inspiration and a wonderful companion on our many trips together; Lawrence and Morag Howell, who supported those trips I made with Ran; Chris Lawrence, who introduced me to the Sahara race and hence to ultra-distance events; and Mary Gadams who first got me involved with the Eco-Challenge.

Lastly, I owe a huge debt to those who were directly involved with the writing of this work: my wife, Thea, who helped me put my thoughts together by discussing things and reading my draft efforts, and who coped with so much while I was busy, and my children – Callan and Tarn – who had 'daddy behind the computer' for weekend after weekend. Tony Colwell, my editor at Jonathan Cape, was patient, critical, methodical and encouraging in turn, and I am sure that without him the book would never have been completed.

PUBLISHER'S NOTE

Mike Stroud was awarded the O.B.E. for his part in the first unsupported crossing of the Antarctic continent from coast to coast, which he accomplished with Sir Ranulph Fiennes in 1993. Dr Stroud's account of that journey, *Shadows on the Wasteland*, was described in the *Sunday Times* as 'fascinating'. Everest climber Stephen Venables called it an 'excellent book' in which 'Stroud was determined to allow the upper lip to wobble, and discuss honestly his and his companion's human weaknesses.'

SALVAGE

SALVAGE

keren david

www.atombooks.net

ATOM

First published in Great Britain in 2014 by Atom

A CIP catalogue record for this book
is available from the British Library.

Hardback ISBN 978-0-349-00137-1
C-format ISBN 978-0-349-00191-3

Typeset in Palatino by M Rules
Printed and bound in Great Britain by
Clays Ltd, St Ives plc

Papers used by Atom are from well-managed forests
and other responsible sources.

MIX
Paper from
responsible sources
FSC
www.fsc.org   FSC® C104740

Atom
An imprint of
Little, Brown Book Group
100 Victoria Embankment
London EC4Y 0DY

An Hachette UK Company
www.hachette.co.uk

www.atombooks.net

For my son, Judah.
A constant inspiration.

1: CASS

Pretty much everyone in town knew my name, although not because there was anything special about me. That was fine, or, at least, I was used to it.

But the day I met Will it wasn't fine at all. It hadn't been fine for exactly twelve days. Dad was still hanging in there as a government minister, but was better known as 'Olly with the Dolly'. Or 'Monty in a Muddle'. Oliver Montgomery MP was having an affair – and a baby – with his intern, and now everyone didn't just know my name: they stared, pointed and whispered, too.

Even before Dad was on the front page of all the papers, day after day, I was always careful. I could never get away with being one of those girls who hitched up their skirts to flirt with the grammar school boys on the bus. We lived in a small town, and all the busybodies who knew my parents from the Conservative Party would certainly gossip if they saw me step out of line. It wasn't worth rebelling. It'd be too embarrassing.

Mostly I didn't bother with the bus, and walked instead. When I did get it (if it was raining, or I was in a hurry to get home before youth choir or county orchestra or a Conservative Future meeting or whatever), I made sure I sat at the back, kept away from the crowd, pulled out a book, plugged in my headphones and enjoyed twenty minutes of peace and quiet.

That day I was sitting near the back next to a large elderly lady, whose cat hissed and spat from a closed basket on her knee. The basket jiggled against my thigh, which was annoying. I'd have preferred to sit next to the window, tracking the raindrops as they dribbled down the glass.

I hardly noticed when Will spoke to me: he touched my shoulder to get my attention, and I jumped and squeaked.

'What? What is it?' My voice sounded high and panicky, and it was all his stupid fault.

'Um, sorry,' he said, whipping his hand away as though I'd burned it. 'It's just … you're Cass Montgomery, aren't you?'

I'd seen this boy around before, but I'd never spoken to him. He wasn't in choir or orchestra, although I'd seen him once or twice at the inter-school debating competitions. He'd done a good job last time, opposing a motion about school uniform promoting discipline.

I knew about him, though. I knew his name was Will Hughes and I knew he was in the sixth form, and I knew he was incredibly popular and supposedly funny and bright. I knew all about which girls had been out with him and which girls wanted to. I knew these things the same way that people knew I was head-girl material and heading for Oxford. He had a reputation: a more interesting one than mine.

2

He was a local landmark, like the art deco cinema. He was massively tall, all gawky limbs, but strangely graceful – he floated like a daddy long legs. His afro sprung loud and proud around his head, and he wore ludicrous, outsized tortoiseshell-rimmed specs.

I knew he went to Bonny's, which is short for Bonnington's School. It's where you go if you don't get into the grammar schools but your parents have enough money to avoid the comprehensive.

My brother, Ben, had started at Bonny's in September. We were all worried because Ben doesn't do disruption very well, which Dad should have thought of before he ... but I didn't want to think about Dad.

'Why do you want to know?' I didn't mean to sound aggressive, but I also didn't want to encourage him. Couldn't he respect my privacy?

'My name's Will. Will Hughes. I want to discuss something with you.'

The back seat was full of Bonny's sixth formers, with Will in the middle. I knew one of them, Jordan Strachan, as he played the bassoon in the county youth orchestra. They were all in hoodies and jeans, but Will had on a tweed jacket, indigo cord trousers and a bright orange shirt. His bag was a neon pink *Dora the Explorer* rucksack, meant for a six-year-old girl. He was like a flamingo among pigeons – attractive, sure, but totally out of place. Except you never saw him without a huge flock of pigeons.

I really wasn't in the mood to talk to exotic, attention-seeking strangers. 'You haven't picked a great location,' I said, in my politest brush-off voice.

He didn't get the message. 'I thought we could go for a coffee.'

'Oh, did you?' I was aiming for sarcastic, but realised too late that I sounded potentially interested.

'You're Ben's sister, aren't you? I mean, I know you are. I wanted to talk about Ben.'

'You've done your homework.' Surely the sarcasm was biting enough this time?

'I'm his form's peer buddy,' he said. 'We're like a sixth-form emergency squad – there to help if kids in the lower years need individual attention, that sort of thing. I've been keeping an eye on Ben.'

Obviously Ben needed special attention. He always did. He struggled with some academic stuff, and he totally lacked any ability to make friends. You'd have thought he would have been used to it by the age of eleven, found coping techniques, accepted his fate as a loner, identified just one person to hang out with, but every day it was as though he'd only just realised that no one liked him. It broke my heart, so I tried not to think about it too much.

'Oh, well, thanks for that,' I said, looking away. Couldn't he see that I didn't want to talk to him? Especially about Ben. Especially right now.

'I just thought, perhaps we could talk about what might help him? I mean, obviously you know him better than I do.'

OK, this was too much. Talk about intrusive. He was using my brother, using my family's situation, to ask me out. On the bus, with Jordan Strachan and all his friends listening. What an insensitive idiot.

Luckily Grace wasn't next to me: she'd have accepted on my behalf as soon as he spoke. She was at the front of the bus, the giggling centre of a group of grammar boys. She'd been late out of school, detained by Miss Graham, head of the sixth form, for one of their ritual chats about skirt length and leggings. Grace thought I was tragic and frigid and sad, and her mission in life was to make me more like her.

I refused to fall into her trap. She might be able to turn out A* essays, do loads of extra-curricular activities and have an active love life, but I didn't think I could. Instead, I purposely kept my make-up minimal and my clothes functional, so I wouldn't attract unwelcome attention. It wasn't just the fear of gossip: I didn't want my future messed up by some stupid teenage romance.

Not that Will Hughes was even a suitable candidate for romance, with me, anyway. I'd seen him on the bus before, holding hands or more with various girls including Daisy Travers-Manning, whom I remembered from Brownies, and Ruby Thomas, who was in my year but none of the top sets. He had a type, and that was blonde and curvy and giggly. I wasn't his type.

I didn't have a type, but if I had, he wouldn't be it.

'What do you think?' he asked, and one of his friends sniggered.

'I'm busy.'

'It needn't take long. We could get off at the high street. We don't have to go to Starbucks. It's just that I think Ben's not very happy right now and maybe I could help.'

How dare he even begin to discuss my family in public? I opened my mouth to tell him what I thought of him. Then I shut it again.

I didn't want some sort of scene. Especially scenes that outsiders could construct as racism, or worse (is there worse?). It wouldn't look good, especially when the entire British media was already running stories on my family.

'Maybe another day,' I told him. He grinned, much more pleased than he had any right to be.

'Look, give me a call, or drop me a text or something,' he said, presenting me with a business card – a business card! What sixth former has a business card? What a poser!

I shoved it into my pocket, looked back at my book. We were nearly at the high street. Usually Grace got off here, headed for Starbucks for coffee and flirting. I never bothered. I actually had no idea how she got any homework done at all.

The bus lurched to a stop, the cat basket crashed onto my knee and I went flying off my seat, cannoning into weirdo Will as he and his mates filed towards the door. For a second I clutched at his scratchy tweed jacket, breathed in his surprisingly sweet smell.

'Oof!' he said, clutching his stomach, as I scrambled back into my seat, hot-faced with embarrassment. His friends howled with laughter. The cat lady glared at us. I wished I could just disappear.

'Bye, Cass!' he yelled back at me, just as he got off the bus. How could he laugh at me in front of everyone? I crumpled his card in my pocket and tossed it onto the seat next to me, but the stupid cat woman said, 'I think you've dropped something,' and handed it back to me. I was too humiliated to disagree, so I shoved it back in my pocket again.

2: CASS

Grace sank into the seat in front of me. 'What was all that about?'

'All what about?' I snapped, and she said, 'Oh, come off it, Cass! Why was that boy shouting goodbye to you? Will Hughes? The one with the hair and the clothes and the bag and the glasses? He's *gorgeous*. Ruby was *devastated* when he dumped her.'

'It wasn't him,' I said. 'It was Jordan Strachan. He plays in the county orchestra. We'd been discussing Mozart.'

Grace grimaced. She thought orchestra was hopelessly boring, but she admitted it would look good on my university application form.

'Why aren't you in Starbucks with all your admirers?' I asked.

'Because I wanted to see that you were all right, of course. I'm going to come back to your house and make you hot chocolate and you can tell me how things are going. I feel so bad for you, and we haven't talked at all, and I want to be there for you, Cass. After all, you're my best friend.'

Grace and I decided we were best friends when we were six. I liked her because she had a pet rabbit and she liked my orange hair. But recently I'd been feeling that our bond was based on a shared history of Brownies and summer camps, parties and prep school and, to be honest, we had hardly anything in common any more. I'd been looking forward to going to university and making new friends more like me. Quieter. Serious. Not given to asking personal questions.

'That's so nice of you,' I told her, 'but you know, I've got that history essay to finish, and then there's youth choir. I haven't really got time tonight.'

Grace pouted, but she didn't argue. She'd done her bit. My friends were being very supportive, but in a competitive way. I put it down to the general ethos of high achievement: we put the 'best' into the phrase, 'Best Friends Forever'.

Grace, Victoria, Megan and Alice kept picking away at my nicely frozen numbness, like Arctic explorers tackling a glacier, and I really didn't want it to start cracking or melting somewhere inappropriate, like a history lesson or during senior orchestra. Or even somewhere appropriate, like my bedroom or the girls' toilet. I needed to maintain a personal Ice Age, and I had to avoid the global-warming effect of concerned sympathy.

I knew I'd made a good decision when I got home and found Mum sobbing over a pile of photo albums on the kitchen table.

'Mum! What are you doing?'

She looked up at me and I was shocked. Her hair lay flat

against her head, limp and greasy. Her skin was blotchy and her eyes red-rimmed. I knew she was upset, obviously she was upset, but seeing my smart, confident Mum unravel like this … well, anything could happen. A hurricane could lift our house and blow it away. An oil tanker could smash into our front garden. Mum always wore make-up; she went to the hair-dresser every week:'An MP's wife has to look the part,'she'd say. Who was this stranger?

I hated Dad for destroying her like this.

'We were a happy family, Cass, weren't we? I'm not making it up?'

I sat down beside her, moved her wineglass out of reach.'Of course we were happy,' I told her. 'We had everything. Of course we were.'

Together we looked at the pictures. Holidays: Tuscany, Vancouver, Devon. The house in France. Mum, wearing a big sunhat, waving at the camera. Dad, in sunglasses, reading the *Daily Telegraph*. My stomach clenched when I saw him. What was he thinking, getting involved with an intern? He was actually old enough to be her father. It was disgusting. He was disgusting.

More photos. Cass with her swimming certificates, her oboe, in the choir. Christmas. Christmas. Christmas. Ben's christening. Baby Ben, big eyed and serious. Me, all dressed up for my first day at prep school, ginger plaits and a straw hat. Thank goodness my hair has darkened a bit since then. I still have the freckles, though.

'Of course we were happy,' I told her.'Look, no one's died. Lots of people's parents split up. We'll be fine.'

'We've let you down,' she said.'It's not the same as if … we took you on, we promised you a proper family ... I feel so bad.'

Oh God. The guilt of an adoptive parent. It's always there, unspoken mostly, but lurking in the background. They're trying to make it up to you, because everyone knows they really would have preferred to make a family the normal way. The home-grown way. Although, mind you, Ben is their biological son and he's as unlike them as it's possible to be. We'd make a fabulous nature versus nurture case study for some scientist.

In my opinion, nurture wins out. What did I inherit from my birth parents? Freckles and red hair. What did I get from Mum and Dad? A whole long list of stuff, everything from religion (C of E), politics, education, aspirations, the way I speak, the way I write, the way I cope with stuff: I put on a brave face and I get on with life. Which is what Mum should have been doing, and what I fully expected her to get on with.

I wondered how Dad's new baby would turn out. The one he'd made with Annabel's nice fresh eggs. Granny Matilda (that's Dad's mum) suggested once that maybe Ben's problems were due to Mum being forty-two when he was conceived. Granny Matilda is famously tactless. When I was ten she told me that it was very fortunate I'd turned out so well, 'because I don't think your real family were nice people at all'. And I've heard her telling Mum they'd have to watch me when I got to sixteen because that's how old my birth mum was when she had her first child; she was eighteen when I came along. 'Don't be silly,' Mum had replied, but Granny Matilda said, 'Blood will out. The mother was clearly promiscuous.' I think Mum agreed with her because I'd had all the lectures about contraception a million times, even though at 16 I'd never even had a serious boyfriend. Or a non-serious boyfriend, come to that.

Anyway, as it turned out, Granny Matilda was the one who brought up a son who wasn't able to be faithful.

Mum gave a huge sniff, ruffled her hair and said, 'I look like a fright. I must pull myself together. So sorry, darling.'

I shoved the tissues at her. 'Mum, none of this is your fault.'

'Oh, Cass, sweetie. Thank you.'

'If Dad hadn't been so ... if the Party Chairman people hadn't made him decide so quickly ...'

'The press had got hold of it,' Mum said, blowing her nose. 'The same thing happened to Rosemary Hayes, do you remember? David Hayes got the call that the press had got wind of his affair, and he announced he was leaving while they were watching *Last Night of the Proms*.'

David Hayes used to be Foreign Secretary. His wife, Rosemary, was one of Mum's friends. She'd been furious when Rosemary didn't even get to come to the Party Conference any more. 'So unfair,' I remember her saying. 'Don't they realise how much work the wives do on the home front? They should get some sort of a redundancy package when their husbands leave.'

Now Mum was in the same position. Annabel would go to Conference. Annabel would open the summer fête. Annabel would have to make friends with all the local Party workers – most of whom were Mum's best friends. I felt an uncomfortable and entirely unwelcome glimmer of sympathy for Annabel, which I stifled right away.

'You should call Rosemary,' I told her.

'I can't bear it. She's become a born-again evangelist of

divorce. "You should have done it years ago," she'll say. "I'm having the time of my life!"'

I didn't see that this would do Mum any harm at all, but I gathered she wasn't ready yet.

'What about, you know, those counsellors you can go to?' One of my biggest shocks had been discovering that Mum hadn't actually been all that surprised about the affair. 'Why didn't you try that?'

'Oh, don't be silly, darling. Have everyone in the world know our business?' Mum pulled out her compact and powdered her nose, with short, sharp stabbing movements. 'I just hoped he'd be, you know ... discreet. After all, he had his own flat in London. There was no need to ... to ...' A big gulp of wine.

'To get her pregnant?'

'He was still deciding what to do about that. It wasn't too late. No, the real mistake was taking the job as minister for families. It made him a sitting target for the press. And once it's out in the open ... well, the spin doctors believe in quick decisions. And he chose her.'

'Do you hate them?' My voice trembled shamefully. 'Do you hate him?'

She managed a tight smile. 'I'm so wet, darling – I can't bring myself to blame him. I mean, you know, an MP's life is difficult. They work such long hours, spend so much time alone in a flat. I suppose ... well, to be honest ... I turned a blind eye. I didn't ask questions. But I thought he'd be satisfied with that. I didn't think he'd ever break up the family.'

All these years I swallowed the lie that my parents were blissfully in love; I never even thought to question it. And actually he was probably having loads of affairs. No wonder he insisted on having his own flat near the House of Commons, even though his constituency is only thirty minutes from central London by train.

If she didn't want to hate him, that left it up to me. I surprised myself how much hate I could generate all by myself.

'Mum, I'm going to make you a coffee. Have you had anything to eat?'

'I didn't really fancy any lunch.'

Mum had lost about a stone in the last few weeks. Normally dropping a dress size would be a reason for great celebration for her. Now, she'd hardly noticed.

I rummaged in the freezer, looking for something to warm up for her. It was full of homemade stuff, packed into foil boxes. Macaroni cheese, Irish stew, shepherd's pie . . . oh God, not shepherd's pie.

Mum had just taken a shepherd's pie out of the oven when it happened. Dad's phone rang, and he accidentally put it on speakerphone. So we all heard his personal assistant say, 'Jesus, Oliver, you're in big trouble. The *Sunday Mirror* has got hold of a story that you've got your intern pregnant.'

Dad dropped his phone onto the red-tiled kitchen floor. The screen smashed, but Gareth kept on crackling into our silence. 'Oliver? Are you there? Sorry to hit you with this, but this is a crisis.'

Dad trod on his phone – his brand new iPhone – and Gareth shut up, but we knew it was true. If it hadn't been true, Dad would have laughed and said, 'Whatever will the journalists think of next?' and started dictating some sort of denial, or calling his lawyer. Instead, he stamped on a perfectly good phone. His face went pink and his mouth opened and closed, goldfish-like

When Gareth's voice finally stopped, all Dad could do was look around the table at us – at all of us, but mostly at Mum, standing in her oven gloves, clutching the steaming dish of pie – and he said, 'I'm sorry. I'm so sorry. I didn't mean it to happen. I didn't mean you to find out like this.'

'But we did,' said Mum, in a strange, far-away, un-Mumlike voice, which jolted me as much as the time in Italy when I drank an iced triple espresso in one, just to see what it was like. Then she dropped the shepherd's pie with an enormous *bang*, globs of mince and mashed potato exploded everywhere, and our family was over, just like that.

We all stared at the mess, the meat and potato all over the floor and the walls and the table. Then Ben made a little noise in his throat and ran from the room, and Mum went after him.

My hand was on my glass of water, and I couldn't let go. My muscles seemed to be in spasm. Just for a moment I imagined lifting the glass, throwing it at Dad. Not just the ice and water, splashing his smooth face, but the heavy glass, too. Would it smash? Would it damage him? Just for that moment I scared myself. Then my hand went limp and I could move again and I got a wad of kitchen roll and started furiously scooping up shepherd's pie from the floor.

Dad tried to talk to me, but the sounds coming out of his mouth weren't anything like his usual smooth fluency. He was blustering and stuttering and I hated seeing him like that. 'Go away!' I hissed at him. 'Go and see that Ben's all right!'

I wiped and scrubbed and tried to eliminate all the random splodges of mashed potato. If the kitchen was clean again, then there was hope that our family could survive. But even if we did stay together, I knew I could never clean away this memory. We were stained and spoiled for ever.

Since then, Dad had told us that he was staying with Annabel, but he hadn't actually moved his stuff out yet. He was everywhere: his papers, *The Economist* – the house even smelled of his expensive aftershave. I kept meaning to buy some air freshener, just to eliminate him from one of my senses.

Right then, bent over the freezer, I felt weary and old and sad. There was no way that I was going to be able to sing that evening. As I warmed up macaroni cheese for Mum and Ben and me, I texted the choir director, told him I wouldn't make it to rehearsal. Two minutes later, I sent another text quitting altogether. I was only doing it for the personal statement that was going to get me a place at Oxford, after all. Singing didn't seem to fit in my life any more.

Over supper, I worked hard to get Mum and Ben talking about neutral things, like television programmes and books, and what to buy Granny for Christmas. I did my economics essay, helped Ben with his geography. Then I kissed Mum goodnight – she was watching a box set of *Downton Abbey* and seemed reasonably stable – went into my room and checked Facebook.

Twenty messages, and a friend request from Will Hughes. It would look snooty to refuse, and I don't put much on Facebook anyway, so I accepted him. Then I scrolled through the messages. Friends and acquaintances checking how I was, sending me love, poking their noses into my business. Fake, fake, fake, I thought as I scanned them.

And then I froze. There was a name I knew and yet I didn't know.

Aidan Jones.

I swear my heart almost jumped out of my chest. I opened the message. I felt sick.

'Cass,' I read. 'I think I might be your brother.'

3: AIDAN

The waiting room is pretty full, which is crap, but normal for our local hospital at midnight on a Monday. I've been here with Rich a few times before. People come and go. Some wait for hours. Some – like Rich – get rushed off on a stretcher. We got here three hours ago and Rich is the only one who got the stretcher treatment, though half of them look nearly dead.

There's nothing to do. Most people have someone to talk to, or a newspaper at least, but I'm just sitting, trying to block out all the times I've seen Rich in a state – pills gone, vodka bottle empty, or once, in the children's home, blood everywhere, like he'd had a bath in it. A blood-bath. Ha-bloody-ha.

I could do with a drink, but you can't get much in the A&E waiting room – just crisps and chocolate and coffee. I haven't got any money on me anyway. I don't know where it all goes. I'll have to walk home, it's only about a mile. I owe Holly a tenner.

An old lady says, 'Call this a health service? It's a disgrace.'

She's talking to me, because the old man with her is snoring. I nod and smile.

'Two hours, we've been waiting.'

The old man doesn't seem hurt like some do – no blood, no bruises – but he leans against her, head on her shoulder, and his breathing makes a rattling noise. If he was a car, his exhaust would be about to drop off.

'How long have you been waiting, love?'

For ever, I want to say. Even being with Holly doesn't stop, it just makes it much, much easier than it used to be. Sometimes I forget that I'm waiting at all.

'Ages,' I say. 'They took my friend in hours ago. I'm just hanging on to see if he's OK.'

'Very bad, was he?' asks the old lady, perking up a bit. 'Must be, if you've been waiting for hours.'

'Pretty bad. He tried to do himself in.'

'Ooh, that's terrible. Overdose, was it?'

'Pills and booze. It's not the first time. He just gets sad.'

'Oh, the poor lad. Did you find him?'

'Yeah.' I shiver, thinking of it. Rich sprawled over his bed. His lips the same colour as his skin, the same grey-white as the sheets. I thought he'd done it right this time, ended it for good, but I'd hit him round the face anyway. When his still body moved, I jumped a mile and yelled, and he gagged and retched and drooled. Then he puked, which probably saved his life.

'Has he done this before?' the paramedic asked, collecting up the empty bottles next to the bed, and when I nodded, she said, 'You'd better come along for the ride.'

A nurse calls 'Albert Brown?' and the old lady says, 'Thank heavens, at last,' and nudges Albert awake.

'I hope your friend is OK,' she says as they stand up. 'Here, have my paper. It's no good sitting here worrying, with nothing to do.'

Normally I'd have made some excuse, but it's quicker and easier to thank her and take her newspaper. It's not as though she's going to watch me read it.

Reading's not my thing, although I can do it if I need to. As long as the words aren't too long and I'm not stressed or tired.

My phone's buzzing in my pocket. You're meant to turn them off in the hospital. I stick the newspaper under my arm and go outside. Walking from the muggy waiting room into the freezing cold car park is like a slap in the face.

'Hey, babe,' I breathe into the phone.

Holly never sounds angry. She's all sweetness and heart, my girl, she's kind and good and she'd never hurt anyone. But I know that anger doesn't always show itself. It can lurk underneath the niceness. It's only a matter of time before Holly kicks me out. I'm waiting for it.

Sometimes I think it'd be easier if she got angry.

'Aidan! Where are you? You forgot to pick up Finn!'

Oh God. This is it.

'I'm sorry, really sorry, Holly.'

'What happened? This is the second time in a fortnight!'

My mind lurches away from the idea of Finn waiting hopefully for me, his eyes teary, his mouth trembling. Not to mention Poppy, our child-minder. We've been trying to keep her sweet. And then there's Holly, worried and disappointed, having to rush out from work because I've let her down.

It's worse because I'd offered to pick Finn up, to make up for the last time that I forgot, which was only last Wednesday. I promised the little guy I'd be there on time today.

'I know ... I'm sorry ...'

Silence from Holly. I know it's the end for us.

'Look, if you want me to move out ...'

'Do you want to move out?' Holly's voice is all wobbly like she might be about to cry.

I did that. I can't bear it.

'Of course not ... I just forgot ... I'm at the hospital.'

'Oh my God, Aidan! Are you all right?'

'Yeah, I'm fine. It's Rich. He's taken an overdose.'

'Not again! Is he OK?'

'I don't know, they're treating him. He can't help it, Holly. He's just programmed to self-destruct.'

'Poor Rich,' she says. 'Poor you. Did you find him?'

'Who else was going to?'

'I know,' she says. 'He needs a proper support network.'

'You sound like a social worker,' I tell her, which is kind of a joke, but also true. Since Holly started working in reception for a GP practice, she's picked up all the jargon. Or maybe it's since she started living with me.

Anyway, when Holly talks about a support network, it makes me want to laugh. Holly has two parents – even if her dad's living in Marbella – a step-mother, two sisters, four grandparents, three uncles, hundreds of friends and aunts and cousins and even one great-grandmother who is still alive aged 101 in a nursing home in Eastbourne. Holly tries hard to understand what it's like being Rich or me. She doesn't always get there.

Rich actually has it worse than me, because both his parents are dead, which is why he ended up in the system. I have a mum and I see her once in a while, even if we don't really get on. And I have siblings, although I never see them at all. Rich has me, and that's about it. I don't count his social worker. She's always off with stress.

Holly's totally forgotten to be annoyed with me. I love how caring she is. 'Are you OK, love? You must have had a shock.'

She doesn't really get that every time I go round to Rich's, I'm half expecting to find him dead. Or expecting to find him half dead. Like today.

'I'm fine. Tell Finn I'm sorry.'

'He's asleep. You'll see him in the morning, you can explain.'

'I'll try.'

'Aidan? You will be back by the morning?'

It's bitter cold, sleet scratching and biting my skin. My hands are numb. But Holly's voice has the power to warm me up, hot and sweet, like the drink of honey and lemon that she made me that time, just because my throat was sore. The best drink ever, I swear.

'I'll be back when I can. I *promise* I will. They're pumping his stomach. I just need to wait and find out whether he's all right.'

'I miss you,' she says. 'It's cold in bed without you.'

'I'll try and be as quick as possible,' I say. 'I will come back, Holly, I promise.'

'Love you,' she says, and I know it's true. I don't deserve her. One day she'll realise.

Back in the waiting room I glance at the paper that the old lady gave me. There's a big picture on the front page. A man and a girl. A girl with dark red hair, the colour of dried blood. A girl with a heart-shaped face and dark eyebrows and eyes that might have been green or they might have been grey.

Something deep inside my memory rustles and stirs, like a bear waking up in its cave, sniffing spring air after a long, cold winter. Finn and I watch nature programmes together sometimes and we saw a bear like that. Finn loved the bear. I thought it looked vicious.

I let my eye move to the black and white columns of whirling letters. I force myself to focus, to try and catch the right ones, squash them into a word, that word I remember. I used to write it on my hand every day until I got sent to Auntie Betty's and she had a thing about hand washing.

Cass.

The curve of the big C, the snaky s's.

She looks like Cass might now, and I'm almost certain that is her name underneath the photo. *Cass*.

My heart is beating fast and loud, and I feel hot and clammy and suddenly dizzy, and I know the signs, so I bend right down, head to my knees, because I don't want to faint in a hospital waiting room. They'd rush me off to some cubicle, when I really need to try and work out the rest of the article, find out why Cass, *my* Cass – it must be – is on the front page of the *Daily Express*.

Letters swim out of the page at me, but I'm too panicky to make them into words. M. D. L. They make no sense. I try to think who can help me. Rich, obviously, but who knew when he'd be in a fit state?

Holly. She'd be happy. She'd understand. Or would she insist that I showed it to my social worker? I'm pretty sure he'd say that I'm not allowed to even try and find Cass. That's what they've all said before.

I could show it to Mum, but that would be like throwing a grenade into an intensive care unit.

'Are you all right?' asks a nurse. She's brought me a drink of water. 'It shouldn't be long now. Your friend is OK. They're going to transfer him to a ward, keep him in overnight. I'll have the details soon, and you can go home, then come back and see him in the morning.'

It's quieter now in the waiting room. I gulp the water and wonder what to do. Normally I wouldn't ask for help. Generally, if people know you find reading difficult, they think you're stupid, and then they take advantage. I can actually read, it's just that the font has to be right, and I need the time to concentrate, and mostly I don't bother. It doesn't make much difference to my life, most of the time, and I can write my name and address and important stuff like that.

I look up at the nurse. She's not that old, about Holly's age, and she's dark and pretty. Girls like her are generally happy to talk to me. I think it's because I've got big brown eyes and I'm good at smiling and, like Holly said – the night we went from being landlady and lodger to something more – 'You look like someone who'd make love like he meant it.'

With Holly, I always mean it. Not with anyone else, though.

I say, 'Thanks for the water. How long are you on duty?' I fix my eyes on hers and I smile my best smile, and I see it reflected in her sudden grin.

We talk a bit about night shifts and being a nurse and missing her family back in the Philippines. Luckily no one interrupts us.

'I know it sounds crazy,' I say, 'but can you just read this article to me? I need glasses to read, and I've forgotten them.'

She looks at the paper. 'It's about the minister. You know. The one who's been having the affair.'

'He's been having an affair with her?' I stare at the photograph. The man – frowning, grey-haired, posh suit – looks way older than Cass. I feel sick.

'Not her, that's his daughter. It says she's called Cass.'

'Read it, please. I think I've found someone I used to know.'

She reads it. The man – Cass's dad – is a government minister. He's been screwing some girl from his office and she's pregnant. He says it was, 'an unfortunate personal error of judgment'. He wants to be left alone to repair the damage he's done.

'Mr and Mrs Montgomery have two children,' says the *Daily Express*. 'Mr Montgomery was photographed yesterday arriving at his constituency home with daughter, Cass.'

'Not nice for the family,' says the nurse, handing me back the paper. 'Not nice for the girl, either.'

'No,' I say, 'not nice.' Then she has to see to some prat who's fallen over drunk and is bleeding from a massive cut on his forehead.

I stare at Cass, and it is her, it is. His daughter. My sister.

And I'm thinking, Thank you, Mr Montgomery, for not keeping your trousers on. Thank you, secretary, for fancying an old man. Thank you, nosy newspapers, stalker photographers. Thank you, kind old lady, nurse with time to talk.

Thank you for helping me find Cass.

4: AIDAN

Rich has to spend the night in hospital but basically he's fine. Out of danger. 'It's a good thing you found him when you did,' says the doctor, 'because otherwise anything could have happened. Do you have any idea why he did it?'

I shrug. All Rich ever wanted was peace and love. Once upon a time he had a nan, who told him all about heaven and angels and forgiveness. Sometimes he tries to go and find her. I didn't blame him, really.

Maybe he'd have been better off if I didn't have a key to his bedsit, didn't know to go and check on him when he rang me, crying his eyes out. But if he'd really wanted his nan and the angels enough, would he have rung at all? This isn't the first time that I've been sitting in a hospital waiting room wondering if I've saved Rich's life or if I've let him down.

They tell me he'll be kept in until he can see a psychologist, so I should come back tomorrow. Not sure how I'll manage that, I'm meant to be at work, but I'll sort something out.

SALVAGE

It's late, so when I get back to the flat Holly and Finn are both asleep. I've gone beyond tired, and I'm starving, so I make myself some toast and switch on the TV with the sound down low.

Is it possible to be in love with a building? I love our flat. It's got incredibly high ceilings, which are decorated with curls and twirls – like a really fancy birthday cake. The floors are shiny and polished. When it's light, the room is bright and friendly. Our furniture doesn't match, it's all bits and pieces, mostly from the shop downstairs, but it's funny how a candy pink sofa, a swimming-pool blue rug and acid green chairs can look good together. At night, we draw the dark, wine-coloured velvet curtains and sometimes Holly lights candles. She buys things from the market too, so we've got old china plates and bent silver forks and a set of green glass tumblers. It's nothing like the plastic plates and tinny cutlery at the children's home.

I've lived in so many places before that they all get muddled in my mind, and sometimes I have nightmares where I'm lost in a house I can only half-remember. Now I wake up every morning and see our stuff and feel at home.

Just saying that I have a home is something new.

I have a job too, and I only have to go downstairs to get to it. Everything's gone right in my life since I started working with Clive Norman, salvage king of London. It's almost like magic. The only thing that spoils it is waiting for the magic to end.

Jon, my social worker, put me in touch with Clive after I left the children's home and he got me settled into a bedsit. 'Further education isn't for you, is it, Aidan?' Jon said. 'And it's not so easy to get onto an apprenticeship now. We could try the construction industry, but I think Clive might be looking for someone. I've placed quite a few care leavers with him. He was a Barnardo's boy himself, believes in giving kids like you a chance.'

I didn't like Jon much. He was one of those young ones who pretend they care but never deliver. Jon had a wife and baby twins and he used to talk about them all the time, like I was interested. He'd been my social worker for two years, and never found me a foster home. It was alright for him, he could go home at night. I couldn't.

Anyway, he left soon enough and now I have some other guy, but we don't have much to do with each other.

I was dead nervous, until I got to the shop where we were meeting Clive. Then I forgot to be scared because I fell in love right away. With the shop, obviously, not Clive.

Imagine a shop for everything, chairs and tables, sofas and rugs, dolls and teddies, games and footballs, plaster angels, huge china vases, gold-framed mirrors, kids' bikes, soft wool blankets, designer jeans, Nike trainers, suede jackets, sunglasses ...

'You'll be working a lot in the shop,' said Clive, who was shorter and wider than me, and wearing a leather jacket that kind of made you notice his grey hairs more than if he'd just worn a jumper. 'Also in the yard, schlepping boxes, unpacking the van. Can you drive?'

'He's only just turned sixteen,' Jon told him.

Clive grunted and said, 'They chuck them out much too young. Mind you, he looks older. If you do OK here, Aidan, I'll pay for some driving lessons when you're old enough, OK?'

I nodded, too dazzled by the shop to say anything, and he said, '"Thank you" wouldn't go amiss. I suppose you don't learn manners in that home of yours.'

'Thank you, Mr Norman,' I said, in a rush, and he said, 'That's more like it. You can call me Clive.'

The yard was just as interesting as the shop: a muddle of metal and stone. Piles of bricks. 'London domestic,' Clive told me. 'People want them for restoration jobs.' Slabs of marble. A mountain of kitchen cupboards. 'Mostly I buy stuff that's been written off by an insurance company. Fire or flood damaged, but still absolutely fine. The original owner can't sell it on, so they sell it to me. Then there's the stuff that comes from building sites. And I do house clearances. Bankrupt stock, too. I take anything that people don't need. Give them a fair price and sell it on. It's 100 per cent above board. I don't take stolen goods, and everyone knows it.'

I'd kind of assumed that some of the stuff must be knock-off, but I believed him because he looked at me hard and said, 'If you do one thing to screw up my reputation, you'll be out on your ear, do we understand each other?'

So I said, 'Yes, Mr Norman, I mean Clive, thank you, Mr Norman', and he said he'd see me on Monday morning, 9 a.m. and don't be late.

I didn't trust him, of course, but then I never trust anyone. It didn't make any difference when Jon told me all about how Clive had made a small fortune and how he lived in a detached house in Hadley Wood with his wife and three daughters. Sometimes the married ones are the worst.

At first I was jittery as hell whenever we had to be alone together, in the van or the yard or the shop. He put his hand on my arm once, and I nearly jumped through the roof. But he's OK, Clive. He didn't mean anything by it. He keeps his hands to himself. He's never asked me to do anything that I didn't want to do. If I turn up on time and keep my fingers out the till, do all the jobs I'm asked and don't complain, then he's happy enough. Mostly that's what I do.

Holly's laptop is on the table, so I switch it on and have a look. I carefully copy Cass's name from the newspaper. And here she is.

Her dad is a Tory MP and a government minister.

He's left home to live with a girl called Annabel who worked in his office. She's having his baby.

It made my head hurt just reading this much, so I switched to Facebook. Maybe I could find her there.

Cass Mont-gom-ery.

I'd looked for 'Cass Jones' before. I'd looked for 'Cass Mackinnon', which was Mac's name. When I turned eighteen, I'd asked my new social worker to find out who had adopted Cass, what her name was. He told me there was nothing he could do. A closed adoption. Cass needed to find me when she was eighteen. If she wanted to.

Cass Montgomery is an unusual name. Only three on Facebook. One from Australia, a grandmother of six. One with no details at all, not even a photo. And then her. Cass. My sister, Cass.

She has 270 friends. She goes to some posh school. She's in a netball team – there's a picture of her holding a silver cup up high, like she's won the Champions' League. She sings in a choir. She's not a party girl – no pictures of her snogging boys or getting wasted or showing off her legs or cleavage. She seems younger than sixteen in her pictures – no make-up, hair long and wavy, pulled back from her face in most of the pictures. Massive grey eyes. Freckles splattered on her nose.

I haven't got any pictures of when we were little. I did have some, once, I used to have a whole Life Book, but I lost it. Maybe it's at Mum's house. I left there in a bit of a hurry, so that's the most likely thing.

There's not much to read on Cass's profile page, not even any pictures of kittens or funny cartoons, all that Facebook crap. But one photo album is of her family at Christmas. The government minister dad. Her mum looks like a TV newsreader, all highlights and lipstick, sleek and scary. I just about remember them.

Cass's parents. The people who took my sister away from me. I get all shivery, just looking at them.

But of course I don't remember the little brother, who you can just tell is a total geek, although maybe I'm judging too much from his glasses and sticky-up hair and chemistry set.

Cass has a brother. Of course she has. How screwed up is it that it's him and not me?

'Aidan?' Holly is standing at the door, eyes half-closed, hair sticking up. 'Aidan? Come to bed, honey, it's really late ...'

'I'm coming in a minute,' I say. She plods back to bed. I open up a Word document. They were good on IT at the referral unit I got sent to when I dropped out of mainstream school, we did a lot of computer stuff.

I write what I want to write, spell-check it two or three times. Then back to Facebook, open message, copy, paste.

Cass. I think I might be your brother.

There doesn't seem to be anything else to say and I'm

dog-tired now, so I press 'send' and shut the computer down and switch off the lights. I walk out of the living room and across the hallway and into our bedroom. So much space! Sometimes I just walk from room to room, enjoying that feeling of having a different area for different things.

Our bedroom has got the same high ceilings as the rest of the flat. I found a chandelier in the shop once, and asked Clive if I could have it. He thought it was a good idea. 'That flat'll have to be sold one day, so it needs to look good. Don't look like that, lad, it's an investment. I'm letting you and Holly live there while she gets on her feet. It's not for ever.'

Clive doesn't really approve of me and Holly being together, partly because she's his niece and partly because we're a bit of an unlikely couple. Not that he put it like that. 'I'm not being funny, but I don't like it when the girl's older. Call me traditional, but it's not natural. Mind you, the baby does need some sort of a dad, even if you're just a kid yourself. Think you're up to it?' When I nodded, he patted me on the shoulder and said, 'Good lad.'

At least he's nice about it, not like Holly's mum.

Finn has a mobile hanging over his cot – a rocket and some stars – and this chandelier is like my version. It looks like diamonds, falling from a golden frame. I just lie and stare at it sometimes, jewels sparkling in the sunlight, hardly believing that I'm lucky enough to live here.

I peel off my clothes and get into bed. Holly cuddles up to me.

'How's Rich?' she murmurs.

I'd forgotten all about him. 'He's going to be fine. Same old.'

'Do you think we should offer to have him to stay?'

We've had Rich to stay quite a few times. He stinks out the flat with his roll-ups. He eats all the food. He cuts up in the living room. Holly spent ages last time trying to get the blood-stains out of the cushion covers.

'No,' I say, and then, 'Maybe. The bedsit gets him down.'

'He should talk to his social worker. Get moved.'

'His social worker's off with stress. He's meant to be getting floating support. I don't think they're going to help.'

Her arms go around me. 'I don't really want him staying here.'

'I know. Nor do I. It's just difficult.'

'You're really good to him.'

I try and explain for the millionth time. 'It's just that for a long time he was all I had and I was all he had.'

'Now you've got me. And Finn.'

I kiss her, slowly, hungrily, happy that I'm her special project. Kissing is easier than talking sometimes. It's easier than explaining that Rich has been my friend for ages – six years now – which is the longest that anyone's ever stuck around for me. Holly and I have been together for nearly two years. That's ages. I'm sure I'm going to mess it up soon.

When we make love, all I want is for her to enjoy it. She talks sometimes about how I'm different from other guys, how selfless I am, how she's never met anyone like me. That's good. She seems to have no idea how bad sex can be, even though she's older than I am, and I never want her to find out.

'You're beautiful,' I tell her. Holly is beautiful, but not in an obvious magazine kind of way. Her figure is kind of lumpy around the middle, and she's got thin, flat, light brown hair, a round face and tiny little blue eyes. When she first came in the shop I hardly noticed her. But now I love the way her body is so soft and comfortable, and I love her smell and her smile and her little snubby nose … I wouldn't change anything about her. I like that she's older than me, too. I like that no one thinks we're together until they find out that we are.

'Oh, Aidan,' she says. 'You're … oh, Aidan.'

Then Finn starts yelling. We freeze in the darkness, hoping it's a blip, hoping he'll go back to sleep.

'Mummy! Aidy!'

He's awake.

5: CASS

I read it again and again. Aidan. Aidan Jones. What was this? Was it a journalist trying to trick me? Trying to find out secrets about our family? Or could it be true?

Aidan Jones was my brother. He was right there in my earliest memory. I remember crying for Aidan, missing him, running through the house searching for him. But I couldn't really remember his face. I couldn't remember talking to him or playing with him. He was just a gap, an absence, a missing person.

That's what it's like, being adopted. It's probably different if it happened when you were just a baby, but I was four when some social workers and a judge decided to reassign me from my useless set of original parents to my bright and shiny new ones.

Some of my memories are just undeveloped ideas of people and places that I can't ask anyone about. They hang around in the corner of my mind like faceless ghosts. Aidan is my brother, but I wouldn't know him if he walked past me on the street. I used to worry about that sometimes. What if we saw each other? We'd never know.

I clicked on the Facebook profile. Aidan Jones joined Facebook last year. He's in a relationship with Holly Norman. He's not very active on Facebook. He only has ten friends. He hadn't bothered with a cover photo. His profile picture shows the back of his head. Dark hair. A white T-shirt. A tattoo on the back of his neck. 'Hope', it said, in swirly writing.

I stare at the photo, as though it's suddenly going to animate itself, so Aidan will turn around and smile at me. It stays stubbornly still.

I click on Holly Norman's profile, but it doesn't say much. There's a cover picture of Camden Market, and a profile picture, but she's wearing sunglasses and a big sunhat. Her pictures, her wall: all private. In a relationship with Aidan Jones. Lives in Camden.

Back to Aidan's message. What should I do? I knew my birth family weren't meant to contact me. There are rules, laws. Social workers say it's best to keep away until you're eighteen. Then it's up to me whether I wanted to trace them or not, and it'd be done with counsellors and support. My birth family had no right to contact me.

I only had a year and a half to wait because I was turning seventeen in April, and I hadn't even thought about whether I'd do it or not. I'd wait, probably, because I wouldn't want to be distracted from my exams.

But Aidan hadn't waited. He hadn't kept the rules. He'd found me on Facebook. He must need me back in his life, I thought. It was extraordinary how happy that made me feel.

I'd looked for Aidan on Facebook a few times, but there are loads of people called Aidan Jones and none of them seemed right, and how can you tell anyway? He probably wasn't living in America, or Ireland – or was he? He definitely wasn't black, although he was darker than me, skin and hair and eyes.

I'd always assumed that he'd got a new family, so he wouldn't be called Jones anyway. But here he was. Aidan Jones. Maybe he'd been adopted by another Jones family: it was a common enough name. Maybe he'd changed his name back when he got to be eighteen. I calculated. He'd be nearly nineteen now.

I clicked 'reply'. I'd be careful, I decided – I wouldn't give anything away. I'd try and find out who this person was, if he was genuine, or if he was really a journalist trying to get the dirt on Dad and his affair.

'Hello,' I wrote, 'why do you think that?'

I wasn't expecting a reply. But the chat box opened up right away. It took ages for him to write something, though.

- yr name is the saym & u hv the saym red hair.

- What do you remember about me?

Long pause. Then:

- u liked red jelly & horsis

- Go on.

- u wear scarred of dark & u had a lite

If this is Aidan, I thought, he can't spell at all. Maybe he was dyslexic.

- Lots of people are scared of the dark.

- the lite had a cat onit.

I thought and thought. I had no memory of a cat light. I did like jelly though, and raspberry was my favourite. And I'd always been scared of the dark. What else could I ask him?

- What's our mum called?

- janette

- What about our dad?

Another long pause. A few minutes. I thought he'd gone. Then:

- mac.

- Do you remember when my birthday was?

- april 15

He *is* Aidan, I thought, or someone who has access to my adoption file. Or someone who knows me – but none of my friends had any idea that I had a brother called Aidan. Some friend of my parents? Someone who thinks this is an appropriate time to mess with my emotions?

- Are you really Aidan?

- yes

- How did you find me?

Another pause. I waited, trying to work out how I felt. A little bit scared, an awful lot excited. One thing was for sure: I wasn't thinking about my parents any more.

- i sor yr foto in a paper

Oh. My excitement faded. Here we were again. Oliver Montgomery, shamed minister, with his daughter Cass outside their £1 million constituency home.

- u had the saym red heyr & the saym naym & i thort it must b u.

- I don't like being in the paper I wrote, and sent it, and then I leaned my head against the screen and wondered why I'd written such a thing.

- i bet. sorry

- That's OK.

- i wish i cd c u becos I mised u.

- I missed you too I wrote, before I thought about whether it was really true. Can you miss someone you can't remember?

- id lik to c u maybe 1 day. He added a confused face to his message, and although I usually despise emoticons, I found myself smiling.

- Maybe. Let me get over the surprise.

- sure.

- Are you in touch with our parents?

A pause. Five whole minutes. Again I think he's gone. Then I see the magic words: 'Aidan is typing.'

- mum a bit not mac.

I wondered, for the first time, whether we actually had the same father. Perhaps Mac was just mine. Perhaps he wasn't really dad to either of us. Aidan could tell me. Did I want to know?

- Sorry. Was Mac not your dad?

- No prob myn ded.

- I'm sorry.

- long tiym ago.

- I'm sorry.

- r u OK with yr dad & al that shit?

OK, this was the first time I really thought he might be a journalist.

- I'd rather not talk about that.

- sure. sorry.

Parents didn't seem to be a great subject.

- Where do you live? Are you at school?

- london camden u no it? I hv a job

Not one that involved writing, I thought, and then felt
mean and nasty. I wondered what sort of school Aidan had
gone to. Not a private grammar school, that was obvious.

- *I go to Camden Market sometimes.*

- *maybe i sor u*

- *I don't think I'd know you. Sorry.*

- *u dont hv 2 say sorry*

This time I sent the confused face emoticon. He responded
with a smile. I sent a smile back. I began to see the attractions
of talking without words.

- *Aidan, how do I know it is you? Have you got a photo or any-thing?*

One minute. Two minutes.

- *maybe i can get a foto*

- *Can you send it to me? A copy?*

- *sure*

- *Thanks.*

It was midnight. I tried to imagine Aidan typing into a
phone or a laptop. Was his bedroom as big as mine, as nice?
What sort of place did he live in?

- *I'd better go. I've got school in the morning.*

- *I hv work*

- *What do you do?*

- *its a shop and a yard 4 salvage*

- *What sort of salvage?*

Mum got the fireplace in our living room from an architec-
tural salvage yard. I imagined Aidan surrounded by Victorian
tiles and original features.

- *stuf from fiyrs and fluds my boss bys it after. he sells on the gud
1s. we get allsorts, toyz, furnytur n stuf*

Aidan's spelling was so random that it would almost be funny if he wasn't my older brother and if every mangled word wasn't making me cringe.

- *Oh, sounds fun.*

- *its OK*

I want to ask him about Holly, but I don't want to look as though I've been nosing around his profile.

- *Do you live with your family?*

- *my girlfrend*

- *That's nice.*

Actually, I think he's incredibly young to be actually living with a partner.

- *shes called Holly. i want u 2 meet her*

Steady on, I thought, and then winced. That was one of Dad's expressions. I was never going to get used to him not living with us any more.

- *Maybe one day.*

- *i hope so*

Mum was moving around outside my room. I quickly minimised the chat box. My door opened.

'Bit late, darling. You've got to be in school early tomorrow.'

I did, too. Every week the headmistress had a meeting with the prefects. As I was on target to be head girl next year, I'd been incredibly conscientious about never missing a week ever since the beginning of sixth form. I was due to take over in May if all went well, as soon as Eleanor Butterworth, the current incumbent, started her exams. I had all sorts of ideas about things I wanted to do; a homework club for Year Sevens, a nutrition advisory committee for school dinners. They all seemed a bit unimportant now.

'I'm just going to bed.'

Mum came over and kissed me on the forehead. She'd brushed her hair, I noticed, put on some lipstick.

'I'm sorry about before. I'm still in shock, I think. I had no idea about this particular girl – and the baby – but you don't need to hear this. I'm here for you, darling Cass. Computer off now, you need your sleep.'

'OK, I'll just finish off,' I say.

'Five minutes! I'm going to check on you.'

She closed the door. I clicked the chat box.

- *Cass?*

- *Cass?*

- *Hv u gon? Cass?*

- *Sorry!* I wrote. *I had to talk to someone!*

But Aidan was off-line.

6: CASS

I couldn't sleep after talking to Aidan, so after I was sure that Mum and Ben were asleep, I turned on my light again. I knew I had photos of Aidan in my Life Book, but I had to climb up to get it – I had so many books that Mum and Dad had fitted floor-to ceiling shelves to one wall of my bedroom.

The album was on the top shelf , and I tried to shake the dust off before dropping the book onto my bed. Then I climbed down and carefully wiped away the rest of the grime and cobwebs, brushing the stray dust off my white duvet cover, before I opened the book.

It was just a normal photo album, big and blue with a label on the front that said 'My Life Book'. I used to look at it with Mum sometimes, when I was really little.

There he was on page three. Aidan and Cass. A wiry boy with dark curls over his eyes, and a big smile. Sweet brown eyes, staring at the baby – me – on his lap. I'm pale and white, his skin is darker. Now it seems obvious that Aidan and I must have different fathers, that we come from different ethnic backgrounds, but there's nothing in the book to say so.

Then the two of us again on the same big, brown sofa. The caption underneath, written in - I assume - a social worker's neat handwriting reads: 'Aidan and Cass in Gran's front room.'

I tried to remember sitting there with Aidan, tried to conjure up the smell and feel of that sofa. Was it lumpy or soft? Was the brown fabric furry or velvety? What did it smell like? There was nothing there. Just the memory of *looking* at the pictures – sitting on our sofa, soft blue with giant cushions, a sunny summer evening, with Mum smelling of the garden and macaroni cheese, saying in her steadiest voice, 'Your gran had been looking after you, but she was getting on a bit and she wasn't coping very well. Aidan was your big brother, and he loved you very much. We wanted to adopt him as well, but the social workers thought he'd be better off if they found him a family who could look after just him, all on his own. So he'd get loads of love and attention.'

'Lucky him,' I'd said once, because I was fed up with Ben crying and crying and getting all the attention, and Mum looked hurt behind her smile, so I had to explain that I didn't really mean it and I couldn't be happier to be adopted by her.

There were lots of people in the Life Book that I didn't remember. Birth mum: Janette Jones, young, pale as an empty exam paper, flaming ginger hair rippling past her shoulders. Birth dad: Mac, no surname, older, bulkier, red nose, moustache. Mac couldn't cope with having a family, according to the captions, and he got into trouble with the police. Janette had gone back to live with her mum: Mrs Anne Jones, grey and fierce and badly in need of an eyebrow shape, but then they argued and Janette got ill, and mixed with unsafe people, and Gran couldn't cope and gave us back to Janette, and she couldn't cope either, and asked social services to find us nice new families who could look after us better than they could. People who could cope.

Then there were the foster parents, Kath and Kevin (six weeks), Jan and Phil (nine months), and then – my favourite pages – Mum and Dad. Lots and lots of pictures. In the garden, on the trampoline. School uniform. Birthday parties. Holidays. Ben. It was the ultimate happy-ever-after book, because I was living it. So I didn't need to read it. Up on the shelf it went.

Looking at it again for the first time for years, I'm surprised to see that Aidan obviously visited me a lot when I first came to live here. There's a picture of him with me and Mum and Dad lighting candles for my birthday. He's the only one not smiling. It must have been hard for him to come and see me so loved and happy in such a nice house. Now I could ask him how he'd felt when I was adopted. Now I could find out what happened to him. Why hadn't I asked right away? I felt uneasy, and turned the page quickly.

All my best friends knew I was adopted, although it wasn't something that we shouted about as a family. 'It's no one's business,' according to Mum, but I think it was because Dad was high profile and they didn't want my birth parents tracing me.

My friends knew because they remembered me arriving. 'You were so quiet in those days,' Grace told me once. 'It was like you were learning how to speak English.' She thought it was romantic and exciting to be adopted, and would ask me if I'd ever thought of tracking down my birth family. 'No, never,' I'd say completely truthfully. My birth family wasn't a family, just a collection of forgotten and useless individuals. Why would I want to meet parents who hadn't wanted me enough to get their act together? All except Aidan, but I couldn't really remember him; he was probably part of another family and must have forgotten all about me.

That's what I used to feel. But now Dad had let me down, and my family was smashed to bits. Aidan needn't be a ghost any more. He could be my brother again.

'Cass?'

I nearly jumped out of my skin. 'What? Oh, Ben! What is it? It's late.'

Ben was all bundled up in his dressing gown, eyes blinking behind his glasses, hair rumpled up. He looked like a small, distressed owl.

'I can't sleep.'

'What's the matter?'

He sat down on my bed, stared at his slippers. 'No one likes me.'

'They're stupid. Ignore them.'

'But they've all got people to laugh with and even if I try and laugh at them, no one laughs with me.'

Over the years, Ben's been to loads of doctors and educational psychologists and special educational needs teachers and so on, and no one has ever really said what it is that makes him different. He's not autistic and he's not dyslexic and he's not dyspraxic and he doesn't have classic Asperger's, although one psychiatrist said there might be similarities. There's nothing anyone seems to be able to do, anyway, to

45

help him catch on to the trick of making friends, which is all he wants in the world.

'Are they being mean to you?' I pretended to consider my options. 'Shall I come and meet you after school? Have a word with the bullies? I could scare them for you.'

Ben screwed up his face. 'I'm not sure that would work,' he said, so polite and thoughtful that I felt terrible for teasing him.

'No, I don't think it would,' I agreed. 'Be fun, though.' Horrible brats. How difficult would it be for them to be nice to him? And how happy would that make him? If I went to Bonny's, I'd probably spend my whole time wading in to defend him. What were the teachers doing there? Why didn't they sort it out?

'They don't hurt me physically,' he said, carefully – Ben's learned over the years to be very precise when explaining to school and family how bad things are. 'But they say things about Dad and they laugh.'

'Oh, Ben, poor you.' I hugged his scrawny little body. 'Isn't there anything you can do? Can you tell a teacher?'

Ben shook his head. 'I tried that already. Not doing it again.'

At primary school, Mum and Dad were in and out of the head teacher's office every time Ben came home crying. Every member of his class had been in trouble for bullying him by the time he was ten. Moving school, he had been determined to change his strategy. It was gutting that it hadn't worked.

'Ben, you've got to tell someone. They can't get away with being horrible to you.'

He shook his head. 'I'll tell Mum I've got a tummy ache tomorrow. Maybe they'll forget to be nasty to me by Monday.'

'Maybe,' I said. To be honest, Ben might as well stay at home. He'd watch his wildlife programmes and read his books, and probably get himself a better education than he would in a day at Bonny's.

'What are you reading?' I asked, and watched his face brighten as he told me about his latest Michael Morpurgo book, about a boy and his wolf, the adventures they have, their search for a home. Ben wanted a dog, but Mum and Dad worried about his allergies, which were many and various and gave him asthma and eczema. But maybe a dog would actually help him? One of those anti-allergy dogs. There was no point talking to Mum about it though. She wasn't fit to consider a cockerpoo. She couldn't even look after herself properly. Maybe in a few months.

We talked about books and animals, until Ben started yawning and I told him it was time to go back to bed.

He wouldn't have to act a tummy ache tomorrow: staying up until 1 a.m. was probably enough to bring on one of his migraines.

'Cass,' he said to me, rubbing his eyes. 'Cass, will we actually have to see this Annabel person?'

'Don't worry about her,' I told him. 'Go to bed.'

'We won't, will we?'

'I'm not sure,' I lied. 'Go to bed.'

'OK.'

He left, but I was still wide awake. I lay in the dark worrying about him. School was bad enough, but how would Ben cope with meeting Dad's new partner? What about when their baby was born? How could I talk to Mum about this when she was such a wreck?

That's when I remembered the boy on the bus. He was actually offering to help me with Ben. He'd noticed that things weren't going well.

I got up again, padded across the room to find my school blazer. I found the crumpled card in my pocket. Fluorescent green writing on black. Will Hughes. Private Investigator.

Private Investigator? What?

7: CASS

Dad moved out officially on a school day. We came home that afternoon and there were big gaps among the pictures on the walls: the long, thin rectangle which was his group photo from college; the three small gaps in a row, original cartoons from the *Daily Telegraph* poking gentle fun at him; the bigger one to the left, his framed front page from the election.

I didn't think he'd be framing the front page of *The Sun on Sunday:* MONTY'S FLING WITH SEXY INTERN.

'Perhaps we should redecorate,' said Mum. 'It looks awful. I can't bear to look in the wardrobe, there's so much empty shelf space.'

'I could move some stuff in there,' I suggested. 'Everything's crammed into my cupboard.' Actually, I had more than enough space, but I could easily rearrange my clothes to make her feel better.

'Oh, thank you, darling Cass. And maybe you could take over his study, too? It'll give you space to spread your books out.'

'I'm OK in my room.' In my room I could pretend that Dad was still living with us. 'Why don't you use it?'

'What would I use it for? Please, darling.'

'Oh, all right.'

'It won't be for long, anyway. We'll have to put the house on the market.'

'We have to sell the house? Why? We still need somewhere to live!'

'Well, we can't afford to buy your dad out of his share, and he'll need a constituency house, anyway. In fact, he was talking about buying me out – giving me a lump sum to buy a new house, and then he can have this one.'

'What? He'd move Annabel in here? That's ... that's obscene!'

'You'd still have your bedroom ... and so would Ben ... It would mean less disruption.'

I felt like she'd punched me in the stomach. 'I can't believe this! I can't believe you'd let him get away with it!'

'If it's that or show lots of strangers around our house ... I mean, there's no good solution here, Cass, is there?'

I wanted to scream and shout, but I didn't want to shout at her, so I marched up to my bedroom and lay on my bed and breathed in and out.

My phone rang. Dad.

'How's it going, sweetheart?'

I growled through my teeth.

'How do you think?'

'I know it's not easy. I'm sorry, darling Cass.'

'You're going to chuck us out of our house, and move Annabel in ... How can you live with yourself?'

'That's just an idea. We'll probably end up selling. Cass, this will all be easier when you have met Annabel, and we've all moved on. It's horrible at the moment, but people do get through the pain. I know it's not easy, sweetheart. I miss you and Ben terribly. How about we all go bowling on Sunday?'

'What, with Annabel?'

'No, just the three of us.'

'I'm busy,' I told him. 'I've got an essay to write.'

'We can talk about it when I see you. You know I love helping you with your homework.'

'You can't just pretend nothing's happened.'

'I'm not. Give me a chance, Cass.'

I knew why he wanted to go bowling: I couldn't lose my temper in a public place. But in some ways that made it safer for me. I really didn't want to lose complete control. I was scared of what I'd unleash.

'OK,' I said. 'For Ben's sake.'

Sunday came and Mum dropped us off at the bowling alley. Dad was waiting by the counter where you change your shoes, talking into his iPhone. A new one, I suppose. We waited patiently while he went on and on about inspection regimes for children in care. Occasionally he waggled his eyebrows at us, the sort of thing I used to find funny a million years ago.

'Cass! Ben!'

I sidestepped his hug.

'I've missed you!' he said, ruffling Ben's hair.

Ben hugged him back. 'I miss you, Dad.'

I wanted to point out that Ben needn't be missing Dad at

all if he'd just refused to leave us, however pregnant Annabel was. Or if he'd just resisted the temptation to sleep with his intern. Instead, I snorted and did up my bowling shoes super-tight.

'Come on, kids,' he said. 'Let's see if you can beat us, Ben.'

I searched for the right size ball, weighed it in my hands, visualised chucking it at Dad's head, knocking him out, smashing his nose. Oh, how the newspapers would love that: DAUGHTER BOWLS MP OVER IN FREAK ATTACK.

'When were you planning to tell us, then? When the baby was born?' I asked.

'Come on, Ben, your turn first,' said Dad, laughing in a completely fake way. I hated him. I despised him.

Ben stepped up to take his turn. Dad took the opportunity to mutter to me: 'Look, being hostile isn't going to help Ben. We've got to support him through this. I accept I haven't handled it well, but I didn't mean to hurt anyone—'

'Yay!' I shouted. 'Ben! Great strike!'

He's only short and his arms and legs are spindly, but this is his sport. He's so proud every time he bowls a strike that I get tears in my eyes watching his capering little celebration dance. Ben doesn't get to be proud often enough.

I took my turn. Dad took his. Then the muttering started again: 'I never meant things to work out like this. I didn't even know she was pregnant until a few weeks ago. It's not easy for me, either.'

'Oh, my heart bleeds.' It was great to get angry. I felt better than I had for weeks.

'How's your mother?'

Neither of us noticed that Ben scored eight and is back with us. 'Mummy's lonely and sad,' he says. 'And she won't even go to the Women's Institute.'

Dad winced. 'She's bound to be … it's a big change for everyone … Your go, Cass.'

I'd never bowled with such power before. My ball sent the pins flying.

'Strike!' shouted Ben. 'Well done, Cass! You're usually rubbish!'

I was usually rubbish on purpose. I couldn't help myself today. It was too satisfying to have the chance to smash something.

'Mum's better off without you,' I said, 'she's enjoying herself. She doesn't have to do all that constituency crap any more. Annabel can do it.'

I was only semi-lying. Mum was completely miserable, but I'd watched her for years, being smiley and gracious, listening to people's problems and views with a patient smile. Perhaps it would be good for her to stop being a supportive partner for once, to be an independent woman, to be free. She'd be a good role model. Finally.

Dad looked relieved to be taking his turn. The ball rolled down the lane, heading for the centre pin, crashing, smashing … two lonely pins were left standing, one either side of the mass destruction. Pregnant Annabel – she was like a bowling ball, I realised: even Dad got crushed by her.

Ben's turn. Dad said, 'You're going to have to meet Annabel. You're going to have to accept her. You'll have another brother or sister before too long.'

Ben scored a nine, turned to us, disappointed.

Dad clapped. 'Well done, Ben! Nine's good! Nine's great!

You can't get a strike every time!' Ben aimed for the single pin and Dad resumed his campaign to persuade me that everything was going to be OK. 'You're going to have to meet her. She's part of my life now. And then there's the—'

'Woo! Another strike! Well done, Ben!'

So we went on. A conversation in bits and pieces whenever Ben was playing.

'I just think for Ben's sake, we ought to prepare him, make Annabel part of his life ...'

'For *Ben*'s sake?' My voice was shaking. 'If you'd been thinking about Ben in the first place—'

'Great strike, Ben!' I knew that voice. I spun around.

Will Hughes, in a neon pink T-shirt and emerald-green jeans, was applauding Ben's strike from three lanes away. Dad and I hastily joined in.

'Great stuff, old man,' said Dad, patting Ben on the back. 'Well done! You're on fire!'

'You weren't watching,' said Ben.

'We certainly were. Didn't respond right away because we were awestruck, weren't we, Cass? Struck dumb with admiration. Come on, give me some pointers on where I'm going wrong.'

I looked over to where Will was playing. He was with three girls and two boys, all completely normally dressed, no one I knew. They were laughing and joking, messing around. I wished I were with a group like that, and not this pathetic little family outing.

Will caught my eye and gave me a smile. I thought he was about to come over, but then Dad and Ben called me to take my turn. And when I'd scored my seven, he was sending his shot rolling down the lane. And scoring a strike, I noticed. One of the girls (pretty, blonde, short) gave him a hug.

Was he with that girl? What did I care? I turned my attention to Ben, chatting to him, cheering his efforts, ignoring Dad's attempts to bring up the subject of Annabel again.

Unfortunately, when the game was finished – Ben won, of course – Dad bribed him with cash for the slot machines. Ben loves them. I had no choice but to stand around with Dad while he droned on and on again.

'She's a nice person, Cass, a lovely, kind person. She didn't mean to hurt anyone. All the fault is mine, you can't blame her. You need to meet her, darling Cass, you have to accept her. Her and the baby. After all, it will be your brother or sister.'

He'd said it again. That was the last straw.

'No, it won't,' I hissed. 'It's nothing to me. It's no relation.'

'Cass, you're my daughter and this baby is mine, too. I know you don't like what's happened, but you just have to accept it—'

'Hi, Cass!'

I had never been so grateful to see anyone in my life. Even though it was the annoying Will Hughes, so-called private investigator.

'Hi, Will!'

Dad looked grumpy and said, 'I don't think I've had the pleasure—'

I would have prolonged the awkward moment, but Will leapt forward, stuck his hand out and said, 'I'm Will Hughes. I'm Ben's peer buddy at school.'

'Jolly good, jolly good,' said Dad, clearly overjoyed to be

seen shaking the hand of an ethnic minority youth in the heart of his constituency. The only disappointment for him was that there was no photographer to capture the moment.

I wondered why I'd never realised before what a big, fat hypocrite he was. All those years thinking he was a hero, and now realising what a fool I'd been.

Will turned his attention to me. 'Hey, Cass,' he said. 'How's it going?'

'Fine, thanks.'

'Think about my suggestion at all?'

I glanced at Dad, who was clearly curious.

'I'll be in touch,' I said.

'Great! Brilliant! I'll just go and say hi to Ben.'

'Seems like a nice lad,' said Dad, in much too loud a voice. I cringed, hoping that Will hadn't heard. 'Friend of yours, Cass?'

'Didn't you hear? He's Ben's peer buddy.'

'He seemed keen on you, though. Anything going on that I should know about?'

I didn't even bother answering that.

'I see, I see,' said Dad, who didn't see anything because there was nothing to see. Well, nothing related to Will, anyway. 'Shall we get a pizza?' he suggested, as Ben finished his game. Ben loves pizza, but I'd had enough.

'I've got an essay to write,' I told him. 'I'm going to get the bus. Ben, you can have a pizza if you want. I'm sure Dad can drop you back at home.'

'Cass—' said Dad, but I was already heading for the bus stop.

*

I enjoyed the peace and quiet of the journey home. I cleared my mind of Mum and Dad and Ben, of Will and his friends, so happy, so relaxed. I thought about me and Aidan, how weird it was to be in touch with him, but how right. He could tell me so much about myself. He could fill in the gaps, tell me who I really was. I'd thought this life was real, solid, secure, but now I saw it was not.

I couldn't wait to meet him. But I wasn't stupid enough to think I could just go waltzing off by myself. I needed an ally. And I had an idea who that could be.

8: CASS

Will Hughes suggested meeting at the outdoor swimming pool, even though it was October and getting chilly. 'Not to swim, unless you want to,' he said. 'But the café there is one of my favourite places.'

'Oh, right,' I said. I hadn't been there for years. It was old and faded and crumbling – most people I knew were members of the big health club on the outskirts of town. I was surprised the lido hadn't been closed down years ago.

We arranged to meet at 2 p.m., and I made sure to get there a little bit late, so he'd be waiting for me. But at 2.15 there was no sign of him. I wondered if I'd been the target of some elaborate joke.

The lido dated from the same era as the art deco cinema. It must have been stylish and fun in the 1920s, but now the paint on the changing room doors was peeling, and the tiles at the side of the pool were overgrown with moss and grass. Amazingly, there were people swimming. They must be crazy. I was freezing, and I had my winter coat on.

The café was just some tables and chairs around a little kiosk. I ordered a hot chocolate, sat down, and pulled out my phone in case Will had texted me to cancel.

'Hey! Cass!'

The shout came from the pool. Will was bobbing in the water, waving at me.

'Oh! I didn't know you were there.' I felt instantly stupid.

'I lost track of time. Sorry! I'll be with you in a minute.' He grimaced. 'Getting out is the worst bit.'

I dipped my head down to my cup as he pulled himself out of the water. His hair sparkled with droplets of water, and his body – well, I wasn't really in the habit of ogling boys. I just got an impression of long limbs, a little scattering of chest hair, before he wrapped himself in a towel and disappeared into a cubicle.

Five minutes later he dropped into the chair next to me. Will Hughes's out-of-school post-swimming outfit turned out to be a fire-engine-red hoody, chocolate-brown tracksuit trousers and a long-sleeved top that was striped orange and black. I mentally christened him Tigger.

'You should have come for a swim. I wasn't sure if you'd want to, because I don't know you, but it's so awesome this time of year.' He took a sip of coffee. Will's beverage of choice was a double shot cappuccino, which would have kept me awake for about a week.

'Isn't it freezing?'

'Only when you come out. The water is incredibly warm, and once you've dried off, then you can sit here and drink your coffee and pretend you're back in 1930.'

I spluttered into my polystyrene cup. 'Why would you want to do that?'

'Don't you ever do that? Try and do time travel? You just have to find the right place. I spent a whole summer holiday once trying to transport myself back to Victorian times. There are loads of places in London which feel almost right – the problem is when you spot a bus stop or a Starbucks or something out of the corner of your eye ...'

I had to laugh. He grinned back. 'That's better. You were looking disapproving before.'

'Was I? I didn't mean to.'

'You've just got one of those faces, then.'

'What? What about my face?'

'When you get to be a headmistress, you'll terrify all the kids without even trying.'

'Oh!' I couldn't believe my ears. A virtual stranger was telling me that I was ugly enough to scare small children.

'I don't mean to be offensive,' he said. 'I just mean you've got the kind of demeanour that commands respect. You might not want to be a headmistress. You could be prime minister, maybe. Or a High Court judge.'

'I don't seem to command much respect from you.' I was cringing inwardly. Naturally he thought I wanted to be a politician, just because of my dad. Some girls at school used to think it was funny to call me Maggie Thatcher. I knew better than to react, and they soon got bored with it but, still, it wasn't exactly the image I wanted to project.

Who was I kidding? I had no idea what image I wanted to project.

'Oh, you'll have to get used to me,' he said, easily. 'I never mean any harm.'

'Good.'

'Blimey,' he said. 'You *are* offended. I'm sorry. Let's start again.'

'Look, I came to talk about Ben. He needs help at school.'

'Poor kid,' he said. 'The problem is that he takes things far too seriously.' His dark brown eyes met mine. 'A bit like his sister.'

I gasped, and then laughed. What else could I do?

'They're being mean to him. The boys in his year, and some of the older ones as well. It hasn't helped, all the stuff with my dad. It's just another thing to tease him about.'

'No point trying to stop them,' he said. 'I think I'll find a few girls in his year who might take him on – keep an eye on him in the playground, that kind of thing. They're a bit more civilised than the boys. That is, they're constantly bitching about each other, but they'll be nice to Ben if I point out the benefits.'

'What benefits?'

'My approval.' He said it in such a self-satisfied tone that I started laughing again. 'What?'

'It's just, you! You're so arrogant! You think they're all in love with you, don't you?'

'It cannot be denied,' he said. 'The whole of Year Seven is in love with me. Boys and girls. And quite a few of Year Eight, as well.'

I'd never heard anyone so big-headed.

'No point looking like that. It's true. Ask Ben.'

I had asked Ben about Will. His little face lit up and his eyes shone and he said, 'He's just the nicest person in the world! And have you seen what he wears!'

Of course I couldn't tell Will that – it would just feed his gigantic ego. So I said, 'Maybe I will,' and tried to keep a straight face.

It didn't work. He crowed, 'I see you have already!' then threw back his head and laughed.

'Maybe your charm only works on naïve eleven year olds.'

'Oh, no, it works very well on naïve sixth formers as well,' he said, putting his hand over mine so suddenly that I blushed and squeaked as I quickly moved it away.

'Look, we're here to talk about Ben,' I said, quite cross, and he nodded and said, 'Yes, ma'am.'

'Can you really keep an eye on him?' I asked. 'Because he is upset about what's going on at home, but he's always been a bit like this. All his life. He's not had an easy time.'

'Must have been difficult.'

'Yes, it's horrible for him.'

'Not so easy for you, either.'

'Oh, well, I can cope.' I said, slamming my cup down on the table.

Will raised his eyebrows. 'Glad to hear it.'

'What's all this about being a private investigator?' I wanted to change the subject. 'You're not really one, are you? It's just a joke?'

'Well, not really a joke. It's my business. I thought it would be more fun than a Saturday job at Waitrose.'

'Do you get any business?'

'Lost cats. Lost dogs. Cheating boyfriends. I trailed one guy all the way to Leicester Square, took pictures of him meeting some girl.'

'Really?'

'His girlfriend was extremely grateful.'

'I bet she was.'

'Of course, she needed some counselling to help her cope with the emotional implications of our investigation. But I provide that as well.' His grin was in danger of becoming a smirk. I raised my eyebrows.

'This job is nothing like that.'

I didn't want to tell Mum about Aidan. She was too fragile, too upset. She couldn't cope. But I really, really wanted to meet him. I was just waiting for him to send me the photo and any other proof of identity he could come up with.

'I'm sure I can satisfy your every need,' said Will. I decided to ignore the innuendo. He was annoying, but I could also see that Will Hughes would be an excellent bodyguard. Something about him - his height maybe, or his bouncy air of confidence – made me feel safe.

'What do you charge?' I asked, cautiously.

'Depends on the job,' Will said. 'And the client. For you, I'd have a special rate.'

'Why?'

'Because you're Ben's sister, obviously, and he's my job anyway. We are talking about Ben, aren't we? Or do you have a boyfriend for me to trail?'

Annoyingly, I blushed again. I sidestepped the boyfriend question. 'I might need a kind of bodyguard. Just someone to come along with me to a meeting. Not actually there at the meeting, but just to be there, in the background. Just in case it goes wrong.'

I knew Will was desperate to know more. He leaned forward on the table. 'You intrigue me, Miss Montgomery.'

'It'd mean going up to London. Camden.'

'What's the meeting about?'

'It's private,' I said. 'Look, if you're not interested I'll ask my friend, Grace. I'm sure she'd come with me.'

'Grace Lilley? As your bodyguard for a top-secret meeting? That sounds like a fantastic plan.'

I couldn't help laughing. It was absolutely true that if I asked Grace to come along then the entire town would know about Aidan within twenty-four hours.

'How do you know Grace?'

'Cass, just because you don't choose to socialise, it doesn't mean that the rest of us plebs have to ignore each other.'

I wasn't sure if he was accusing me of being snooty or teasing me for being shy. So I got to the point. 'Are you up for it?'

His huge hand thrust forward over the table. 'I'll do it for the special rate of twenty quid. Let's shake on it, Cass Montgomery.'

9: AIDAN

I swing my legs out of bed, shivering in the cold. I'd like to switch on the central heating, but we've decided to keep it for emergencies. Clive doesn't charge us much rent, but the heating bills are a killer.

'Aw, Aidan, love, why're you getting up so early?' mumbles Holly, and Finn makes himself comfortable on my side of the bed.

It's so tempting to lie back down next to her. All I want to do is be with her, so warm and inviting and safe. But Finn is there and he's a bit smelly and besides, I'm on a mission.

'I'm going to see Rich,' I tell her. 'Just check he's OK. You know. Now he's out of hospital, he might be a bit lonely in that bedsit.'

'Can you be back by midday? I wanted to go down the market today, to get some stuff for Finn's birthday. Mum said she'd watch him, but she's only free for a few hours.'

'I'll be back as soon as I can, I promise. Truly, Holly, I do have to see him. I don't know how long I'll be. Enfield, you know. It's a long way.'

Holly snuggles up to Finn. 'I just thought you'd want to come shopping. You know, get him some little presents and stuff. He's so excited.'

'But I'm busy,' I say, before I can stop myself, and then, immediately, 'Sorry. Sorry, Finn. Sorry, Holly. Look, I'll do my best.'

She sighs and says, 'Never mind. We've got a few weeks. Maybe I'll take advantage of Mum babysitting and go and have a swim, eh? You make sure Rich is OK.'

'I don't deserve you,' I say, and I bend down and kiss her. I'm so tempted to get back into bed. But I need to get on.

Holly knows that I won't be back by midday. I'm just incapable. Once I leave the flat, time doesn't seem to exist any more. I'm back in my old life, mooching around here and there, catching buses just to see where they go, talking to random people.

Sometimes I get into situations, and it takes ages to get back.

That's why Holly is so hot on arrangements and timetables and me getting to work on time. Sometimes I find Holly's nagging annoying and sometimes it makes me angry, but mostly I understand why she's doing it. I can't just carry on the way I used to, now I'm with her. I've got responsibilities. Mostly I remember that.

Anyway, I'm feeling bad now, because I don't like lying to Holly. But I have to see my mum, and if Holly knew that, she'd have felt sad and left out, and I hate that. So it's better not to tell her.

My mum's been working in a hairdresser's in Southgate, and I think she's still there. I've visited her there a few times, and she would have told me if she was planning to leave. At least, I think she would.

But when I get to the salon there's a sign on the door saying 'Under New Management', which might mean the staff has changed as well.

'Can I help you?' asks the bored girl on the reception desk, and I grin at her and say, 'Can I just have a quick word with Janette? It won't take long.'

Phew, she doesn't say, 'Who's Janette?' I'm expecting her to call Mum over, but she gives me a long look up and down instead and stops chewing her gum just long enough to jerk her head towards the line of mirrors. 'She's there, at the end.'

Mum is cutting some woman's hair, scissors flying, talking about a wedding in *Hello!* 'That dress is too young for her,' I hear her saying. 'I wouldn't bet on it lasting.'

'Hey, Mum,' I say.

'Aidan!' she says, blades still for a moment. 'Why don't you ever warn me? You can see I'm busy.'

'Sorry.' Of course I didn't warn her. What if she said no? Anyway, I haven't even got her number.

'You'll have to wait,' she snaps, pointing with her scissors at the sofa in reception. 'Ellie-Mae will get you a coffee if you ask nicely.'

I back away.

'He's gorgeous,' says Mum's client, who must be sixty.

'That's my son,' says Mum. 'He's nothing but trouble. Ought to be his middle name.'

'Lovely looking boy. I like the mixed-race ones.'

'You should've seen his dad,' says Mum. 'A soldier. Died in Kosovo. A hero. Course, the British army never gave me a penny because we weren't married.'

I back off more quickly. I can't stand it when she starts her stories. My dad, as far as I can work out, was a soldier. She never got anything from the military or anyone else, because when he died he was married with four other kids.

He wasn't much of a hero, as far as I'm concerned.

Mum's still complaining about me to the client. I can hear her right across the room. 'Yes, he looks like a lovely boy, but you can't judge a book by its cover.'

I sit on the sofa and look at the pictures in the magazines and wipe my clammy hands on my jeans. I only see my mum every now and again, and each time I've forgotten what it's like.

'Coffee? Tea? Hobnobs?'

Ellie-Mae is the gum-chewing girl at reception. She tells me she's the daughter of the new manager. His name is Brian and he's got a chain of hair and beauty places in Essex, Hertfordshire and North London.

'Impressive,' I say, and she says, 'You should come for model night. Free cut, and your picture taken by a professional.'

Holly would love that. She's always going on about how expensive it is to get your hair done properly. But that'd mean Holly meeting Mum, which would be a terrible idea.

'Here you are.' Mum's finished with her client and she's standing in front of me. 'I've got five minutes before my next one. What do you want?'

'Can we go outside or something?'

'I'm taking a fag break, Ellie-Mae,' says Mum, and we go out into the high street. She fishes in her bag for her packet of Silk Cut and offers it to me. I'm meant to be giving up – Holly's really strict about never smoking in the flat – but I take one.

'How are you?' she asks.

'I'm fine. Good. I'm working and I'm living in a nice place and I'm going to learn to drive.'

'That's good,' she says. 'Make sure you don't cock it all up, as usual.'

I've found Cass! That's what I want to tell her. I imagine the look of joy and surprise on her face, the love and pride. I imagine her flinging her arms around me. *Aidan, I'm so happy! You've found my little girl!*

But I can't tell her. I haven't even met Cass yet. Maybe she won't want Mum to know.

Since that first time, we've swapped messages a few times. I'm slow because writing's not my thing, but I can do it, just about. Cass wants to see a photo from when we were small before she'll meet me. She's sensible. She thinks about things. She's completely different to me.

'Mum, I just wondered, have you got any pictures? From when I was a kid?'

'What for?'

'Just, you know. To remind me of the good times.'

She snorts. 'I'd have thought you'd want to forget.' Then she rummages around in her handbag, opens her purse. 'How about that?'

Cass, red-haired and smiling. Me, dark and scowling. Gran's scratchy brown sofa. I can almost smell it, the unique mix of cigarette smoke and stale cat food.

'Wow, Mum. You keep it on you?'

Mum thaws, just a little bit. A tiny smile. 'I want to remember the good times, too,' she says, so sad all of a sudden that it reminds me just how bad those times were. She crushes her cigarette under her kitten heel. 'Look at the two of you. My beautiful children.'

'Can I have a copy?'

'I suppose so. You can get it photocopied at the newsagent. I want it straight back, mind you.'

'Thanks, Mum.'

She reaches up and pecks my cheek. I breathe in her smell, cigarettes, musky perfume, hair products.

'I'm glad to see you, Aidan. I hope the good times last.'

'Thanks, Mum.' This is as good as it gets with her. It's pretty special.

Back in the salon a man is waiting for her. 'What's going on? Janette?'

He must be her new boss. He's all spray tan and sideburns.

Mum steps backwards. 'Brian! Sorry! This is my son, Aidan.'

Brian shakes my hand, bares his dazzling white teeth. 'Nice to meet you, Aidan. I'm sure you'll understand that Janette's got clients to see, and Saturday is our busiest day. Let's try and keep family visits to break times, eh, Jan? And fag breaks out the back, not in the high street. Doesn't look good, seeing the staff puffing away, right outside.'

'I didn't know my son was coming,' says my mum, 'and he's going now, aren't you, Aidan? It's OK. He doesn't come and see me often, do you, Aidan?'

I smile at Brian, grin at Mum. 'I'll just get it photocopied,' I tell her, 'and drop it back to you.'

Brian follows her as she clip-clops over to her next client, and Ellie-Mae asks me if I want her to copy the picture. She does it on their photocopier in the tiny back office, and she leans right in towards me as she's doing it. Her arm acciden-tally-on-purpose brushes against mine as she hands me the photo, and somehow I can see more of her bra and cleavage than I could before.

I don't have to do anything, though. A photocopy isn't much of a favour. Anyway, her dad might walk in on us.

'Shall I take your number? So I can tell you about the next model night?' asks Ellie-Mae, and I smile at her again and say, 'Tell you what, why don't you get it from my mum,' although I'm not sure that Mum's ever saved my number.

Then I go and wait for the bus to Enfield and it takes ages to get there, and it's past two already, and Holly's rung and left voicemails, but I don't reply because I know I'll feel bad to lie. In the end, I text her a smiley face and a love heart, and I hope that'll keep her happy. She texts back a confused face; I send her a kiss.

I knock at Rich's door, steeling myself for what I might find if I have to use my spare key. I'm relieved to hear him open the door.

Rich is looking a hell of a lot healthier than the last time I saw him. His hair is newly dyed – a bright, showy red – and his eyes are focused on me, and he's smiling. I tell him he's looking good. I know the new look means he's trying to put it all behind him.

'How's it going, mate?' I ask him.

'Oh, you know. Not too bad. My social worker came round, helped me clean up a bit.'

The bedsit looks a hell of a lot better than the last time I saw it. Someone's sprayed air freshener to get rid of the smell of vomit.

'That's good.'

'They're going to look at my meds again. See if they can find a better anti-depressant.'

'Excellent.'

'They might find me a counsellor.'

Rich has had counselling before; he's been on happy pills for years. It helps a bit, I suppose, but it would help him more if he could find someone like Holly. Someone who felt like home. Every bloke he meets he thinks is The One, but no one ever thinks that about him.

'Rich, something's got to change. You can't keep doing this.'

He shrugged. 'I don't want to live like this. Nothing ever changes.'

'Come on. Let's get out of here.'

We walk to the pub. We drink some beer. Then we argue a bit about where to go next. Nando's? Into town?

And then I find myself just blurting it out: 'Rich,' I say, 'I've found Cass. I've talked to her on Facebook. I've really found her.'

I couldn't have found a better person to tell. Rich knows, he really understands, how much I missed Cass. How losing her was like punching a massive hole in the middle of my life. He laughs, he cries, he hugs me. 'That's the best news ever!' he says. 'Let's celebrate!'

I forget about Mum. I forget about Holly. It's like the old times with Rich. We drink a lot, we laugh a lot. He's all right, is Rich.

By the time I get on the bus, it's already late.

And in the dark and in my joy it's hard to remember where I'm meant to be getting off. I'm at King's Cross before I realise it. The night crackles with possibilities. I want to dance. I want to meet new people. I need a drink.

My phone is heavy with messages. It stays deep in my pocket.

10: CASS

The picture that Aidan sent me was almost exactly the same as the one in my Life Book. For a minute I felt as though someone had crept into my room and taken the book away from me.

This must be Aidan, I thought. Who else would have the picture?

But the quality was awful – it looked like a copy of a copy. Maybe the picture was on my social work records? Maybe some old foster family had a copy? A photo can't really prove anything, can it?

He sent me a picture of him all grown up. I stared at it for ages, trying to match up the curly-haired boy from then with the stubbly guy from now. Dark hair, check. Sweet smile, check. Big, dark eyes – well, they weren't as big in his adult face, but they were still staring right out of the photo as though he could look into me. He was wearing a check shirt and a white T-shirt and jeans. He was drinking beer out of a can, raising it in a toast to the camera.

He looked OK. If I had seen him in the street, I'd have thought he was good looking.

I used to worry about that sometimes. Growing up and meeting a guy and falling in love and then finding out we were related. Not just Aidan, but a cousin or something. I used to think I'd have to go somewhere far, far away if I were ever going to find someone to marry. But since Dad had left that didn't worry me any more, mainly because I doubted I'd ever trust anyone enough to marry them.

I had a history essay to write, and three books to read for English literature. But I was not doing any of it. I was on Facebook, just flicking from picture to picture, from my wall to chat, waiting to see if Aidan messaged, waiting to hear when we could meet . . .

A message! But it was just Will Hughes, bugging me.

Hi Cass,

Do you think your Mystery Meeting will be during half-term? Because I've got a lot on at the moment and I want to make sure I'm free.

Will xxx

Grrr. I was seriously regretting involving Will in this. Not only was he arrogant and irritating, he also never stopped asking me about the arrangements for a meeting that might never happen if Aidan didn't get a move on and suggest something. If Aidan was Aidan. Back to the picture. Did this man . . . boy . . . man . . . look like the brother I couldn't really remember?

I'd actually cracked and asked Grace how she knew Will. What a stupid mistake. Her eyes nearly exploded and she did a totally fake double take.

'Will Hughes? Oh my God! Why are you asking if I know him?'

'No reason. Ben mentioned him.'

'Oh my God, it *was* him, wasn't it? Shouting to you on the bus? What's going on?'

'Absolutely nothing. He's Ben's peer buddy or something. I just wondered what he was like.'

'He's a legend. Have you seen his clothes? And he's so tall and gorgeous.'

'He's not all that special,' I said, completely accurately. 'He's just an attention-seeker. The clothes are ridiculous.'

'They'd be ridiculous on anyone else, but on Will – he looks great! And he's so clever and so funny. He'd be perfect for you! Why didn't I think of it before?'

'We've got nothing in common,' I said, before I realised that I'd inadvertently and totally accidentally made it sound as though I was actually taking seriously her mad idea that there could be anything between me and Will.

'You'd complement each other! You'd be the serious one! So, has he asked you out?'

'Grace, you're crazy. There's nothing going on between me and Will.' I paused. 'I bet he's got a girlfriend, anyway.'

'As far as I know he's not seeing anyone. He broke up with Ruby over the summer. I can find out, of course.'

'Look, I'm not interested. I just wondered if you knew him. Ben thinks a lot of him and I wondered if he was a nice person.'

'He's amazing! He's, like, incredibly nice! He's got this way of looking in your eyes and asking how you are, like he really, really cares. And he's so cool! He plays the sax and he's in a band and his sister's been on *Holby City*.'

Luckily, the bell rang just then and I was able to escape to my economics class. We had a test for which I'd totally forgotten to study. I chewed my pen and tried to cobble together some answers, and cursed Will Hughes for the bad mark I was going to get.

Back to Facebook. I replied to Will:

- *It might be at half-term. I suggested Monday, waiting for confirmation. Don't worry if you can't spare time in your busy schedule because I'm sure I'll be fine on my own.*

He wrote straight back:

- *No way are you going without me! As your security consultant, I specifically forbid it.*

OK, fine.

- *I'll let you know then.*

- *Yes, ma'am, at your service xx*

I opened the history essay document and tried to concentrate. Russian history. 'Compare nineteenth-century Tsarist rule with communist rule under Lenin.' I knew about this. I really did. I just had to focus.

My phone rang. Will Hughes.

'What do you want?' I probably sounded rude, but surely texting was a perfectly adequate way of communicating?

'Have I offended you?' His voice was amused rather than contrite.

'No, of course not. I'm just working on a history essay.'

'You'll have to get used to me. People usually stop finding me annoying after a while.'

'You surprise me.'

'I knew you had a sense of humour buried away under your Ice Queen exterior. I like it.'

He was *incredibly* annoying. But the thought of finding someone else to come with me to see Aidan – someone who didn't know my parents, someone who wouldn't gossip, was too exhausting.

'Sorry, I haven't got time for this. I have an essay to write.'

'So have I. Tsars versus communists, who were the biggest bastards to the Russian people. Discuss.'

'That's what I'm doing. Although not exactly that title.'

'I like to adapt the titles. Liven things up a bit. Make the teacher smile.'

'Good luck with that.'

'We should get together to study sometime. I hear you're the cleverest girl at your school.'

'I just work hard.'

I'm not really clever at all. I'm just very conscientious and I do lots of work and I listen to what the teachers say. When I got my GCSE grades, I was actually shocked. I knew I was in line for 10 A*s but, seeing them written down, I felt as if I'd pulled off a successful con trick.

Sometimes I feel like I've pulled off the same trick just to be accepted as a Montgomery. When I'm with all the relatives and everyone's so effortlessly clever and confident and knowledgeable ... what am I doing there amongst them?

Would I feel different with Aidan?

'I got 8 A*s and 3 As,' said Will. 'Best in my school. Not sure if I'll make it to Oxford, though.'

'You want to go to Oxford?'

'No need to sound so surprised. They do let black people in.'

'Are you calling me racist?'

'Of course not. Just joking. Don't be so touchy.'

'I'm not touchy!'

'That's a shame.' Will's laugh was more of a chuckle. I ground my teeth.

'Are you trying to annoy me?'

'Am I succeeding? Anyway, Lenin's much more annoying.'

'That's true, and I'd better get on with writing about him. I'll let you know when I have a date for the meeting.'

I went back to my essay. But I couldn't stop thinking about Will and how irritating he was.

Mum stuck her head round the door. 'How's it going?'

'OK. Lenin. Not very inspiring.'

'Do you like having more space?'

I was working in Dad's old study, trying to forget it used to be his. 'I like the space. Not the reason.'

'Make the most of it.' Mum sighed. 'Keep this to yourself, but I talked to the estate agent today.'

'Mum, no! Why is Dad doing this?'

'This is what happens. We're not the first family to go through this.'

'But you put so much work into this house!'

Mum and Dad bought our house about a year before they adopted me. It was an old farmhouse, and it needed completely modernising. Dad wasn't an MP then: he worked for a bank in the City of London, while Mum worked at home as a freelance editor and supervised the builders. She found the furniture at auctions and antique shops. She picked the wall colours. She co-ordinated the architect, the plumbers, the builders, the electrician. She did less and less editing and more and more home-building. She was making the house ready for me. That's how the story went.

'It was a full-time job,' she told me once. 'To make up for the emptiness I was feeling about not being able to have kids.'

When I came, Mum gave up the editing altogether. The social workers advised her to dedicate herself to me 100 per cent in the first year. Essential bonding time, apparently. Then Dad got elected and she got pregnant with Ben.

'It was like a miracle! After all those years of trying! I'm sure it was down to you, Cass. I had my wonderful daughter and somehow that kick-started my system.'

Dad bought the flat in London just after he was elected, because of the late-night sittings at the House of Commons. It worked out well financially to have a second home, and he didn't want to spend his life catching the last train or driving for hours late at night. We stayed there sometimes, but it never felt like home. This house was full of antiques, family history: the portrait of Dad's ancestor who fought at the Battle of Waterloo; the dining table that belonged to Mum's great-grandparents. The London flat was bland, beige, modern.

Even so, I hated the idea of Annabel being there. I suppose selling this house was better than letting her invade and colonise it.

'I'll keep everything tidy,' I told Mum, and she kissed my forehead.

'Thanks, darling.'

Back to the computer. Facebook again. Two messages.

One from Will:

- *Lenin said: Give us the child for eight years and it will be a Bolshevik for ever. Agree?*

And one from Aidan.

- *monday 2pm? camden markit? let me no?*

11: CASS

Camden is one of those places that teenagers are meant to like – full of noise and people, shops and stalls, clothes and music, and cheap stuff to make you feel like you're very individual and different.

I didn't like it. It was too crowded and dirty, too smelly and unpredictable. It made me feel small and nervous. I could easily get lost somewhere like this. I wondered why Aidan chose to live here.

I was actually grateful to have Will with me.

'I can't believe you won't tell me what this is all about,' he complained. 'I mean, what could it be? Are you a spy?'

'No. I am not a spy.'

'You could be, with your government connections.'

'I promise you, I am not a spy.'

'Is it an internet date?'

'No, it is not an internet date. Will, I'm not going to tell you, all right?'

'How far in the background do you want me to stay?'

'About five miles. Seriously, I don't want him to know you're there.'

'Oh, it's a he, is it? And not an internet date? Are we talking secret lover?'

'No!'

'You'll tell me eventually,' he said, and I vowed silently that I would never tell Will Hughes anything about Aidan. Ever.

Half an hour to go. I was calm, in control, absolutely fine – on the surface, anyway. Underneath, I felt sick, scared, and like I could quite easily forget how to breathe. Will being there helped, because there was no way I was going to let him see how nervous I was.

We arrived at the café Aidan had suggested. My plan was to find two tables – one for Will right at the back, and one for me at the front.

Disaster. The whole place was completely packed with tourists.

'Oh no!' I moaned. 'What can we do?'

'You'll have to head him off out here,' said Will. 'I'll be behind you. Completely invisible. Don't worry.'

'I suppose so ...'

Quite unexpectedly, he grabbed my hand, squeezed it. 'Cass. I don't know what this is all about, but I won't let anything bad happen to you.'

I took my hand back, hoping I wasn't blushing. 'Thanks.'

I'd tied my hair back as usual that morning, and dressed in jeans and a pale blue T-shirt and a grey cardigan. Then I looked at myself in the mirror. Somehow, my normal ponytail made me look severe, off-putting; somehow my ordinary clothes seemed drab and boring. So I'd let my hair loose and I'd found a bright blue shirt that Grace had bought me for my birthday and I'd never worn before.

I wanted to look different for Aidan. I wanted to look more grown up than usual. I even took more care over my make-up than usual – well, I put on mascara, which I only do for special occasions.

Even Mum had noticed when I left the house. 'You look lovely, Cass. Going anywhere special?'

I had my story all planned out. 'I'm meeting some friends and we're going up to London. There's an exhibition at the Imperial War Museum. Our history teacher recommended it.'

'Is Grace going? Megan?'

'No, she's away all week. Pony-trekking in Andalucía. And Megan doesn't do history. It's just people. No one you know.'

'Oh, well, have a good time. When will you be back?'

Mum always wants to know where I am and who I am with. Normally, I hardly notice that she's asked. That Monday I was lying and lying. 'I'm not sure. We might see a film or something.'

'Oh, well, ring me if you're going to be late. I'll come and get you from the station if you need me to.'

'It's OK. I've got taxi money.'

Will looked at his watch. 'Five minutes to *rendezvous*. I'm off.'

'Where will you be?' Suddenly, I didn't want him to go.

'In the shadows, round the corners ... don't worry. I can blend into the crowd.'

Off he went. Six foot two – so he claims – and his hair adds another two inches at least. A dark indigo hoody and green skinny jeans.

I'd asked him if he had the right clothes for surveillance when we met at the station. 'What do you mean?' he replied. 'This is completely neutral for Camden! I go there all the time!'

Will was completely unblendable ... What was I doing ... ?

'Cass?'

I whirled around. That must be him. Aidan. Dark hair, a slight frown on his face. He looked as confused and nervous as I felt.

'Aidan?' I had thought I'd remember right away. But seeing his picture had confused me. What if I only felt as though I knew him because of the photograph he'd sent?

Then he smiled. His eyes crinkled at the sides, and I could see his teeth – a gap between his front two, just like mine before I spent two years in a brace.

I knew him! I remembered him!

'I thought you'd be inside the café,' said Aidan. 'I was right at the back. Then I came out looking for you.'

'I'm sorry! I was going to come in but it looked so busy. I didn't know if I'd find you.'

I was burbling, because I'd never actually talked to anyone who looked like Aidan. He had three piercings in each ear – a line of studs in the lobe – and a ring in his left eyebrow. The tattoo on his neck wasn't the only one: where his lumberjack shirt was unbuttoned at the top, I could see blue ink and a tangle of dark hair.

I didn't know what to do. I stuck my hands behind me and said – too posh, too formal – 'It's very nice to meet you.'

'Don't I get a hug?' he asked. 'It's been a long time.'

'Oh! OK!' We had the briefest of hugs. I could feel my whole body go tense and my shoulders hunch up to my ears.

He released me. 'It's so good to see you,' he said.

'Where shall we go?'

'Maybe the park? I know it's cold, but it'd be good to get away from all the people. We get loads of tourists around here.' He pointed the way and we started walking.

Momentary panic. Was he just some internet nutter? I glanced over my shoulder to try and see if Will was following us. No sign of him. Oh, Jesus, I was breaking every rule of personal safety that my parents had drummed into me: walking somewhere with someone I didn't know; someone who was older than any of the boys I'd ever socialised with; someone who was more of a man than a boy, in fact.

But I *did* know him. He was Aidan. He was my brother. I was almost certain I remembered him.

We walked away from the crowds, past huge houses with pillars at the front. We crossed a road. 'Here we go,' said Aidan, and we were in a park. Regent's Park. Mum and Dad had brought us here before, to have a picnic and watch a cricket match. Press versus parliamentarians. I wondered if the journalist who broke the story about Dad and Annabel had been playing. Did he even know what damage he'd done? Did he care? Or was it just a game for him, the journalistic equivalent of hitting a six?

'There's a café at the top of the hill,' said Aidan, and we kept on walking.

The hill wasn't much more than a gentle slope, and the café had tables inside and out.

I asked for a cappuccino, he ordered two. The bill came to £5.80. Aidan dug into his pocket for change, spent ages counting it in his hand.

'I can get this,' I said, hastily. He couldn't earn much, working in some salvage yard.

'I can get my sister a coffee,' he said, handing the change over the counter.

'Oh. Sorry.' I hoped I hadn't offended him.

We went and sat outside. You could see right across the park. No sign of Will anywhere.

'So,' said Aidan. 'Here we are.'

'I can't believe it.'

'It was so easy. Thank God you're on Facebook.'

'I shouldn't be, really. After all the stuff with the papers, Dad's press person said I ought to make my profile more private.'

'You've not got much there. No party pictures.'

'That's what I thought. And I don't like being told what to do by Party workers.'

He frowned, just slightly.

'I mean, people who work for the Conservative Party ... you know ...'

'Oh, right,' he said politely, stirring his coffee. I didn't know if he understood me or he didn't, and whether he thought I was an idiot or the most patronising person on the planet.

'Um, so, how's your job?' I asked.

'Yeah, it's good. I got a half-day off today. Clive was pissed off with me, but I told him I had to see my social worker.'

'Who's Clive?'

'He's my boss. He's OK. 100 per cent straight.'

'Oh. OK.'

'People kind of assume he isn't because of the shop. You know. But it's all kosher.'

I wasn't sure if he was talking about his boss's religion or his sexuality or what, so I just nodded and smiled and said, 'Oh, that's good.' I felt like a member of the royal family, visiting some scheme for young offenders.

He leaned forward. 'Are they good to you? Your parents?'

How could I answer that question? It felt as though I didn't really have parents any more – just two individuals. 'Well, you know, it's been a bit difficult lately. They've just split up.'

'So ... you're not happy?'

'Yes, well, I thought I had the perfect family and everything was fine. So you can imagine what we're going through.'

He smiled. 'I never had much of a perfect family. Not until now, anyway.'

I ran my hands through my hair. 'Oh, I'm sorry. I didn't mean ...'

'Don't worry about me. Everything's good for me at the moment. Especially finding you.'

I'm embarrassed. 'That's really nice of you to say so.'

'Tell me what it's like, growing up in a perfect family.'

I had no idea what to say. 'I mean, we live in a nice house and I go to a good school and we have holidays in lovely places. All the middle-class trimmings.'

He laughed. 'Pony riding? Ballet? That kind of thing?'

'Everything.'

'My girlfriend grew up a bit like that. She didn't like it. Too much pressure. Everyone pushing her to go to university.'

'There is that, yes. I always feel ... well ...'

'Go on.'

'Like I have to do the best I can. To thank them ... sort of.'

'For rescuing you?'

'That sounds awful. I don't mean it like that.'

'No, I get it.'

'Aidan? Do you think it was better for me to be adopted?'

He shrugged. 'I don't know. It was hard, that's for sure.'

I'm not certain if he meant it was hard to lose me, or hard to be left behind. Either could be true.

'I don't know much about our background,' I say, tentatively. 'They gave me a Life Book with photos and things but it didn't have many details.'

'I had one of those. It must have got lost along the way.'

'Aidan, why didn't you get adopted, too? Mum – I mean *my* mum – said that's what she thought had happened to you.'

He looked away, staring over at a football game between two teams of little boys. I glanced over my shoulder to see if I could spot Will, but there was no sign of him.

'That's what she said? She never talked to you about me?' Aidan was still looking into the distance.

'They said the social workers thought you needed a family of your own, to give you lots of attention.'

'Oh, right. Well, I think that was their plan. But no one wanted me.' It sounded bleak, but he made a funny face as he said it, so we laughed together. 'The older you get, the more difficult it is. You were lucky. One more year and you'd have been like me. Unadoptable.'

'Oh, Aidan, that sounds awful.'

He shrugged. 'I had some brilliant foster parents. There was one couple – Betty and Jim – Auntie Betty and Uncle Jim, I called them. They lived out near Epping Forest.'

'Did you stay with them for long?' I'm thinking of my Life Book, how I stayed with one couple for six weeks, another for nine months. Was that what Aidan's whole life was like? No wonder he'd lost his Life Book.

'I stayed with Betty and Jim for four years. From when I was eight, to just after my twelfth birthday. They were talking about adopting me, too.'

What went wrong? That's what I want to ask, but I can't think of a tactful way to phrase it. So, instead, I ask if he still sees them at all and he shakes his head and says no, he hasn't seen Betty and Jim since he moved out. 'I didn't have a mobile.'

Not having a mobile seemed to me to be a pretty lame reason to lose touch with people who might have adopted you. What about landlines? What about letters? But Aidan was looking into the distance again, and something told me that it might be better not to ask any more. After all, these people changed their minds about adopting him. I couldn't begin to imagine how much that must have hurt.

'That's a shame.'

He switched on that smile again. 'That's how it goes. But, anyway, tell me about you. You're at school? You like it?'

'I like some things about it. History. Netball. English.'

'English was my worst subject. They hated me, the English teachers. I couldn't do it at all.'

'Do what?'

'Anything they asked me to do. I didn't get on with school. Dropped out when I was 14 or so.'

'At least you thought about it. I never think about whether I actually want to be going to school or not, I just do whatever's expected of me. I'm meant to be applying to Oxford next year, and I'm not even sure it's what I want to do ... But I'll end up doing it ...'

Aidan whistled. 'Oxford University?'

'If I get in. I probably won't.'

I was lying. I was almost guaranteed a place at Oxford. As long as I got the grades, I'd get in. But how to explain that to Aidan? He couldn't even spell.

'You'll get in,' he said. 'Why wouldn't you?'

'You have to be really good.'

'I bet you are.'

As we talked, I was trying all the time to match his face with the boy in the photos. Straining after tiny flashes of memory.

I remembered Mum plaiting my hair. I remembered her tying pale blue ribbons into it. I remembered her telling me that I'm going to see Aidan, to say goodbye.

'Do you remember the last time we saw each other?' I asked him.

He paused, scratched his head, looked away into the distance. 'Do you?'

'No. I can't remember anything much.'

He shook his head. 'I don't think so,' he said. 'It's all muddled up in my memory.'

12: AIDAN

It's one of the most difficult conversations I've ever had, and I've had some shockers in my time.

She hasn't got a clue. No idea what foster care is like. No idea about children's homes. She knows nothing about Mac or Mum, or why she was adopted. She knows nothing about me. It's like the social workers waved a wand and sent her to frigging fairyland.

She's probably never met anyone who doesn't live in a big house, goes on holiday to swanky hotels, or gets top marks at school.

That makes me sound bitter and jealous and full of hate, but it couldn't be less true. I think she's great. It's extraordinary just looking at her. She has shiny red hair, like those elegant dogs with long floppy ears. There are a few freckles on her nose, which is straight, a smaller version of my big beak. She looks nothing like me – of course, why should she, we have different dads for a start – but the way she holds her head, the curve of her jaw, her neat little ears …

It's like looking in a strange, mixed-up mirror. She's a little bit of me and Mum and Gran – and him, of course, he's there, too. You wouldn't think it was possible, but it is.

Mac. The monster who hurt everyone he came across. I'm looking at Mac in my sister's pretty face and it chills me inside, it's actually scaring me, just that twist of her mouth, the way her eyebrows arch.

'Aidan?' Her voice is nothing like mine or Mum's or – thank God – Mac's. Her voice is posh and confident and completely her own.

We've been chitchatting about this and that, her school, my job, her parents. I don't say much about Holly because I haven't told her anything about Cass, so it feels disloyal to talk about her.

Obviously, I should have told Holly this morning, but I sort of felt a bit awkward about the whole thing, plus she might not approve of this. She might think it was against the rules, because it is, and although Holly doesn't like rules much, she might go for this one, she might agree that you shouldn't see your family before you're eighteen and only with your hand held by social workers and counsellors, because Holly believes in all that stuff a lot more than I do.

I don't think the rule is fair, because the new family have had Cass for 99 per cent of her entire life, and I only want a few hours or whatever.

Anyway, Cass just looked at her watch and I have to go and pick up Finn soon, so maybe we're going to avoid the most difficult questions of all.

'So, do you see much of her? Our mum?'

OK, we're not going to avoid them all.

'She's ... well, she's got her life together now. She's got a husband and two kids – two more kids, I mean – and she's a hairdresser.'

Cass raises an eyebrow just slightly. Is she too posh to have a mother who cuts hair for a living? What about a brother who unpacks boxes of salvaged sleeping bags or stacks bricks in a yard?

'Two more kids? We have more brothers? Sisters?'

Oh, shit.

'One of each. The boy's about eight, the girl's younger.'

'Wow! What are their names? Have you got a photo?'

'Louis, the boy's called. Louis. I think the girl is Scarlett. I'm not part of their life.'

She's shocked.

'But why not? I thought you were in touch with her?'

'Yes, but she, kind of, she keeps things separate. Them and me. I just see her now and again. She's not really like a mum to me.'

'What is she then?'

'She's ... I don't know ... someone I used to know. I see her, you know, sometimes.'

I'm sweating. This is going wrong.

'Oh! That's sad. I'm sorry, Aidan. I'm probably saying all the wrong things.'

'No, don't worry about me. My life is good! I have a great place to live – a beautiful flat. I have a girlfriend, a job. It's really something. It makes a big difference.'

'But you've had a hard time in the past.'

What does she expect? Not everyone's life is all holidays and riding lessons.

'Well, everyone has, you know, challenges. I'm fine now. It's all behind me.'

'Are you sure? I mean you're only, what, eighteen?'

'You grow up really quick in care. They chuck you out when you're sixteen, and you have to cope on your own.'

'Really? I can't imagine that.'

'They put me in a bedsit. All on my own. I thought I was going to go crazy.'

'I wouldn't cope for a week!'

She would, actually. She's clever and she's pretty, so she'd probably get a job easily. She knows lots of people with money, so she'd be able to get loans, sleep on their sofas. She's got connections. I've got them too, but only through Holly. Good connections through Holly. I've got some bad connections of my own.

'I was lucky. I met my girlfriend and she offered me a room. I was her lodger at first, but then we got together. Except we have to keep it a bit quiet because you know ... it might affect our benefits ...'

Just a moment too late I remember that her dad's a government minister. The sort that thinks that anyone on benefits is a scrounger.

'You won't tell, will you?'

'Of course not.' She seems offended. 'So, can you tell me anything else about our mum? And my dad?'

'He went away a long time ago,' I say. 'No idea what happened. I can't remember much about him. Mum's OK most of the time. A bit touchy. She seems happy with her bloke.'

'Do you think she'd want to meet me? Not now, but maybe one day?'

I know she would. I'm not sure I'm ready, though.

'Maybe. She's unpredictable.'

'She must be if she doesn't want to let you know her other kids. Maybe she'd feel the same about me.'

'Maybe,' I lie. 'Maybe.' I'm sure Mum would love to introduce Cass to Louis and Scarlett. She'd be their new big sister.

My phone beeps at me. Time to pick up Finn. I'm proud that I've remembered.

'I'm sorry,' I say. 'I've got to be somewhere.'

'Aidan, has this been OK? Shall we ... can we ...'

'Yes! Of course! We'll do it again!' I mean it. It's just that her questions have opened up all kinds of memories, and she's hardly started. She doesn't even know the questions to ask.

'I'd like to meet your girlfriend.'

'She'd like that, too.'

I'll have to tell her about meeting you first, though.

We're making our way back through the market, standing there, saying goodbye, trying to get over the awkwardness, when it happens. A shout of excitement, and then something grabs me round the knees.

'Aidy!' shouts Finn. 'Aidy! Aidy!' He rubs his runny nose on my jeans. 'I sawed you!'

What the hell? I look around for Holly. Why is Finn running around Camden Market on his own?

'Finn! Where's Mum? Why aren't you at Poppy's?'

'Granny!' he shouts. 'Go park!'

That's all I need, the Wicked Witch of Primrose Hill.

I hoist him up into my arms. 'Finn,' I say, 'meet Auntie Cass.'

13: CASS

'Daddy, Addy!' the little boy shouted. 'Addy!'

I shouldn't have been so shocked, I suppose. I knew people did have children young; in fact, Dad was always talking about teenage mums and how to discourage them. Janette, our mum, she'd been really young when she had Aidan and then me. Of course it was natural that he'd have done the same thing.

Anyway, Aidan might only have been eighteen but he seemed a whole lot older than the boys I was used to seeing on the bus and in Starbucks. Boys like Will, for instance.

What I was surprised about was that he hadn't mentioned having a little boy all the time we'd been together, and he hadn't said anything in his messages, either. He'd talked about the flat where he lived, and a bit about his girlfriend, Holly. Nothing at all about being a dad. And this wasn't a tiny baby. Aiden must have started really young.

An older woman pushing a buggy came rushing up to us.

'Finn!' she said. 'You mustn't go running off like that! Naughty boy!'

Finn screwed his face into Aidan's neck, but Aidan put him back into the buggy, which made Finn cry and kick his legs. 'He could've been hurt or lost or anything,' he told the woman.

'Don't blame me!' she snapped back. 'The buckle doesn't do up properly. Tell Holly I'll buy her a new buggy. Finn, you're a naughty boy to run off. Say sorry to Granny.'

Granny! I thought for a minute that I was going to faint. My knees wobbled and I felt hot and cold and clammy. But then I looked at her again. She was blonde and smart and she wore nice clothes – a grey wool coat and a soft blue scarf – but she was easily sixty. She was far too old to be my birth mother.

'No!' Finn kicked and arched his back to break free of the buggy's straps. The buckle sprung open, and Aidan held Finn down in the seat and forced the clasp back together again. I could see it wouldn't last long. Aidan stayed bent over the buggy, trying to work out how to mend it.

Granny didn't give up. 'Say sorry!' she insisted.

'Sorry, *borry*!' screamed Finn, clearly intending the 'borry' to cancel out the 'sorry', lashing out with his fists and catching Aidan in the eye. Aidan swore, and the woman made a tutting noise.

Strangely enough, Finn fell silent. His legs gave up kicking. His thumb went into his mouth and he gazed thoughtfully from Aidan to his granny. He and Aidan looked so alike, with their dark curls and light brown skin, their dark eyes and worried frowns, I felt almost jealous. Aidan was my brother, but we couldn't have looked more different.

'Umm,' said Aidan. 'Sorry, Juliet. Sorry, Cass. Sorry, er, Finn. It just slipped out.'

'I'm sorry, Aidan,' said Finn's granny. 'Of course he shouldn't have been out of the buggy. He just wriggled out ... nearly gave me a heart attack.'

'Good thing I caught him,' said Aidan.

'I don't think he'd have run off if he hadn't seen you.'

Her voice didn't quite fit the way she looked. Cockney upgrade, my dad would call it. Nouveau riche, I found myself thinking, and hated my parents for making me a snob.

'I didn't know you had Finn. I thought he was at the child minder,' Aidan mumbled. 'Holly asked me to pick him up. I was just on my way.'

'I waited there for you. But you were half an hour late, and Poppy needed to get on with other things. You must learn to be more reliable.'

Aidan didn't bother to answer. He hung his head, like a sulky child.

I was feeling incredibly awkward, just standing there. Surely he should have introduced me? She must have been wondering who I was. I held out my hand. 'I'm Cass,' I said. 'Cass Montgomery.'

Juliet frowned and took my hand with as much enthusiasm as if I had offered her a dead rat. 'Hello, Cass,' she said, and then to Aidan, 'Holly says he's been a bit unsettled at the child minder's, poor little Finn. She said she'd been trying you, your phone was switched off, and she wasn't sure you'd remember you were meant to do the pick up.'

Aidan puffed out his cheeks, jiggled from foot to foot. 'He's

OK there,' he said. 'He likes it. Poppy's got loads more toys and stuff than we do. And I was on my way.'

'We're going to have a little walk in the park before meeting Mummy for tea, aren't we, Finn?'

Finn nodded and removed his thumb. 'Duck!' he said.

Aidan winced, but Juliet said firmly, 'Yes, darling, we're going to feed the ducks.'

'I didn't realise you had Finn,' I said to Aidan, and then I wished I hadn't, because Juliet's thin eyebrows went up and her mouth tightened into a smirk.

'Yeah, I should've . . . there wasn't really . . . it's complicated,' he said, staring at his feet.

'I'd better go,' I told him. 'It's been great. Really great. I'll give you a call.'

'I'll walk with you to the tube,' he said. 'Bye, Finn. I'll see you later.'

But Finn gave out a huge shriek, and screamed, 'No! No! Addy come ducks! Addy come tea!' so I patted Aidan's arm and said, 'Don't worry. You go and have a nice time with Finn.'

'Cass,' he said, glaring at Juliet, 'Cass – be in touch, OK? It was so good to see you.'

I smiled, although I suddenly felt like crying. I wanted to hug him, kiss him, something, but with Juliet watching and Finn kicking, I didn't feel that would be appropriate. I didn't know what to do. So I stuck out my hand and Aidan stared at it, and then grabbed it and we shook hands. Which was totally wrong, but the best I could do.

I nodded politely to Juliet, who was still maintaining absolute permafrost towards me. 'Goodbye,' I said. And then

I was backing away and walking blindly through the crowds, in the vague direction of the tube station.

As I walked away, I heard Juliet say, 'Well! A new friend, Aidan?' but I couldn't hear Aidan's response.

Will caught up with me just before I got to the tube station and just after I'd given up hope of finding him. He emerged out of the crowd and grabbed my arm. I thought he was a mugger, and hugged my handbag to my chest, letting out a tiny, embryonic scream.

'It's me, you idiot!' he said.

'Oh! You didn't have to grab me like that!'

'You were about to rush into the tube station! What were you thinking – perhaps I'll bump into Will on the Northern Line?'

'I'm sorry. It's just, I thought, I was ages and I thought you must have gone hours ago.'

Will rolled his eyes. 'No faith at all! I've been hiding in corners spying on you. A bit difficult, mind you, in Regent's Park where there *are* no corners, but I found suitable trees.'

It was sunny but cold, and I felt bad to think of Will hanging around for more than an hour just so I could feel safe talking to my own brother. How stupid I was, thinking that I needed a bodyguard.

'You must be freezing, I'm sorry.'

'I am freezing. And starving. Shall we get something to eat?'

He didn't wait for an answer, but opened the door to a Turkish café and ushered me in. The place was completely empty.

Will headed for a table in the corner. I hesitated. I didn't really want to have a meal with someone I hardly knew, when

I could have been analysing my meeting with Aidan, but I couldn't think of an excuse. Will looked at the menu. The café was warm and quiet, it smelled of cumin and lamb stew, and the owner brought us cups of mint tea without asking.

'Mezze ... are you OK if I order ... anything you hate? Vegetarian? No ... OK ...'

The waiter brought us hummus and pitta bread. Will took a bite and chewed.

'So,' he said, mouth half-full. 'The tall, dark stranger. The half-hearted hug. The heart to heart. The little boy. The older woman – she really doesn't like him, does she?'

'I have no idea,' I said, frostily. The waiter put down six more little plates of food.

'Yum!' said Will. 'I love Turkish food, don't you? Here, try this one. You're going to have to tell me, you know. Do you want me to die of curiosity?'

'Yes.'

He clutched his chest, made a choking noise and slid under the table. I ignored him and helped myself to a kebab. I did like Turkish food, I discovered. After a moment he was back in his chair.

'You are made of ice. What if I'd really had a heart attack? I thought you'd at least try and give me the kiss of life.'

'You picked the wrong person.'

'I was sure you'd know how to do it. You must have been a Girl Guide at some point. You're just the type.'

Stung, I retorted, 'What type is that?'

'Do you really want to know?'

'Probably not, but I'm going to hear anyway.'

'You're a joiner. You like being involved. Orchestra, choir, debating society … That's why you were definitely a Girl Guide. Am I right?'

'I don't like being involved. But I was a Girl Guide. Well done, Sherlock. That's how I know a real heart attack from a completely fake one.'

'And that's why I'm desperate to know what the story is! Girl Guides do not run around London with guys who look like that! Earrings! Metalwork in his eyebrow! A tattoo on his neck! Hasn't shaved for a day at least! Who is he? Is he an ex-boyfriend?'

'No!'

Will drummed his fingers on the table. 'Now you're the annoying one.'

'Sorry.'

'Who will you tell?'

'What?'

'Are you going to tell Grace? Because that's incredibly unfair, what with me coming all this way and lurking so excellently that I bet you never saw me all afternoon.'

'I didn't see you, that's true. You are a superb lurker. Surprising in one so attention-seeking. And I'm not going to tell Grace anything. She's great, but she's got a big mouth, as you know.'

'I haven't got a big mouth. I'm incredibly discreet. You can totally trust me. Besides, this is a professional relationship. You can add a confidentiality clause.'

'To what?'

'To our contract, which I will draw up this week for us both to sign.'

'Oh, well, I'll wait until my lawyer's checked it out.'

'But you will tell me then?'

I didn't answer. He leaned forward. 'I mean it – you can trust me. If you need someone to talk to, give me a try. I'm here, that's all. Underneath all my messing around, I'm actually a nice person.'

'Well, thanks,' I said. 'I'll remember that.'

'What are you up to for the rest of half-term?' he asked. 'I thought I'd go to Oxford one day. Check it out for next year. Maybe at the weekend. Do you fancy coming to have a look?'

'I'm busy next weekend.'

'Oh. OK. No problem. I've got a few other friends who are thinking of applying. I just thought you might like to come, that's all.'

Why was I like this? Why was it so hard to just relax and say yes to someone who was trying to be friendly?

'Ben and I have got to go and see our dad next weekend. We've got to start, you know, weekend access. I'm a bit worried about Ben. It's going to be confusing for him.'

'So you feel you've got to be all strong and selfless for him?'

God, Will Hughes was annoying! I was trying to talk about Ben; he was completely missing the point.

'There's not really an alternative.'

Will traced a pattern in the leftover hummus.

'How's everyone going to cope when you leave home?'

I'd been deliberately avoiding thinking about that.

'Oh, well, it's not for ages yet. Anyway, I might take a gap year.'

'I think you should come and look at Oxford,' he said, decisively. 'Spend a day thinking about your future. You don't need to come with me if you hate the idea so much, but it'd be good for you to get away from all this family stuff. And if you did decide to come with me, it'll be fun, I promise. If you can bear the idea of spending a day with someone as irritating as me.'

There were so many things I could have said to this. About how my family and my future were no business of his. About how I knew Oxford really well, because my granny and aunt lived there. About how I had a load of homework to get finished. About how he had no idea what I'd find fun.

But I was kind of exhausted by meeting Aidan and I fell into his trap. 'It's not *you* ... I don't *hate* the idea ... it's just that next weekend is no good.'

'Oh, well, that's easy, then. How about Thursday?'

He was grinning at me, very smug and pleased with himself, and I was panicking because I didn't know if this was a date or a friendly gesture or a wind-up or what.

And then I thought about my mum. My birth mum, Janette, teenage mum. And Aidan, not even nineteen yet, with a son ... If I'd grown up with them, I'd know how to handle situations like this. I'd be confident about boys and dates and sex and all sorts of stuff. Grace was right. I was a frigid freak, sweet sixteen and never been kissed, and somehow that seemed much more awkward and difficult and wrong than it ever had before. It wasn't that I was interested in Will Hughes, or that anything was going to happen. He was definitely a flirt, though, and a day with him would be good practise for when I found someone more suitable.

'Thursday?' I said. 'I think that's OK. As long as I've finished all my half-term homework, then I could probably do Thursday.'

14: AIDAN

Don't ask me why Holly's mum hates me so much.

Maybe it's because I'm not posh like her, or perhaps because I didn't finish school, or maybe it's something to do with Holly being a few years older. Juliet was OK when I was just the lodger. But she came round one morning to surprise Holly, drop off some toys for Finn's first birthday, and she let herself in with her key, and I was in the wrong bed. With Holly. She's never got over it.

Clive says it's because she wants to see Holly married to someone who's older and richer, with a proper job. 'She's got big ideas, my ex sister-in-law. Never stopped nagging Bernie. No wonder he buggered off to Marbella and hooked up with that tart, Monica. Not your fault, Aidan, but you're not what she had in mind for Holly. And she's never going to let you forget it.'

Juliet winds me up all the way to the park. 'So, not working this afternoon, Aidan?' and, 'Was that a new friend, Aidan?

She's a very pretty girl. Can't be more than about seventeen?' then, 'Do you think Finn's got a cold? They do pick up germs at the child minder's, don't they?'

This one offered me an escape. I crouched down, took a good look at Finn. 'Maybe you're right,' I say, 'he's looking a bit under the weather. Tell you what, I'll take him home. You go and have a cup of tea with Holly, and tell her I'll see her at the flat.'

'Oh!' I've nutmegged her and she doesn't like it. 'Are you sure? But he's looking forward to feeding the ducks! And I promised him a toasted tea cake.'

'Can't be too careful,' I say, grasping the buggy and wheeling it round. Finn's asleep, luckily, so we don't have to say any goodbyes.

On the way home, I call Rich and tell him about Cass, how great she was, how we knew each other right away. 'I owe you a favour,' I tell him. 'If I hadn't seen that newspaper in the hospital that day . . .'

Finn only wakes up when we're five minutes from home. He spots a pigeon. 'Duck!' he says, loud and clear. 'Finn go feed ducks!'

We're at our flat door. I undo Finn's straps, tell him to go up the stairs. He doesn't want to. 'Ducks!' he yells. 'Ducks!'

So I have to carry him up three flights of stairs. He's getting heavy, and he's kicking me in the chest. I'm still talking to Rich on the phone, but only half listening to what he's saying.

'Go park!' screams Finn. 'Go park with Granny!'

'Is that Finn?' asks Rich. 'Jesus, Aidan, you've got the patience of a saint.'

'Call you later,' I say, although I know he's got more to tell me about his counsellor, and the new guy he's met, and the college course he's thinking of signing up for.

'Aidy!' shouts Finn, as I open the front door and put him down. My back is aching – I'm not as fit as I should be.

Sod's law, the first thing Finn sees are his toy ducks, which ought to be in the bathroom. 'Ducks!' he cries, and then the penny drops that there's not going to be any duck-feeding or special tea with Granny, or treats for Finn today.

His face goes red and then white. His eyes screw into little slits and squeeze trickling tears. The volume of his crying goes up and up and he kicks his legs at me, at the furniture, at the telly . . .

'Not the telly!' I yell at him and I pull him away, because although I don't really think he could reach up to smash it, he needs to know now that telly-kicking's a bad idea because if he goes on like this, he's going to cost us a fortune by the time he turns three.

So he attacks my shins instead, and I'm hopping around saying all the words that a toddler's not meant to hear.

And I just don't need this right now. I don't. I don't want Finn bringing me down.

'Stop it! Shut up!' I'm screaming at him now. He's still crying, raging, kicking out at me.

Holly and I have talked about this. She's really hot on parenting stuff, reads all the books and magazines. I can hear her voice in my head: 'Pick him up. Take him into his room and put him in the cot. Shut the door. Walk away. Calm down.'

My mouth is clamped shut so I don't shout at him as I take him into his room. Finn's room used to be my room. But when I moved into Holly's bed, Finn's cot got moved in here. He doesn't remember, but I still feel a bit guilty. And I feel bad that he's still in a cot when he's too big for it really, but we've been waiting to see if Clive gets some children's beds in the shop.

He's screaming and yelling and struggling. 'I hate you, Aidy! I hate you!'

My whole body tenses.

'I don't want you! Go away!'

'Bedtime,' I tell him.

'No! No bedtime!' He's wriggling and gasping like a fish.

'Yes!' I shout. 'Bedtime!'

He lurches backwards, and I lose my grip on him. Just for a moment. But it's enough for him to bang his head on the side of the cot. His screams get even louder. I bet Clive can hear them downstairs. What if he thinks I've hurt Finn on purpose?

His scream is higher, more piercing, unbearable.

'Finn! I didn't mean to hurt you! Finn!'

'Bad Aidy! Bad!'

I'm confused. I ought to walk away, get calm. But what if I've really hurt him? My stomach clenches. I'd never hurt Finn, never. I know how it feels.

'Finn, let me see your head.'

Finn flails like an old boxer who can't land his punches any more. 'No! Want Mummy! No Aidy!'

I stumble out of the room, shut the door, slide to the floor. My heart is racing, my breathing shallow. Finn's still screaming. He hates me. He doesn't want me. I stuff my fingers in my ears.

'*Do you remember the last time we saw each other?*'

I lied to her. I do remember. It was one of those places where they have supervised visits. Toys that smell of bleach.

Bright red sofas and shiny floors. I hadn't seen her for six months: I'd been living with some foster parents. I can't even remember their names.

Cass looked different to how I'd ever seen her before. She was wearing clothes that I'd never seen. Her face was thinner. Her hair was tied into pigtails with pale blue bows.

She wasn't my sister any more. She belonged to them. She was their little girl.

The social worker had explained it all beforehand: 'It's not Cass's fault, Aidan,' she had told me. 'It's just the way it worked out.'

All I remember is that as soon as I saw Cass I was so full of hatred that I thought my head would explode. 'You look stupid,' I told her. 'I hate your hair like that.'

'Aidan!' said the social worker. 'Don't talk like that to Cass! Come on, let's see what's in the toy box.'

Cass just stared at me. She held onto the social worker's hand.

'Look, here's a jigsaw! Shall we do the jigsaw?'

Cass and her social worker started doing the jigsaw. I hid behind the sofa, watched her from there.

She was the same and she was different. She didn't know who I was. She'd forgotten me. She didn't care about me or Mum at all.

'Come on, Aidan,' my foster mum said. 'Come and join in.'

They got me to come to the table. I looked down at the jigsaw. It was a picture of a house: a little white cottage with roses climbing up its walls. A blue sky, a shining sun, a cat on the doorstep. They'd done about half of it.

I picked up a piece, tried to shove it into place.

'That's wrong,' Cass told me. 'It doesn't go there.'

'It does,' I insisted, pushing it so hard that half the pieces sprang into the air.

'You broke it!'

I shoved her then, hard, in the chest. 'Go away!' I told her.

Then she cried and the social worker took her away and they tried to talk to me, but I knocked the table over and scattered all the jigsaw pieces and then I kicked my foster mum and I was in trouble again.

That was the last time I saw Cass. I don't think she remembers it at all. I've carried it around, all those years, how angry I was, how mean and scary, and she had no idea.

I don't know whether to laugh or cry. Crying seems most likely, so I go into the living room where it's harder to hear Finn. A big fat tear rolls down my cheek. I need a drink. I need to go to the pub. But they won't let Finn in there. I take deep breaths and whimper pathetically into my fists. I'm not even sure why I'm crying. Only a few hours ago I was the happiest I've ever been, just at the idea of seeing Cass again. But seeing her again hurt more than I'd thought it would, and now Finn hates me, and doesn't want me.

I need a drink. I need to make myself numb.

Mac used to drink whisky. Whisky didn't numb him. It made him meaner. He reigned by fear, did Mac. He'd shut me in my room with no supper. He'd whack Mum round the head for answering back. He loved Cass, though. That was the joke. He hated the rest of the world, but he loved his daughter.

Mac played with your head. He'd make out everything was your fault. He'd make me confess to every crime, beg him for punishment. I hate him worse than anyone else who ever hurt me, because he was the first person who put it into my head that I don't deserve much. He opened the door to anyone who ever wanted to do bad things to me, and there were plenty of them, just waiting.

Finn's stopped crying now, and he's calling for me, so I stagger into his bedroom and pick him up, and carry him back into the living room. He traces the tears on my face and says, 'Poor Aidy,' and that makes me cry some more, which scares him I think, because he goes completely quiet.

I switch on the telly and pick our favourite DVD and we watch David Attenborough in silence. It's the one about the penguin chicks: the parents go away and leave the chicks, and they all huddle together in a big group, but the parent knows the chick and the chick knows its parent by the way they whistle and call. The parent makes the chick wait for its food, makes it cry for it. Eventually the mum or dad penguin coughs it up.

We fall asleep together on the sofa. I only wake up when Holly bangs the front door shut.

'Hello, gorgeous,' I say, lifting my head from Finn's warm curls.

'Is he all right? Mum said you thought he had a cold.'

'He's fine. He knocked his head on the cot, but it was only a bump.'

'Oh, Finn, sweetie, let me have a look.' She lifts him out of my arms. It's cold without him. 'There's a bruise. I'll be glad when he has a proper bed. Maybe we should just buy one.'

'They're expensive.'

'Mum would help out.'

She's avoiding my eye as she cuddles Finn to her. What

have I done wrong now? It wasn't my fault he was running around the market – or was it? Was I meant to have mended the buggy?

'When were you going to tell me?' Her voice is calm and even, and I know she's hiding something. She's angry or upset. I wish she'd show it.

'Tell you what?'

'That you've found her, Aidan. You've found Cass.'

15: AIDAN

I'd been planning to tell Holly about Cass, eventually. When I'd got my head around it, and I'd met Cass and I'd made sure that there wasn't any reason to keep her and Holly separate. You never know. But of course Holly's mum had to turn up at the wrong time, nose into my life, give everything away.

'Why didn't you tell me before?' She's trying not to look hurt but I know Holly better than that. 'How long have you been in touch with her? Is this where you were the other day?'

'What day?'

'You know ... when you didn't come home all night and I thought you were dead.'

Whoa! She never said that at the time. I thought she'd swallowed my story about Rich being in a state, needing me to stay over.

'I didn't ... Rich ...'

'Oh, come on, Aidan. I called Rich at midnight. He said he'd seen you onto the night bus.'

'You never said ...'

'What would have been the point? I didn't want to fight with you. So, were you meeting Cass?'

I'd gone to a club, was the truth, and I'd drunk a lot and I'd woken up in some girl's bedroom and sneaked out without asking any questions. I don't know if anything happened and I'm leaving it that way.

'No – today was the first time we'd met. I've been in touch online.' I can't help boasting a bit. 'I messaged her and she messaged me and we had a whole chat. In writing. A long one.'

'And you won't even send me a text.' Holly looks exhausted. She sits down next to me on the sofa, stroking Finn's head, rocking him, checking his forehead.

'It's not just a bruise,' she says. 'Look, it's swollen.'

I take a look. I'm sure it wasn't like that before.

'He just knocked himself when he went into the cot. He was wriggling around, you know how he does ...'

'How long has he been sleeping? Was he sick? He might have concussion!'

'He's fine! He was yelling and screaming, going off on one.'

'Why? What was the matter?'

She's talking about a little kid. There doesn't need to be anything actually wrong. Finn's always making a huge fuss about nothing. When I think about what was happening when I was his age, well, I don't think he's got much to complain about.

'He was upset about not feeding the ducks.'

'I was upset about that! I'd arranged it all, called Mum, got her to pick him up so I could meet them ...'

'You never told me! I was on my way to pick him up!' This comes out much more sulky than I expected.

'Take a look at your phone sometime.'

I pull it out, switch it on. Four missed calls. Five texts.

'Honestly, Aidan.'

'I didn't know you were going to change everything. Mostly you're cross because I don't pick him up.'

'You need to keep it switched on! I need to be able to get hold of you!'

He's your kid, not mine. I'm doing you a favour.

Of course I don't say this. And, after all, where was my dad when I was Finn's age? You have to take what you've got when it comes to parents. And Finn's got me right now.

I lean my head back against the sofa. 'Holly, meeting Cass was huge for me. I was going to tell you about it afterwards. I wasn't sure how it would go, whether she'd even turn up.'

Holly's forehead is all pinched together, like her mother's. 'But she did. Was she nice?'

'She is nice. She's very nice. She's clever. She might go to Oxford University.'

'That's good,' says Holly. She's looking at the bruise on Finn's forehead again. I remember too late that she's a bit weird about the whole university thing. She went to Cardiff for one term to study psychology before she decided it didn't suit her.

'She's incredibly posh. Posher than your mum, even.' I'm thinking of how Cass stuck out her hand and introduced herself to Juliet.

'Aidan, how did Cass find you? Was it through your social worker?'

'He doesn't know about it. Cass and me, we just found each other. Through Facebook. It was easy.'

'She found you through that profile I set up for you?'

'Yes ... sort of ...'

Finn's waking up. He looks at Holly and back to me. I hold my breath. If Finn decides that I'm to blame for the bump on his head then it could be disastrous.

He doesn't. He flaps his hand at the telly. 'Where penguins, Aidy? Where?'

'We were watching David Attenborough,' I explain to Holly and she's so happy to see that Finn's OK, that there's no concussion, no brain damage, that she kisses him on the head and goes to make us a cup of tea.

I think everything's going to be all right. But then Holly starts again.

'You never even look at Facebook.'

'I know, I know, but ...'

'So how did you even know she'd contacted you?'

'Well, it was me, I looked for her. I worked out her name.'

I told her the whole story. She doesn't say how clever I was, all that reading and writing, finding Cass from a picture. No, she gets jealous. 'So Rich knew – you told Rich before me?'

'Only because he knew about Cass, from when we were in the children's home.'

'But I knew about Cass. You told me about her.'

'I know, but Rich, he understands what it's like ...' I trail off.

'I'm meant to be your partner, Aidan. You should tell me first. I understand too. Rich has met your mum, hasn't he?'

'Yes, but ages ago.'

Mum came to visit me in the children's home, three or four

times. No, three times. After the first time, I made sure Rich came with me. It was easier. Mum talked to him and he talked to Mum and admired the baby and they got on OK, and I sat in the corner of the room not looking at them. Especially not the baby.

'So, what was she like, Cass?'

'Yeah, she was nice.'

'You said that before. Can I meet her?'

'Well, you know ... If she wants to meet me again, I can talk to her about it. She's only sixteen, Holly.' Cass seemed younger than sixteen in some ways, and older in others. She spoke like a teacher or something, but she had no idea about quite basic stuff.

'And what about your mum?'

'What about her?'

'Is Cass going to meet her?'

'I dunno.'

'Can I meet your mum?'

Holly's asked this so often that it's boring.

'This is about Cass, not Mum. It's different. I'm not even thinking about Mum right now.'

Thinking about Mum tends to spoil my day, so I'd rather not.

Holly pouts. 'I don't understand why you won't introduce me.'

I swear, Holly's obsessed with my mum.

'She's not part of my life, Holly. It's easier that way.'

'Do you think she wouldn't approve of me?'

'What, like your mum doesn't approve of me?'

'She doesn't mean it. She just thinks you're a bit young.'

'Yeah, well, one disapproving mother is enough.' I mean it as a joke, but she doesn't take it like that.

'There's nothing actually wrong with our age gap. If it was the other way around, if you were older than me, no one would care.'

I'm not actually sure about that because when Denise at the children's home was seeing an older bloke, we did all think it was a bit dodgy. The gap was bigger though, about fifteen years, and Denise was only fourteen.

'It's nothing to do with you. Why would my mum disapprove of you? It's not about you, Holly.'

That came out all wrong. I try again. 'Look, it's complicated. Don't bring my mum into it. It'll be great when you meet Cass.'

'But you hardly know her. Do you really think you've got that much in common?'

She doesn't say it, but the words 'a girl who's going to Oxford University' are floating in the air.

'That's different. Cass and me, we went through a lot together. I suppose it was the best thing for her, being adopted. I mean, obviously it was, as she wouldn't be going to Oxford University if she'd stayed in care or with Mum or whatever.'

'You don't know that. She might be one of those people who's just naturally brilliant, whatever her circumstances.'

So, Holly thinks I'm just naturally thick. 'She might, but the thing is that because she was adopted we never got to see each other. We lost each other. And now ... now ...'

'So you're not actually officially meant to see each other?'

Just for a minute I think it's a threat. That Holly's going to call the social worker, or the police or Cass's parents; that she's going to stop us meeting again. Without even knowing it, my hands ball into fists.

'I don't know.'

'It's so unfair! They really treated you badly. They didn't find anyone to adopt you, and they banned you from seeing your sister!'

I let myself breathe again. My hands relax.

'They did find someone – Jim and Betty! They said they wanted to ... but Mum contested it ...'

'I know, that was awful, but you were eleven then. Why didn't they find someone before? Like they did for Cass?'

She asks these questions without thinking how much it might hurt to tell the truth. That no one wanted to take me on. That I was unadoptable. I'm not even certain that I've got it right about Jim and Betty, or whether it's a story that I made up to make myself feel better... If they'd wanted me that much, wouldn't they have taken me back when Mum and Keith chucked me out? I asked the social worker I had then and she said it wasn't possible, but I never understood why. Maybe Betty and Jim were glad to be rid of me, too. They just put on a good show of hiding it.

Anyway, after I did what I did, maybe they didn't want me any more.

I shrug. 'There aren't enough people out there. I mean, who wants to take on some screwed up kid?'

Finn's DVD is coming to an end.

'I'll make supper,' says Holly. 'Can you put Finn in the bath?' Her hand strokes Finn's bump again. 'Take care, monster. Don't wriggle too much. You don't want to hurt yourself again.'

'It wasn't my fault,' I say, quickly.

Holly looks at me, startled. 'I never said it was.' A beat of silence. Then she asks, 'Spaghetti, OK?' and I say, 'Time for Finn's bath,' and life goes on. Normal, ordinary family life.

How long can it last?

16: CASS

Will stood behind me, hands over my eyes. I could feel his breath on the back of my neck, his fingers soft on my skin.

I was completely distracted from what he was saying.

'Just turn a little bit to the left ... and then wait ... go away, tourists ... OK ... now!' He took his hands away. I was looking over one of the quads at Trinity College, all honey-gold buildings and velvety grass.

'Imagine you're wearing a silk dress. A corset thingy underneath. OK, and so it's 1850. Just try and put yourself there.'

Just for a moment I do it. I hear the rustle of the dress. I imagine another Cass, a Victorian girl, hair in a bun, petticoats swishing.

'I wouldn't even be here. I don't think they took women students until later on.'

He laughed. 'You got it, though, didn't you? You time travelled. I could tell.'

'How could you tell?'

'There was something about the way you were standing. You changed your posture.' He imitated me, dropping his shoulders, lifting his head. 'You looked like a swan.'

'More like an ugly duckling!'

'Are you fishing for compliments? You're looking very good today.'

I blushed – I hated how he teased me. 'Of course I'm not. Fishing for compliments, I mean.'

'You always wear such boring clothes,' he said, mournfully. 'It's such a waste. With all that hair. Why did you tie it back again?'

'It's neater.'

'Neater!' said Will, in such a disgusted voice that I had to laugh. He reached out for my hand. 'Come!'

We walked across the quad, pushed at a door.

'We can go in!' Will was as thrilled as a little boy. I was just grateful he'd picked this particular college: one I didn't know all that well, one where there was no danger of ...

'Cass!'

Oh, no. No, no, no. Just what I'd worried about.

'Darling! Why didn't you tell me you were coming?'

She was walking down the corridor of the wrong college.

'Auntie Kate! I didn't think ... this isn't your college—'

'I've just been meeting a colleague. We are allowed to cross each other's thresholds, you do realise?'

'Oh, yes, well, obviously.'

'How lovely to see you, darling.' My aunt planted a cool, crisp kiss on my cheek. 'Aren't you going to introduce me?'

'Oh, yes, sorry, this is my ... my friend, Will. Will Hughes. Will, this is my aunt. Kate Fielding. She's ... she works here ...'

'What a shame I didn't know you were coming,' said Aunt Kate, looking at Will and raising her eyebrows slightly, just so that only someone who knew her really well would notice.

'We only decided yesterday,' Will said, shaking her hand. 'We thought we'd come and look around. Get a feel for the place, you know, before we apply next year.'

Aunt Kate looked faintly amused. 'Well, I'd have thought Cass had a good feel for the place already,' she said.

'I'm just showing Will around,' I said, squirming inside.

This was naturally the cue for her to start quizzing Will about what he planned to study – 'History! Excellent!' – and start advising him about college applications and interview technique.

Will was all enthusiasm. 'Wow, that's great, thank you so much.'

'I'd take you both to lunch, but I have a faculty meeting. How long are you staying?'

'Not long,' I said. 'Next time, I promise I'll give you more warning.'

'Do, Cass darling. We're all so concerned about you. How is Ben coping? Your mother seems to be bearing up.'

'Ben's fine,' I told her.

'I'm his peer buddy at school,' Will offered. 'I've been introducing him to some kids, helping him make friends, that kind of thing.'

'Excellent,' said Aunt Kate. 'I'd love to hear more about it, but I'm afraid I have to get going. Do give my love to everyone. Tell Susie I'll give her a call. Good luck with the applications, Will.'

'Bye,' I said, and off she went.

Will stared after her. 'Did she just give me the brush off? Or am I being super sensitive?'

'No, she's always like that.' I was so used to Kate's briskness that I hardly noticed it. 'All my mum's family are like that. Not my mum so much, but Granny and Kate and Mum's other sister.'

'Blimey,' said Will. 'And she's what – a professor?'

'She's not a professor. She's a Fellow in History at Corpus Christi. That's not quite as senior as a professor.'

'Yes, I do actually know that. What's her specialism?'

'France in the eighteenth century. The enlightenment. Maybe you've heard of it?'

I thought he'd call me out on my sarcasm, but he ignored it.

'Wow,' he said. 'Just wow. Well, no wonder you're applying here.'

'What's that meant to mean?'

'Well, you know. You virtually live here already ...'

'I'm not applying to Corpus Christi. I'm not looking for special favours.'

'No, but it's got to help ...'

'Well, I thought I'd get in on my own merit,' I said, stung. 'I don't want to get in because of my family. But I'm happy to ask for inside information on your behalf if you want me to.'

'Yes, do,' he said. 'We ordinary people need as much help as we can get. Actually, Corpus Christi was on my list.'

I turned and walked towards the door, went out into the quad, headed for the gate. Will followed me.

'Oh, Cass, don't go off in a strop. I'm very impressed, that's all. And kind of daunted. The nearest I've got to a Fellow at Corpus Christi is my cousin, Derek, who's the head teacher of a school in Croydon.'

'Yeah, well, it's not so amazing having family here. It means that everyone expects ... they just take it for granted ... '

There were tears in my eyes. I turned away so he couldn't see.

'Well, it'll be nice for you to have family around. I don't know anyone in Oxford, supposing I get in. I'll be all alone in a strange city. I just hope you'll be here too, to keep me company. Show me the ropes.'

'It'll be just as strange for me as it is for you.'

'Oh, come on, Cass! Don't be so touchy! You'll fit right in at Oxford. It's in your genes.'

That did it. 'Look, if I get to Oxford I'll probably be the first one in my family to go to university. My biological family, anyway. I'm adopted. My birth mum's a hairdresser. I'm just as ordinary as you are.'

Will stared at me. For about five minutes. It was so unlike him to be quiet for so long that I began to feel self-conscious and worry that I'd accidentally said something really insulting.

'What? What's the matter?'

He grabbed my hand. 'Come on.'

'What! Get off!'

He took no notice, but increased his speed. I ran with him, through the gate, back out onto the street, along the pavement, into a café. 'Sit!' he ordered.

'You're incredibly bossy!'

'I thought girls liked it when men are masterful.'

'You're not a man and no, we don't.'

'Well, tough. And who qualified you to judge who's a man or not?'

I was nearly certain I was blushing again. 'You're at school. Wait until you've left to call yourself a man.'

He deepened his voice. 'I can see I'm going to have to prove my manhood to you, Miss Montgomery.'

'No thanks!'

'Oh, well, at least I've got you out of your strop. Although you are magnificent when you're angry.'

'Where do you get your lines from? Old movies? Mills and Boon?'

'Old movies!' A waitress appeared by our table. 'I'll have a tuna melt and a cappuccino, thanks.'

I hadn't even looked at the menu. 'Oh, I'll have the same.'

Will smiled. 'Good. I thought you might walk out on me there.'

'I'm hungry.'

'So practical. I can just imagine you at Guide camp, counting out the sausages, working out how many baked beans everyone should have ...'

'I could always get a sandwich on the train ...'

'No, don't. I want to talk to you. I didn't mean to upset you.'

'I'm fine. I'm not upset.'

'Good, because I think I've worked it out.'

'Worked what out?'

'That's who it was. That guy in Camden. He must be your brother – am I right? Your real brother.'

17: CASS

'Not "real",' I said, flustered. 'Ben's just as real—'

'Your birth brother. Am I right? I'm right!'

'Half-brother, actually.' There was no point in lying. 'You're right.'

'Oh!' He was bouncing with excitement. 'I love it when I work things out! Sherlock Holmes, you are so *rinsed*!'

'"So rinsed"?'

'I'm reviving it. Vintage slang.'

I wanted to hate him but I couldn't. And it was good to tell someone about Aidan. I'd been bursting to talk about him.

'Tell!' he ordered.

'Why should I tell?'

'Because I was so clever and guessed. I deserve a reward. Or shall I pick something different?'

'No,' I said, hurriedly, 'I'll tell you about it ... I've been keeping it to myself because I don't want my parents to know. I would talk to Grace or Megan or someone, but they all gossip so much and all their mothers know mine ...'

'No danger of that with me,' he said. 'Do tell me, though. I've been dying of curiosity.'

'I hadn't seen him since I was four. That's when I was adopted. He's two years older than me. We weren't in touch at all. Then he contacted me, out of the blue, a few weeks ago.'

'You haven't seen him for twelve years? But you knew him right away, didn't you? The way you two looked at each other ...'

'I can't explain it. I knew him.'

'Well,' he said. 'Well. This is something else. What's he like? Does he have a name?'

'Aidan. Aidan Jones. He's nice. He's really keen to include me in his life. He works in a shop, selling second-hand stuff. He's got a girlfriend, he lives with her, and he's got a little boy. You saw him, right at the end.'

'He started early.'

'Yeah, well, I suppose some people do.'

'Did Aidan get adopted as well?'

'No, he was in foster care most of the time, I think. Maybe a children's home as well.'

'How old is he?'

'He's nineteen in a few months.'

'You're going to meet him again?'

'I hope so. We talked about meeting his girlfriend.'

'Maybe I could come with you? To lurk? You don't have to pay me this time.'

'Oh, God, I'm so sorry! I never paid you.' I pull out my purse. 'I've actually only got twenty pounds on me.'

'Whenever you want to pay,' he said. 'I'm not bothered.'

'I'll get lunch.'

'No, my treat. After all, you're showing me around.'

I felt bad that I hadn't told him how well I knew Oxford. I'd meant to. But when we were on the train and he started telling me about the research he'd done, showing me places on his map that I'd known for ever, well, I just didn't have the heart.

And now I was here and it was strange being in Oxford without my family. In fact, it was making me wonder why I'd always been so sure that Oxford was my goal. Was it because I couldn't imagine doing anything that wasn't connected with my family? What would it feel like to plan a future based on exactly what I wanted?

I had no idea at all where I would even start.

'I love Oxford,' I told him. 'We could go punting this afternoon, if you want. It's a bit late in the year, but it'll be quiet, and the sun's shining.'

'Would it be good for time travelling?'

'Very good. It could be any time.'

'Cass, I didn't mean to upset you before. I was just teasing. I'm sorry.'

'It's OK.'

'I had no idea you were adopted. How about Ben?'

'No, my mum got pregnant after I was adopted. Apparently it happens quite often. People try for ages for a baby, nothing happens and then they fall pregnant as soon as they adopt.'

'Do you have any memories of before you were adopted?'

'Not really. I have some photos and things. That's how I knew about Aidan.'

'Four years, though. That's a long time without memories.'

'How many memories have you got from before you were four?'

'Loads! When I was two and a half my oldest sister took me shopping and lost me. I remember asking a lady if she knew where Layla had gone. When I was three I fell off a climbing frame and broke my wrist. When I was three and a half, I got a toy car with mobile controls for Christmas. When I started nursery school, a boy asked if my skin colour would wash off.'

'No!'

'Yes! Quite confusing for both of us.'

'That's awful.'

'Not really. He was just a kid. That's what happens when you live in a town where you look different from 80 per cent of people.'

'Try being ginger,' I told him. 'When I was six, I emptied a pot of black paint on my hair. I thought I could dye it.'

'Now, that is tragic. You must stop hiding it away. You've got to embrace the ginger.'

'It's not really ginger any more,' I said, defensively.

'Ginger rocks. Ginger is cool. Ginger is the new black.'

'It's got a lot darker.'

'Dark ginger,' he said. 'Burnt ginger. Raw ginger. Stem ginger. Crystallised ginger.'

I was so sure that he was going to say something about chocolate-covered ginger that I completely lost the ability to think of anything witty. I was sure I must be blushing. Luckily our food arrived, and I was able to look at my sandwich and avoid his gaze.

'Oh, shut up.'

'"Oh, shut up" is not worthy of you. Especially as a leading light of the interschool debating competition team. Not that I'm bitter or anything.'

'I only summed up. The others did the hard bits.'

'We went out in the second round.'

'That wasn't your fault, though.'

'A compliment? From the County interschool debating champion team captain?' He pretended to faint. I chewed my sandwich and ignored him.

'So, where next? Any more secret relatives you forgot to mention? Head librarian at the Bodleian? Professor of Mathematics at Christ Church?'

'Actually, my grandfather was a professor,' I told him, in the interests of full disclosure. 'Of English literature. My mum studied Old and Medieval English. She gave it all up to be an MP's wife, and mother. Now, of course, she's been made redundant.'

'She's still your mother.'

'Yeah, true. She's going to have to work out what else to do with her time, though. At the moment, she's looking at smaller houses and seeing her solicitor.'

'That can't take up all her time.'

'Having her hair done and going to Pilates.'

'Do you think she'll get a job?'

'Maybe. I don't know. She's got to work out a financial settlement with Dad, sell the house and all that.'

'It's rough for you and Ben.'

'It's OK. I'm fine. Ben's getting there.'

'You are so tough,' he said. 'The iron lady.'

I made a face at him.

He bounced up from his seat. 'We're wasting our time. I've

always wanted to go punting. I looked it up. This is the last week you can do it. Come on! It'll be fun.'

The route to the river took us through the covered market, and he stopped by one of the stalls. 'Just what I was looking for.' He picked through a heap of scarves.

'Oh, come on,' I said. 'You're already wearing orange and purple. You can't possibly add a pink scarf.'

'Come here,' he said. He held up the scarf. It was a deep, dark pink, the colour of beetroot soup. 'Perfect.'

Before I could stop him, he unwound my grey, woolly scarf and tossed it into a rubbish bin.

'What are you doing?'

He wound the pink scarf around my neck. 'So much better. Never wear that dishcloth colour again.'

'That was my favourite scarf! This is going to clash with my hair.'

Will paid the stallholder five pounds.

'The clash is the point. Take a look.'

There was a full-length mirror propped up against the stall. I looked. Will, tall and dark and laughing; me, pale and stressed and only brightened up a little bit by the scarf.

'Take the ponytail thing off,' he said, bossing me around again.

I let my hair loose. Watched as the girl in the mirror instantly looked more relaxed, and even a bit older.

I was nearly certain that this proved Will thought we were on a date. Do you buy a present for a random girl? Do you have days out with random girls? I didn't think so. I wished I could be sure. I was not completely hating the idea, to tell the truth, but I needed clarity.

'Better, eh?' he said. 'Right. Show me what punting's all about.'

Punting is not all that easy, and I was secretly hoping that Will would find it impossible, so I'd have to take over. It's a matter of shifting your weight to lever the pole most efficiently. Dad taught me a few years ago. But Will seemed to get the hang of it really quickly, and insisted that he did all the work. 'You relax,' he said. 'I want to practise.'

Sitting in the punt, moving through the water, it was easy to travel in time. I was a medieval princess. I was a Victorian lady. I was someone who didn't have to take exams, worry about university, try and be perfect all the time.

'I wish we could actually time travel,' I said. 'We could time travel instead of doing our A levels. Just skip the next few years.'

'A levels are a lot of work, aren't they? That's why you have to make time to do other things. Oof! This is hard!'

'Let me take over,' I said, but he mopped his brow theatrically and said, 'I can cope.'

'I still haven't finished my essay. I really shouldn't have come out today.'

'You're not back at school until Monday.'

'I know, but I've got to go away at the weekend.' I should have finished the essay easily, to be honest: I'd had two full days. But between thinking back to my meeting with Aidan and worrying about this day with Will, let alone the weekend with Dad and Annabel . . . I couldn't concentrate.

'Why do you want to apply to Oxford?' I asked him. 'Why not somewhere else?'

'Because it's there,' he said. 'Or, rather, because it's here. Because I can. And because I failed the exam to get into the sodding grammar school and my parents have paid a small fortune to send me to Bonny's, and I'd like to show the grammar school that they were wrong about me and my parents that they were right. Also, I like history and there's a lot of it here.'

'That's a much better reason than mine.' Will seemed so sure of himself. Next to him I was a marionette. Wooden and controlled and hollow.

'What did I say? You look sad.'

That was the thing about Will. Most of the time he was bouncy and annoying and I couldn't tell if he was serious or winding me up. And then he'd stun me by being so kind and nice that all I wanted was to carry on talking to him forever.

'I'm not sad.'

He sighed. 'No, you're fine. I know you are. You're always fine.' The punt lurched as he shoved the pole deep into the river.

We made our way to the station as the sun set, moving closer together as the light faded, his arm bumped against mine.

'Hey,' he said. 'What's that down there?'

'What?'

He led the way down a little alleyway.

'What are you doing? The station is that way.'

'We've got fifteen minutes,' he said, 'and I want to ask you something first.'

'Why can't you ask me on the train?'

'I don't think you'd say yes. You don't like people looking at you.'

'Why would they be looking at me?'

'Because I'd like to kiss you. But I wanted to warn you first, or you'll do that thing you do whenever I touch you.'

My mouth went dry. 'What? What thing?'

'The blushing and screaming thing.'

'I don't scream!'

'Squeaking, then. The blushing and squeaking thing.'

'Oh! You are so annoying!'

'I know, but can I kiss you anyway?'

'I don't know, I'm not sure, I don't think … why do you want to? What does it mean? Does it mean we're going out?'

'What do you want it to mean?' He was getting closer.

'I don't know what I want it to mean. I won't know unless you do it. Look, OK, kiss me, but just on an experimental basis. It doesn't commit us to anything.'

'Purely experimental. Do you want me to write down any scientific data? Time of day, location, length of experiment, anything like that?'

'No, you don't need to do that.'

'You realise that just one experimental kiss won't be enough? We'll have to do some under different conditions to compare?'

'Only if I think I want to,' I said, totally aware that I'd shown myself up as a completely inexperienced idiot. 'Go on then.'

'"Go on then"? How romantic.'

He slid his arm around my shoulders, leant down and brushed his lips against mine.

'That wasn't much of a—' I said.

And then he kissed me. For about two minutes. His hand stroking the side of my face. His lips moving against mine.

Our first experimental kiss.

I decided right away that it wouldn't be the last.

18: AIDAN

Holly's tired, so I'm giving her a lie in and taking Finn to work with me.

Finn loves the shop. There are toys and cardboard boxes and corners for hide and seek. It's like an adventure playground for him. We've got cheese rolls for a snack and orange juice in cartons.

Clive doesn't really like me babysitting Finn in the shop, but he doesn't often come in Saturday, and there's nothing much to do except unpack some bedding and tidy up the storeroom. Finn and I play Hide in the Box, which he loves, and then we build piles of boxes and topple them down. I find a load of old bubble wrap and we stamp on it to make the bubbles pop. And then we – less successfully – sweep the floor until he starts grizzling and rubbing his eyes and I put him in the buggy and wheel him around the shop a bit and he's crying but it's because he's tired, and soon he falls asleep.

AIDAN

When he was a baby – when Holly was just my landlady – I used to babysit for her a lot. I didn't like it much, to be honest, because I couldn't bear him crying – it made me a nervous wreck because I didn't know what was wrong with him. But then Holly taught me the difference between hungry cries and tired cries and angry cries and it was much easier to cope with. I could be with him without panicking most of the time.

With Finn asleep, there's actually nothing for me to do, so I put my feet up on the desk and prepare for a little snooze myself. But then the door opens.

A customer. Except that I recognise him.

'Well, if it isn't Aidan Jones.'

I tense right away. Glance over to Finn. Can I nudge the buggy out of sight in time?

'Hey, Neil,' I say. 'Long time, no see.'

'You being funny?' Neil's acquired some new tattoos since I last saw him: a spider on his hand and some letters on his knuckles. It's probably good that I can't read them.

'No, just saying.' It's a bit much if Neil's going to start being all sensitive about spending time inside. He shouldn't have got involved if he wasn't prepared to serve his sentence. But then logic was never Neil's thing.

'How's the girlfriend?'

I'm confused. I'm absolutely certain that Neil's never met Holly. We've been together for just over a year, and Neil went down for eighteen months, if I remember right. Armed robbery.

'What girlfriend?'

'That little runt. What was his name?'

'Rich,' I mutter. 'C'mon, Neil.'

Neil laughs. 'Just joking, Aidan. I heard you had a lady, though. Nice one.' He jerks his head towards Finn. 'He's your boy?'

'No.'

'Ah. Like that, is it? Well, well. And she knows the whole story, does she? Knows all about you? Well, what it is to be loved.'

There are lots of things I could have said, like, How would you know? Like, What do you mean? But I don't. I might not be able to read and write very well, but I do know a bit about how you talk to someone like Neil.

He's looking round the shop, picking things up, examining them. A stand for earrings. A plaster angel. A big cut-glass vase.

'Funny sort of shop, this. What's the idea?'

'It's salvage stock. Stuff that was in floods and fires. It's all good stuff but the insurance has written it off.'

'I see,' he says. 'Bits and bobs that no one wants. Damaged goods, eh, Aidan? Damaged goods. I bet you feel right at home.'

I haven't seen Neil for years. I'd forgotten how much I hated him. I'd forgotten how he used to run that home, the things he did, the way he got other kids to fight his battles.

There's stuff I'd never forget, though. How he punishes people. How he enjoys hurting people. What he did to Rich. What he did to me.

Every time he moves near to Finn I want to shout at him, tell him to get out. I don't. I know better. I'm still scared. But if he – or anyone – ever does to Finn what was done to me, then they're dead.

'Nice kid,' he says. 'Nice face. Pity if anything happened to it.' He's still holding the big vase, and he pretends to drop it, catching it as it slips out of his hand.

'Nothing's going to happen to his face.' I try and keep my voice steady. I can't show him that I'm so scared that I could easily hurl.

'It's hell looking after kids, isn't it? Got to watch them every minute. It's so easy for them to have an accident. Run out into the road, stick their fingers in the power socket, get their hands on a knife.'

We stock knife blocks, but we keep them locked away in a glass cabinet. I'm trying very hard not to look at them.

'Your girl must trust you a lot if she's leaving you watching the baby. It'd be terrible if something happened to him. Left him with a scar or something worse.'

Bang! Neil drops the vase onto the wooden floor and Finn wakes with a shock. He starts screaming and I try to get to him, but Neil's in the way.

'Hang on a minute, Aidan. I want to ask you something.'

Finn screams for me. I'm rooted to the spot.

'What?' I bark at Neil. 'What is it?'

'I want you to do me a favour.'

'What favour?'

'Well,' he says. 'I've got something that needs keeping safe. And somewhere like this ... it's just what I've been looking for.'

'No,' I say, and then I see that he's picked up a piece of glass from the floor.

He's looking from Finn to me, and back again.

'I mean, yes. Anything you want, Neil. Just tell me what I can do.'

19: CASS

When I first saw Annabel, sitting on the edge of the sofa – our sofa – in Dad's flat – our flat – she looked so glossy and perfect that I knew she'd be easy to hate.

She had a degree in Politics, Philosophy and Economics from Oxford University and a post-graduate something or other in International Relations from Cambridge. Dad gave us the history in the car after he'd picked us up from the station, but the *Daily Mail* had told us everything weeks ago.

Her skin was flawless ivory, her hair shiny chestnut, she wore over-sized glasses with black frames – all the better to show off her big, brown eyes – and a stylish blue dress that clung to her chest and flowed over her body, so that you could tell she was pregnant, but not in a flaunting-her-belly kind of way. She was twenty-three. Only six years older than me.

Her glossiness reminded me that my hair was frizzy (the

weather suited the occasion – all wind and drizzle), my unmade-up face was pale and freckly, and my jeans and sweatshirt made me look like a thirteen-year-old tomboy.

I'd been thinking of slightly changing my style, but I hadn't had a chance to go shopping. I was wearing my dark pink scarf though. It smelled ever so slightly of Will.

'Come in, come in,' said Dad, right behind us. 'Come and sit down. This is Annabel, obviously. Annabel, this is Cass and this is Ben.'

Annabel made a brave effort to make eye contact. I stared at the wall behind her head. I might have to be there physically, but I could remove myself mentally. I tried to focus on key dates in twentieth-century European history, but found my mind wandering to Will, and how we'd kissed in Oxford, and then sat on the train and kissed some more, and then it'd been quite awkward on the tube and the train home because I was nervous someone might see us, even though I didn't recognise anyone, and he'd kept on leaning right over to me and teasing me when I pulled away.

'What are you laughing at, Cass?' asked Dad now, and I quickly rearranged my face and said, 'Nothing.'

Will and I had met up the day after Oxford – I felt it was important to set down some rules about experimental kissing, just in case it got out of hand. We went to the park, which was nice and quiet, thanks to the weather.

'Look, I just want to make it clear that I'm not in the market for a relationship,' I told him.

'Oh, you made that clear enough. We're experimenting, yes? I don't suppose I could come under your umbrella? I'm getting soaked.'

'OK,' I said, a bit annoyed because once he was holding the umbrella there was no choice but to walk really close together, which made us look like a couple when, of course, we weren't. Also, he was better at stopping the wind turning it inside-out than I was, moving it skilfully to avoid the gusts of wind that were battering us.

'Have you got any ground rules? Or shall we see where our experiments take us?'

'No coupley stuff, no talking about boyfriends and girl-friends, no being seen together in public, no Facebook statuses or cuddling in Starbucks, or on the bus ... you know ... '

Will laughed at me. 'That wasn't really what I meant—'

What else could he mean?

'I mean ... oh, look, the bomb shelter ... '

There's a brick bomb shelter from World War II at the edge of the park that smells of wee and is littered with broken bottles. Really the council ought to clean it up – in other circumstances, I'd have taken a picture on my phone for Mum to mention to the local councillors. Anyway, once we were in the bomb shelter, we experimented with kissing again. Longer kissing this time, with additional neck stroking and a small amount of hip grinding.

After about half an hour, the rain stopped and Will said, 'Actually, I was thinking more of guidelines about this ... you know ... '

'I don't have any guidelines about this. That's why we're experimenting.'

'Oh.' Will blinked a few times. 'You are quite unlike anyone I've ever kissed before.'

I was going to reply to this but then I spotted the deputy

head walking her standard poodle, retrieved my umbrella and insisted that we took separate routes home.

The loathsome Annabel was talking to me. Reluctantly I stopped thinking about Will's kisses and focused on her impossibly perfect face.

'This is horrible, isn't it? I'm sure you completely despise me. What can I say? I understand totally if you don't want anything to do with me.'

Oh God, she was pretending to be nice. Just when I was successfully managing to forget all about her.

'I don't think Ben or Cass feel like that,' said Dad, in his fakest politician's voice.

'I can't answer for Ben,' I said, making absolutely sure that I was not smiling or moving or doing anything at all to make the situation less awkward than it was, 'but you've got it just about right as far as I'm concerned.'

Ben had a *Star Wars* action figure in his hand – Yoda – and he stared intently at the little guy's pointed ears and baldy head.

Way to go, Ben, I thought. Let's look as crazy as possible. And then I felt horrible and disloyal and mean. No wonder Dad had abandoned us. Ben was strange and I wasn't a nice person.

'I'm so sorry,' said Annabel.

'Don't be silly, Annabel,' said Dad. 'You've got nothing to apologise for.'

'Yes, she has,' I said. 'She's broken up your marriage. She's hurt all of us. At least she's got the guts to admit it.'

Annabel tried to give me a sympathetic smile. I kept my face stony.

'Oliver, darling,' she said. 'Why don't you go and pick up some stuff from the deli? We'll be fine.'

Dad huffed a bit, but agreed. We watched him go in silence.

'That's better,' said Annabel. 'No good having him here. He feels so guilty that he'll just get upset.'

'He should get upset.'

'He is. He truly is. Give him a chance to make it up to you. He loves you both so much.'

'He should have thought of that before.'

'I know. But what's done is done. And we will be friends eventually. At least we've got the scary first meeting over with. It only gets better from now on. I promise you.'

'Don't bother promising anything,' I snarled, 'because this is not going to happen often. Eh, Ben?'

'What do you mean it gets better?' asked Ben, talking to Yoda.

Annabel wasn't in the slightest bit fazed. 'Well, I've been in your position myself, you see. Daddy's been married twice – but with plenty of girlfriends in the meantime, and Mummy's on her third. I know all about meeting the new partner. Ghastly business.'

'If you know what it's like, then why did you do it to us?'

'I fell in love,' she said.

How convenient it was to have a catch-all excuse. I promised myself I would never lose control over my emotions in a selfish way that hurt other people.

'I know how much you must resent me. But it all works out in the end, truly. I've got loads of lovely half- and step-siblings now. And step-parents, and, actually ex-step-parents – in fact, one of Daddy's ex-girlfriends is my best mate, and I honestly wouldn't want it any other way. I feel as though my family's got bigger and bigger.'

That's all right for you, I thought, furiously. You are never going to be my best mate. And I don't need you to provide a bigger family. I have Mum and Ben, and now I have Aidan and pretty soon I'm going to meet Holly and Finn and maybe my mum and Scarlett and Louis and perhaps she'll tell me more about my real dad. My family is growing and growing.

'Are you really all friends?' asked Ben.

'All of us,' she said, firmly. 'I'd like to be your friend, too.'

'Can I be friends with all your sisters and brothers?' Ben asked, and I wanted to hit him. Just because you haven't got any friends, Ben, you don't have to like her! You can do better than that!

If Ben was going to like her, then I'd have to hate her all the more.

Dad came back with a bag full of deli treats – stuff he knew we liked: crackers and tapenade and olives. He and I sat in silence. I was remembering family holidays, eating olives in Spain, rosemary flatbread in Tuscany. I don't know what he was thinking about.

Annabel talked to Ben. I have to admit she was good at it. She knew a lot about *Star Wars*, and she liked *Dr Who*. She'd probably been reading up, watching DVDs, studying for meeting her lover's children as though she was taking an exam. Pathetic.

My exam preparation wasn't going so well. The history essay had been my perfect excuse for getting out of coming to London for the weekend. I'd told Dad about it on the phone that morning: 'I don't think I can spare the time,' I'd said. 'They've really put the pressure on and I'm a bit behind.'

'That's not like you, Cass,' boomed Dad. 'Well, you'd better bring it along. I mean, I'm sure Ben would prefer it if you were there.'

I might have known he could manipulate me into doing what he wanted. He is a politician, after all.

'Did you get that essay finished?' he asked me now. I told him I hadn't and he started telling me what to write.

'I can do it,' I snapped.

Annabel and Ben went into the kitchen so she could show him how the new coffee machine worked. I wondered how much Dad had spent on it.

'Can you try and be nice?' Dad was frowning at me.

'I've got a headache. I'm a bit stressed.'

'Mum says you've been out all the time this holiday. Are you sure you're spending enough time studying? This is a crucial year if you want to get into Oxford.'

'It's my business, how I spend my time. And why were you speaking to Mum, anyway?'

'We're still your parents. We're determined to carry on doing our best in that role. Together.'

'Oh, well, good luck with that.'

'Don't be cheeky. Cass, this isn't like you.'

'A lot has changed,' I said. 'Haven't you noticed?'

'Cass, you're going to have to accept this, whether you like it or not.'

'I don't like it,' I said. 'Look, I'm not up to this right now. I'm going out. I'll see you later.'

'Cass – no, Cass, wait—'

I stuck my head round the kitchen door. I didn't want to, but I couldn't just abandon Ben.

'I'm going out,' I told him. 'See you later.'

'Look, Cass, I can froth milk!' he told me.

'Excellent,' I told him, trying not to show him my disappointment that he could be bought off so easily.

'See you later, Cass,' said Annabel, and I nodded in her direction. I'd have preferred to ignore her, but I had no inten-tion of acting like a stroppy teen. Dignity and disdain, that was the aim.

I walked down the stairs and out of the apartment block. Down the street, across the road, into the tube station. I dug around in my bag, found my Oyster card. I was down the escalator before I knew where I was going.

Camden Town.

My other family.

Aidan.

20: AIDAN

The doorbell jangles again as Neil's inspecting the store room, deciding where he wants to stash whatever it is he wants to hide. It can't be good. I'm betting weapons, but it could be drugs, or even money. Or stolen goods. As the whole yard looks like a fence's paradise anyway, we do have occasional visits from the police who want to see all the paperwork. Could I use that to put Neil off?

I look up to see who's come in. A group of customers would be good. Something that might scare Neil away. A policeman would be best.

'Aidan!'

It's her! It's Cass! What the hell?

Neil mustn't realise. If he knows I've got a sister then she could be in danger.

Finn's nearly back to sleep again, grumbling in his buggy. So I jerk my head towards him and put my finger over my mouth. 'Shhh.'

She whispers. 'Oh, sorry! I just thought ... I was in London and I thought ... I asked one of the stallholders and he told me how to find the shop. He knew all about it.'

'Well, hello!' Neil appears from behind a pile of boxes. 'I don't think we've met. Friend of yours, Aidan?'

I mutter something, and Neil grins at Cass and says, 'The boy's got no manners. Must've been dragged up. I'm Neil. Me and Aidan go way back. I don't think we've met.'

Cass has been too bloody well brought up for her own good. She sticks out her hand. 'I'm Cass,' she says.

'Cass, great to meet you. You do look familiar. Are you from round here?'

'No, I don't live in London.'

Shut up, don't talk to him, don't give anything away ...

'Come far, have you?'

'No, not far.'

'So how do you two know each other?' says Neil. 'I'm sure I remember you from somewhere ... hmm ... not the children's home.'

'Were you two together in a children's home?' Cass's eyes are wide. She's probably thinking that a children's home is a bit like an orphanage in the olden days: all of us sitting with our bowls of gruel, wondering who'll dare to ask for more. Or maybe she thinks it was like some boarding school, with midnight feasts and practical jokes.

Well, it wasn't like either of those things.

'Oh, yes, we knew each other very well, didn't we, Aidan? Shared a lot. All them memories, eh?'

I have a very clear memory of Neil breaking my finger because I wouldn't deal drugs for him at my school . Quite soon after that, I stopped going to school altogether. Safer.

'Yeah, yeah, loads of memories. All them parties, eh?' I smile weakly. 'You want to buy anything, Neil, or were you just going?'

'I've seen exactly what I'm looking for,' says Neil, looking Cass up and down.

Her face turns pink.

'Bye, then, Aidan and Cass. And the lad, of course. I'll be back soon, Aidan. Later today, maybe.'

'My boss'll be back soon,' I say.

'No worries. We'll keep an eye out. Cass, good to meet you.'

Jesus. Jesus! Now he's going to be trying to find out who Cass is. What if he uses her to get at me? At least she didn't tell him her surname.

At the door, Neil turns and comes back. 'I think I've got it!' he says to Cass. 'How I know you! What did you say your surname was?'

'You never met her, Neil. She doesn't live round here,' I say.

'No, I'm sure of it. Some big party. I'm right, aren't I?'

'No,' says Cass. Good for her. She sees through him.

'You're an actress? On telly? I know I've seen you.'

Cass hates this. Her cheeks are pink and she's looking away. I hate Neil more than I ever did before.

'Leave her alone. You don't know her,' I say.

'It's killing me!' says Neil. 'Hey, didn't your mum have red hair, Aidan? I remember her visits. Blimey, the fuss you made about having to see her. Understandable, I suppose. In the circumstances.'

'Bye, Neil,' I say, as firmly as I dare.

'See you around, Aidan.'

The door slams behind him. Finn wakes up again and

screams. I lift him up, cuddle him close. 'It's OK, Finn, everything's OK.'

'You're really good with him.' I think Cass means Neil and I'm confused, but she doesn't. She's talking about Finn.

'Oh, well, you know. I've been like a dad to him. Holly and me, we've been together for more than a year now. Holly doesn't have any contact with Finn's real dad.'

'I thought he was yours.' Cass seems to have forgotten Neil the minute he walked through the door. I wish I could. 'He looks like you,' she adds, although it's not really true. It's our skin colour that's the same, and our dark hair. Finn's hair is a lot frizzier than mine, and his face is round and cute like Holly's. But, judged by colour, Finn and I are more alike than Holly and Finn, or me and Cass. Blood isn't everything.

'Why are you here?' I ask. 'I mean, it's great to see you, but kind of unexpected.'

'I was at my dad's, to meet his new girlfriend, and I just felt suffocated. I needed to escape. So I came to see if I could find you.'

I look at my phone. 'It's almost time for lunch. Let's shut up early, eh? I need to deliver Finn to Holly, anyway.'

'Oh, can I meet her? I'd love that.'

I'm suddenly shy. What if it all goes horribly wrong?

'She'd like that, too.' It's true. I'll have to hope for the best.

I lock up the shop, put up the 'Closed' sign, bump Finn's buggy down the front step. Clive doesn't like it when I close up during the day ('Just bring some sodding sandwiches,' he said last time, when I explained that I needed to go for lunch at 1 p.m., 'and besides, no one gets a three-hour lunch break.

I should sack you, really I should') but he'll have to lump it. I'll tell him Holly was ill, that I had to check up on her.

Cass follows us up the stairs, making faces at Finn so he laughs all the way. I'm kind of excited about showing her the flat. I bet she'll be surprised when she sees how nice it is. I don't know anyone else who lives in a flat like this. She's got to be impressed.

'My boss is Holly's uncle. He's converted the flat and he lets us live here. He'll sell it eventually, but until then we take care of it for him.'

'So one day you'll have to move out?'

'Yeah, but not for ages, probably. Holly's a single mum, and she works part-time at a GP's surgery and she's not got much money. Clive wanted her to have somewhere nice to bring up Finn '

'He sounds nice.'

'Holly's lucky. She's got a big family and they all help out.' I open the door. 'Holly! We've got a visitor!'

Holly's not looking her best. She's still in her dressing gown and she's got no make-up on and she hasn't brushed her hair so it's all over the place. She's pale and her forehead is all pinched up again.

'Aidan! Why didn't you warn me?'

Cass steps forward. 'I'm so sorry, it's all my fault. I just turned up out of the blue. Aidan didn't know. I'm Cass. You must be Holly.'

'It's lovely to meet you.' Holly's putting on a brave face. 'I've just got up. I'll go and get dressed. Aidan, why don't you get Cass a drink? Oh, Finn darling, are you hungry?'

Finn's bellowing for food. I pick him up, steer Cass towards the living room.

'Here we go!'

She steps in and stumbles over one of Finn's toys. 'Sit down!' I say. 'I'll just tidy up a bit.' Obviously, if I'd known she was coming, I'd have done this already. I sweep all the toys into the toy box, ignoring Finn's shouts of fury, pick up the plates and mugs on the floor where we'd left them last night, and straighten the rug. 'I'll just get Finn some food,' I say, scooping him up. 'We've got a kitchen as well as this room. And two bedrooms!'

Holly hasn't cleaned up the kitchen, either. She made spaghetti last night and there are splashes of tomato sauce all over the worktop. I plonk Finn in the high chair and give him a pitta bread, wipe the tops, do the washing up.

Cass stays on the sofa where I left her. She's talking on her phone. I can hear her voice but not what she's saying.

Holly comes into the kitchen. She's looking a lot better. A pretty top – pink flowers, a bit see-through – and jeans. A bit of a muffin-top spilling over the top of her jeans, which makes me want to touch her right there. I don't though.

'I'm sorry!' she hisses. 'I didn't know you were going to come home! I thought you were working all day!'

'I couldn't keep Finn at the shop all day! You know that.'

She yawned. 'I'm just so tired, Aidan.'

'Are you all right?' Sometimes Holly loses energy and stays in bed a lot. She's trying all sorts of things to fight it. Yoga is one, and vitamins, and she's been thinking about taking up tai chi. Her mum is always nagging her to go back onto anti-depressants, but Holly kind of reacts badly to being told what to do by her mum. You'd have thought Juliet would have realised by now but, no, she keeps on with her, 'Just consider seeing the doctor', and, 'Perhaps a counsellor would help, sweetie.'

Holly puts it down to glandular fever, says she had it when she was seventeen and never got over it 100 per cent. She was seventeen when her dad ran off to Marbella, so that was a pretty bad year. And then the next year she had a go at university, but that didn't work out, and then she had a few bad relationships. And then she got pregnant, but the dad wasn't interested. And the next boyfriend ... well ... Holly's not had it 100 per cent easy, which actually helps us get along.

Cass comes into the kitchen. Holly smiles and gives her a hug. 'I can't believe you and Aidan have found each other! It's so exciting!'

'I can't believe it, either,' says Cass. 'That was my dad on the phone. He's really angry with me.'

'Why's he angry with you, sweetie?'

'I'm meant to be staying with him, meeting his new girlfriend. I just walked out.'

Holly shook her head. 'It's the worst! When my parents split up, Dad wanted us to meet the new girl right away. She was only young, a year older than my oldest sister. I can't stand her—'

'It's a shame, because otherwise we could go and stay with them and it'd be a free holiday in Spain. Finn would love it.' I've pointed it out before, and I get the same response as usual.

Holly shakes her head and says, 'Not a good idea, Aidan.'

'It might be better than you think,' I say.

'I hadn't even thought about holidays,' says Cass. She pulls a face. 'He won't want us tagging along. Not with Annabel and a new baby.'

'Make sure he doesn't use you as a free babysitter,' I tell her.

Holly starts clattering around with the kettle. 'Cup of tea?' She sounds pissed off. I'm not sure why.

Cass is looking at us curiously. I wish she hadn't seen Holly without make-up. She's lovely whatever, but without it you can totally see that she's a bit older than me. I feel like Cass is judging us. I don't want her disapproving, like Holly's mum.

Holly makes tea and cheese sandwiches and we go and sit in the living room. I can see bits of dust dancing in the sunshine. Finn and I start putting together his train set and Cass and Holly sit on the sofa. And then Holly starts with the questions.

'How do your parents feel about you meeting Aidan again? I've seen that programme on ITV – you know, the one where they reunite birth parents and children – and I sometimes wonder what the adoptive parents think. They keep them off screen mostly. Don't even talk about them a lot.'

'I haven't actually told them,' says Cass. 'I suppose I will, eventually. It's so stupid. You're meant to wait until you're eighteen and do it all with a social worker holding your hand. Well, I don't need that. Meeting Aidan is one of the best things I've ever done.'

'I suppose it'll be your mum, next,' says Holly. 'I tell you what, why don't we do it here? I can make a little party, you can invite your mum, Aidan, and we can get all the meetings over together.'

I drive Finn's train over the bridge and let it roll into a line of trucks. Finn squeals with laughter.

'I don't think so,' I say.

'Maybe it would be easier,' Cass says, 'if there are other people around. And she might want to bring her kids.'

Holly's mouth drops open – if she was in a cartoon, her entire lower jaw would have dropped to the floor with a huge crash. 'Her kids? She's got other kids? Aidan! You never told me!'

I concentrate on threading train track under Finn's bridge. It's a really cool train set. Naturally Juliet bought it for him. 'I've only met them once or twice. I don't think she'll want to bring them.'

'Well, just her then. Just her and us and Finn and Cass. Do you have a boyfriend, Cass? Anyone you'd like to bring?'

'No,' says Cass. Then, 'Well, sort of. Maybe. Probably not.'

'We'd better do it soon. Before the Christmas rush gets started.'

'I'm at school,' says Cass. 'It would have to be a weekend.'

'Oh, that's no problem.' Holly's all excited. 'It'll be wonderful to reunite a family!'

'It might be,' I say, cautiously. 'Mum might not want to, though. She's a bit funny.'

'What mother won't want to see her daughter? You're so lovely too, she'll be thrilled.'

Cass is looking so pretty today. She's got a pinky purple scarf wound round her neck which ought to look bad with her hair, and somehow doesn't. Her hair is wild and curly. She's glowing, a golden girl. She doesn't need any slap to look good.

'You're really kind,' she says to Holly.

'It's nothing. Anything for you and Aidan. Tell you what, why don't I open a bottle of wine? We should toast the occasion.'

They talk about Holly's job at the GP's surgery, how people can be really aggressive and nasty when they can't see a doctor right away, and how Cass wants to study history.

'How did you and Aidan meet?' Cass asks.

'Hasn't he told you?'

I freeze. What's she going to say?

'He was working in the shop. He was very shy, weren't you, Aidan? Wouldn't talk to me. I had to make all the running.'

'I wasn't shy. Just, things were different, then.'

'I had no idea that you were only sixteen. He looked a lot older, Cass. Do you think I'm awful, a cradle-snatcher?'

'Of course not,' said Cass. 'You can't be much older, anyway.'

'I'm twenty-one.'

Cass hardly blinks. I do, though, because actually Holly will be twenty-three in two months' time.

'He told me about himself. No family. Coming out of care. Sixteen years old, living in a bedsit ... Can you imagine? Anyway, I offered him a room in the flat.'

This version of events is kind of irritating, because it leaves out a lot. Holly's obviously got her reasons though, so I stay quiet and concentrate on the train track. 'Look, Finnster, let's line the engines up here. Then they can have a race.'

'That was nice of you,' says Cass. 'Where were you living before, Aidan?'

'In a bedsit. But the ceiling fell in.'

I'd got back after a night out and found my room covered in soggy plaster and my bed soaking wet. Someone had left a tap on upstairs in a room that'd been empty for a few weeks.

It was too late to call the landlord or a social worker or anyone. I didn't think they'd let me back into the children's home – I'd been gone for six months by then. The room stank and everything was wet and I couldn't stay there. The pubs were closed and, anyway, I was under age. The other people in the building – well, I wouldn't want to draw their attention at night when I had nowhere to sleep. The caretaker knocked off at seven. I was all alone.

I'd sat there for a bit looking through the handbook the council gave me when I left care, looking for a page which told you who to call in a night-time emergency. There were pages about claiming benefits and what to do when you got into trouble with the police. Details of religious holidays. A list of words you might not know the meaning of. Words like correspondence, solicitor and subsistence. Pathway plan, amenity, eligible, accountable. Entitlements, eligibility and fundamental. Martyrdom. Words to tie you up in knots and keep you prisoner. Useless words. Scary words.

My head was spinning and the words were dancing. The choice was the street or the shop. I picked the shop.

Clive had given me keys a few weeks before to open up in the mornings, and so it was easy to let myself in. I knew we had a stack of sleeping bags, and I hadn't even needed to switch the light on to find them. I climbed up the stairs to the upper floor and curled up on one of the beds. I even set two alarm clocks – we had about twenty. I'd be up and out by eight, all ready to come back again for Clive at nine. I wasn't due to be opening up the next day, and I didn't want him to get suspicious.

The bed was a hundred times more comfortable than the one in the bedsit, and I felt cosy and comfortable, lying there among all Clive's treasures. I was yawning and sleepy and wondering if I could do this every night, and never have to go back to the bedsit. There aren't many places that are improved by a whacking great hole in the ceiling, but my room was one of them.

Then I heard them. The baby crying. Shouting. A huge thump, a crash.

And Holly's scream.

21: AIDAN

I ran down the stairs, stumbled through the shop, opened Clive's desk drawer and found his bunch of keys. Then I let myself out at the front, opened the door to the flat and ran up the three flights of stairs to its front door. I was panting by the time I got to the top, but I kept going.

They didn't hear me at first because Finn was crying so hard. They were in the front room, screaming at each other: Holly and her boyfriend. Some loser called Don. I'd seen him a few times, carrying Finn, pushing the buggy. A big, muscley bloke. I'd disliked him on sight for no real reason.

Now, I had a reason. He had his hands on Holly's shoulders and he was shaking her, and there was blood trickling from her snubby little nose.

'Get your stinking hands off her!' I shouted, running across the room and grabbing Don's arm. He let go of her and swiped at me. I hit him, hard in the jaw. He staggered backwards.

'Christ! Where did you come from?'

'Call the police!' I told Holly, but she went running to pick up Finn, and stood rocking him, crying, shivering.

'You've got no right coming in here!' Don told me. 'I'm calling the police!'

'You were beating her up!'

'She's not going to say anything.'

I kicked him in the shin and he doubled over.

'You get out of here, now.'

Holly was shushing Finn, rocking and shushing, and I saw her nose and the blue shadow of an old bruise on her jaw and I remembered my mum. All those years ago. Shushing us and rocking us and telling us that everything would be all right. And it wasn't all right. It was never all right.

'You're never going to hit her again,' I told Don. 'You're going to pack your bags and get out.'

'You can't throw me out!' He was taller and bigger than me, but I was more used to fighting, and I wasn't drunk.

'You want to bet?'

Don spat abuse at her, ugly names, names I'd heard from Mac's mouth when I wasn't much bigger than Finn. Holly held Finn tighter, kept her head down.

I hit Don again, straight in the mouth, enjoying shutting him up. He flailed around, trying and failing to land a punch on me. He roared like a toothless lion.

'Don't call her names. OK, don't pack a bag. Get going.'

Now the blood was coming from Don's mouth. 'I'm going! You can have her! She's good for nothing, anyway!'

He stomped out of the room. We heard the flat door slam, and then the distant sound of the front door as well.

Finn was still wailing, but Holly's tears had stopped. Finn's cries turned to whimpers, and then stopped altogether. The only noise I could hear was a faint buzz of traffic and calling sirens, the sound of London at night. I waited, suddenly embarrassed. How could I explain where I'd come from?

'Thank you,' said Holly. 'I'm sorry you had to see that.' She rocked Finn in her arms. It made my heart ache watching them.

'What were you doing here?' She was puzzled. 'The shop can't be open, can it?'

'I was sleeping there,' I told her. 'Don't tell Clive ... please ...'

'I won't.' She blinked a few times, trying so hard not to cry. 'I'm so embarrassed. Don, he's just, he was drunk, that's all. I made him cross. It wasn't really his fault.'

'Don't take him back,' I told her. 'For you, for the baby's sake, don't do it.'

'I love him. How can I just finish it?'

'What's to love about that?'

'You don't understand. I'm all alone.' She glanced down at Finn. 'It scares me, Aidan, having to do this all on my own.'

'He's not the dad, is he?' I'd guessed that because Don was big and white, and Holly was little and white, and Finn was small and definitely half-black. Plus Clive had filled me in on how worried the family was because Holly seemed to go from one bad bet to another.

'No, but it's not so easy to find someone when you're ... when ...' She couldn't talk any more. She held her fists to her mouth.

'Call a locksmith in the morning. Pack up Don's things – he

can get them from the shop if he's got the balls to come back, which I doubt.'

'What if he comes back before I get the locks changed?'

'I'll stay here on your sofa,' I told her, even though I knew that meant Clive would find out that I'd slept in the shop. My sleeping bag and the alarm clocks would give me away. I'd have to hope that he understood it was a crisis.

So that was the first night I spent in the flat. Lying awake, listening for Don. Worrying about Clive. Remembering Mum and Mac and Cass and me. And full up with love for Holly. Nothing to do with sex – I just wanted to care for her, protect her, make sure that her smile never left her face.

Clive was pissed off with me, but when I told him what had happened, he changed track right away. 'Thank God you were there,' he told me. 'I knew he was bad news. I'll call the lock-smith right away. Call my solicitor too, get a restraining order on him. Holly's so soft she'll end up taking him back if we don't act now. You can sleep in the shop for now. Call your social worker, get him to sort out the bedsit.'

Two days later, Holly asked me to come up to the flat. She opened a bottle of wine. We sat on the sofa, side by side.

'I never really thanked you.'

'That's OK.'

'I was just so shocked. You came out of nowhere.'

'Sorry. I heard the noise and I had to do something.'

'You could have phoned the police.'

'I'm sorry. I didn't mean anything wrong. I was worried about you and Finn.'

'I know. I mean, lots of people wouldn't have got involved like you did. I appreciate it.'

'I'll be downstairs,' I told Holly. 'For a few weeks, at least. So if Don comes back, just call me.'

She sighed. 'He was great when he was sober.'

I didn't say anything. I'd heard it all before.

'Aidan, I was talking to Uncle Clive. I've got a spare room and you've got nowhere to stay. Why don't you move in? As my lodger? Got to be better than a bedsit.'

'Really?' I couldn't believe it. The flat was beautiful. She was nice, and kind, and smiling. This was the best thing that had ever happened to me. 'Are you sure? Really sure?'

'I'm totally sure,' she'd said.

It was kind of awkward at first. I spent most of my time in my room, because I wasn't sure she wanted me to be in the kitchen or the living room. Then she started asking if I wanted to watch stuff on telly with her. Offering to share her supper – 'I've made a curry and there's loads of it' – and asking if I could watch Finn for her. We got used to each other.

After two months, I ended up in her bed. I was sixteen, nearly seventeen, and she was twenty. Best thing that ever happened to me.

I can understand why she doesn't tell Cass any of this. It's personal. But it also kind of pisses me off that I don't get the credit for being a hero.

'My uncle's really pleased with Aidan. He's done a great job in the shop.'

OK, now Holly is talking about me like I'm about eight and she's my mum. I'm more than irritated. 'Look, Finn, the train's going round and round and then ... *smash!*' Finn thinks this is hilarious. He loves accidents.

'Aidan needs to pass his driving test, though – the theory exam is holding him up. You find it difficult, don't you, love? I'm sure he's dyslexic.'

'I'm not dyslexic.' I try and make a joke of it. 'I'm just thick.'

'I'm almost certain you are dyslexic. I wish you could get properly tested.'

I shrug. 'It's too late.'

'My mum knows some people,' says Cass. 'My little brother, Ben, he's had loads of tests.'

I don't really see what good that is to me.

'Crash it again, Aidy, crash it again!' says Finn.

So I do.

22: CASS

I loved seeing Aidan. I loved meeting Holly. It couldn't have been more different from everything I'd ever known.

Their flat was so small – one tiny bedroom and then Finn's room, which was barely more than a cupboard – and nothing matched, not the furniture or even the plates. It all seemed to come from the shop where Aidan worked. Second-hand junk. And it was really messy – they obviously don't do much tidying up. It felt young and free and unstuffy.

They opened a bottle of wine mid-afternoon. Mum and Dad always wait until 6 p.m., which they call wine o'clock. I had a few sips of wine, but not much. I was in enough trouble with Dad as it was.

I got a bit of a shock at first when I realised that Holly was loads older than Aidan – maybe ten years, I thought, although she looked younger when she put on some make-up and tight jeans. Actually, she's nearly the same age as Annabel. How strange is that?

They were always touching. She put her hand on his shoulder, or he stroked her leg, or they sneaked a kiss when they thought no one was looking. It was cute. It made me wish I had a real boyfriend, someone who loved me, not just an experimental kisser. I'd have to move on from Will pretty quickly, I thought. But only after my A levels, and only if I could find the right person. I could go on one of those volunteer schemes in my gap year and meet someone in Costa Rica who was into helping the environment.

I was lucky really, that Will seemed fine about keeping everything private and experimental and non-involved. Let's face it, he was probably relieved that no one had to know he had anything to do with me, given that I wasn't like the sort of girl he usually went for. Just the idea of people gossiping about us made me cringe. If people knew, they'd try and define us – not that there was 'an us' – they'd talk about me, judge me. The pressure would be too much. I didn't mind experimenting with kisses. I knew I would never want to fall in love.

Did that make me abnormal and cold? Quite probably. Being with Aidan and Holly made me sad, in a way. Being brought up in care obviously hadn't closed Aidan off to love and partnership; in fact, he was living as part of a family before he was out of his teens. It must just be me that was somehow unloving. I couldn't blame it on genes or upbringing.

I got back to Dad's flat, and managed to be polite to everyone, and he didn't tell me off for running away because (I assumed) he was so grateful that I'd come back. They were taking Ben out for a pizza, but I stayed at home, using my essay as an excuse.

I didn't write the essay. I poured myself a glass of wine from a bottle I found in the fridge, lay on the sofa, closed my eyes and imagined being grown up and independent and free. And sexually experienced. I imagined that as well. I tried to imagine a fantasy boyfriend for when I was Holly or Annabel's age. I wouldn't go out with some older man or younger boy. He'd be my age, handsome and rich. He'd have a starred first from Harvard or Cambridge in economics or possibly philosophy.

He'd be extremely serious and never make stupid jokes. He'd very possibly be American or Canadian. I'd be working in a museum or an archive or something and he'd take me out to smart restaurants and then back to his stylish flat. He'd think I was wonderful ... respect my intellect ... He'd never make demands on me, but he'd be effortlessly charming and impressive. If we got married, we'd have a very traditional white wedding, and my dress would be ivory raw silk with a fishtail ... *urgh*. I was like a six year old, imagining her dream wedding. No, we'd get married in a registrar's office and I'd wear a suit. No, we wouldn't get married at all, I'd live in London and he'd live in New York and we'd Skype a lot.

My phone rang. Will.

'Hey, just wondered how it was going with the Big Bad Girlfriend.'

'Oh, it's OK. I'm here on my own. They've taken Ben out for a pizza, but I'm meant to be doing my essay.'

'Meant to be? I thought you were definitely writing it today. Are you bunking off?'

'I just can't be bothered. Who cares anyway? It's all happened, and everyone's had their say.'

'Hmm,' he said. 'If you say so. So what's she like?'

'Much older than him, but really nice. She wants to invite my mum round.'

'I thought he'd run off with a younger woman?'

'Oh! You mean Dad! I went to see Aidan today.'

'You went without your bodyguard?' Will made tutting noises.

'I went to see him in his shop – actually, there was someone a bit dodgy there – and then I saw his flat. And I met Holly. Finn's not his baby, she had another boyfriend then.'

'So she's an older woman? OK, figures. And what's Annabel like?'

'Perfect. Fake. She's charmed Ben.'

'But she won't charm you, eh?'

'Never.'

'No, because you take a whole load of charming.'

'That is true.'

'Have I charmed you yet?'

'No, you're going to have to work a lot harder.'

'I'll write your essay for you,' he said, 'if it'll win me any points.'

'I don't approve of cheating,' I said, thinking how great it would be if I could just magically create an essay out of thin air.

'Well, hurry up and finish it,' he said, 'and then we can continue with our very interesting experiments.'

'It's so long though, and such hard work.'

'I thought you were quite enjoying it so far. Enough to continue, anyway.'

'Not the experiments, the essay.'

'I have offered to take it off your hands. Think about it. Then we could go to the cinema tomorrow night. Experiment in the back row.'

'No,' I said. 'No cheating and no public experimentation. This is a secret project. I told you that on the train.'

'But that's so unscientific! You might find that experiment-ing in public is the key factor that makes up your mind.'

'Makes up my mind about what?'

'Well, I don't know. You were the one going on about working out what it all meant and the need for scientific analysis of the situation. I'm quite happy to be a little more spontaneous.'

Spontaneity seemed a bit dangerous to me. We'd so nearly been spotted on the train, and in the park.

'We could just forget the whole thing,' I said.

For a second I had him. There was an uncharacteristic pause.

'OK, if that's what you really want.'

'I just don't want to be gossiped about.'

'That's it? The sum total of your concerns?'

It suddenly occurred to me that perhaps Will was slightly insulted that I wasn't parading him in front of the crowds in Starbucks. That he thought I was ashamed of him. That my natural cautiousness was coming over as ... no, it couldn't be ... racism?

'I just think we might find it easier to get to know each other better without loads of attention. I know that you live and breathe by your social life, but I'm not like that.'

'I understand. You've been in the spotlight a lot recently, with your dad and all that.'

This was a much better excuse than the truth. But I couldn't tell Will that I both wanted him and didn't want him at the same time.

'And we hardly know each other.' I told him, as briskly as possible. 'We might get fed up with each other after a week or so, and then it's better if no one else knows.'

'You say the sweetest things.'

I laughed. And laughed some more.

'Miss Montgomery, have you been drinking? You sound remarkably cheerful for someone who is having a miserable weekend meeting her father's hateful new girlfriend.'

'Only a glass of wine,' I confessed.

He sighed. 'What a shame I'm not there. This would be a perfect experimental opportunity. We haven't even *begun* with artificial stimulants.'

'Shh,' I said, trying to stop giggling. 'I think they're back. I'll ring you tomorrow.'

I rushed to the kitchen, rinsed the wine glass as they trooped through the front door. Ben was yawning, Annabel chatting and making him laugh. I kept my face unamused and unfriendly.

Dad asked how I'd got on with the essay. 'Fine!' I said, leaping up to collect my books together before he could demand to read it. 'I've nearly finished! I'm really tired. I think I'll go to bed.'

Dad's flat has two bedrooms, and Ben and I are used to sharing when we're here. Annabel tactfully disappeared to the kitchen to make tea, so we wouldn't have to see her going into the room that had always been Mum and Dad's. I tried very hard to forget that she'd be using Mum's en suite bathroom, sleeping in Mum's bed with Mum's husband.

'She's nice,' said Ben. 'I like her. She understands all about how I feel.'

'Ben, she caused how you feel.'

'No, she understands about school. She said she never had any friends, either. Everyone thought she was a geek.'

Annabel was a big, fat liar in my opinion. Except she's obviously never been fat in her life. Even her baby bump is slender and toned.

'She's going to be my friend. And she's going to introduce me to her little brother and her sister, and her big brother and her half-sister, and they're all going to be my friends.'

All his life I'd loved Ben, felt every hurt he suffered. I never minded that he was the 'real' son, the biological one. I stood up for him, tried to help him … and he'd let me down. Betrayed me. Allowed himself to be manipulated by the offer of some extra friends.

Well, two can play at that game.

'Ben,' I said. 'Can you keep a secret?'

'What secret?'

'You have to promise.'

'I promise.'

'I'm going to introduce you to more friends, too. A friend called Aidan and another friend called Holly. And a little boy called Finn. And maybe one day some children called Louis and Scarlett.'

'Who's Aidan? Who's Scarlett?'

'Aidan is my brother, just like you're my brother,' I began.

'How can he be?'

'You know how I was adopted? I came from a different family? Aidan comes from the same one.'

It took him a while but he got there. 'But why don't I know him? If he's your brother, too?'

'You will meet him, and his girlfriend Holly and her baby, Finn. You mustn't tell, Ben. Mum and Dad don't know.'

'Who are the others? Louis and Scarlett? How old are they?'

'A bit younger, I think.' Ben does tend to go around with people younger than him. 'They're also from the same mum. I haven't met them yet.'

He sat up in bed, squeezing his knees. 'How come we don't know them now?'

'That's how adoption works. You join a new family and you leave the old one behind. You don't have contact any more.'

That's not true, actually. Mum explained it to me once. She said there are all sorts of levels of adoptions, and some people have open adoptions where they exchange letters and meet up with the birth parents occasionally. That wasn't deemed appropriate in my case; mainly, I think, because Dad was an MP.

Ben yawned. 'I want to meet them,' he said. 'I want more friends.'

'You will, Ben, one day,' I promised. I loved him so much. He'd been my brother all his life, and Annabel had only just met him. If Ben was getting any more family, it was going to come from me.

23: CASS

'Will Hughes!' said Grace at school on Monday morning. 'Will Hughes! *Will Hughes!*'

'What about him?' I tried to look unconcerned, but inside I was furious. Will had obviously been talking.

Grace was as smug as a little girl with a secret. 'You! And Will! On the train! Together!'

'Who's gossiping about me?'

'You were spotted,' said Grace. She tapped her nose. 'I've promised to protect my source.'

'Well, anyone who spotted us can tell you that there was nothing going on. Will is a friend, that's all. He's Ben's peer buddy at Bonny's. I happened to bump into him on the way back from London.' I was pretty certain that this was the train she was talking about, not the one from Oxford to London.

Grace's face made it obvious she didn't believe me.

'Ben's not happy at Bonny's. He's Will's special project.'

'Cass, Will Hughes is a legend. Please tell me you're going to go for it this time.'

'I don't know what you mean.'

'Duh! Any boy who's ever plucked up the courage to approach you gets crushed and sneered at, and treated like dirt.'

'I do not treat anyone like dirt!'

'Declan? When he dared to ask you to the Year Eleven prom at Holy Virgin High School?'

'Oh, honestly. Declan! I mean, why did he even bother?'

'Simon Potter? Darius Mainwaring?'

'Who?'

'Will Hughes is in a different league!'

We'd reached our history class. I winked at Grace. 'Talk later,' I said.

History is my favourite subject and Mrs Harper is a good teacher, but my mind wandered as she went on and on about people fighting in faraway places a long time ago. In fact, I started to wonder if history was really what I wanted to study at university. What was the point, after all? Reaching back into the past to theorise about what could have, should have, might have happened. Maybe I should study something more relevant to today? Maybe I should think harder about what I actually wanted to do when I left uni. At the moment, I had a list of things I didn't want to do: no teaching, no politics, and vague clichéd ideas about museums or theatres.

Would it be easier to be someone like Aidan, who just left school and found a job and then got on with it? Or someone like Holly, who dropped out and got pregnant and got a job? Actually, that sounded really dreary. I'd have to grit my teeth and concentrate.

What if I hadn't been adopted? What if I'd stayed with Aidan and his mum, my real mum? What if we'd been a real family and I'd grown up in north London somewhere. Would I be a hairdresser?

'Cass!' said Mrs Harper. 'I asked you a question!'

I had absolutely no idea what she was talking about. I blushed, apologised. 'I don't know, I'm sorry, Mrs Harper.'

She looked disappointed. 'Well, perhaps Grace can tell us something about the cult of personality as it applies to Stalin.'

Grace could and did.

When the lesson was finished and we went our separate ways (English for me, biology for her), Grace nudged me and said, 'Day-dreaming about Will? See you at lunchtime.'

But at lunch I had a prefect's meeting and she was rushing off to play hockey and we didn't get another chance to talk until the end of the day, when we got on to the bus.

'So!' she started, but I glared at her and said, 'Let's go for a coffee if you insist on interrogating me.'

'Oh, I think you'll want to interrogate *me*,' she said.

No Will on the bus, I was relieved to see, and no sign of him in Starbucks either, although there was a big group of Bonny's sixth formers. I found us two seats in a relatively quiet corner; Grace queued up to buy us lattes. I kept my head down, checked my phone.

'Well! Fancy seeing you here. I didn't think you deigned to mix with the rabble.'

I might have known it. Will was towering over me, coffee in hand, on his way – I hoped – to the huge group of Bonny's kids in the corner.

'I don't. I mean, I don't usually come here, but I'm having a quick coffee with Grace.'

'You got your essay finished, then? Shame you couldn't come to the movies.'

He'd kept his voice soft enough that no one would over-hear, but I still frowned at him. 'Shame,' he said again, 'it was good. Oh, well, see you around.'

I wondered if Will had other girlfriends to take to the cinema. After all, there was nothing to stop him: we hadn't made any promises to each other – quite the opposite, really.

Grace sat down. 'He was talking to you!'

'About Ben, actually.'

'Oh, yes. Very likely. Cass, you have to tell me everything.'

'Nothing to tell. No, really, nothing. He's just an acquain-tance. He's looking out for Ben in the playground. You know my brother has a difficult time making friends.'

'Well, make the most of it. Will just broke up with Ruby, remember. He won't be on the market for long.'

Ruby was in our year, but we'd never been in the same form or had the same friends. She was blonde and girly – not the sort to be a prefect, or get involved in debating or anything like that. I couldn't imagine what she'd put on her UCAS personal statement.

'So he's on the rebound?'

'So you *are* interested? Oh God, Cass, you're smiling!'

I rearranged my face. 'I'm allowed to smile.'

'I've never seen you smile like that.'

'Oh, come on. Can you see us together?'

'You need to scrub up more, Cass. If you'd make a bit more effort with your clothes and make-up ... you're really pretty when you can be bothered.'

I was actually wearing mascara that day, which proved her wrong.

'That's not what I mean,' I snapped. 'He's completely weird, him with his stupid Dora bag and tweed jackets. He looks ridiculous. Why on earth would I be interested in him?'

'Aw, Cass, he's lovely. Everyone at Bonny's is in love with him.'

'It's a cult of personality. Like Stalin.'

I glanced over to the Bonny's kids. Stalin was right at the centre, telling some long story that involved silly voices and big, sweeping gestures. The group was full of girls, laughing, staring, admiring. One was the blonde girl I'd seen him with at the bowling alley. Just his type.

'You'd better get on with it, if you are interested,' said Grace. 'Looks like he's got a long line of admirers.'

'I'm so not interested,' I told her, and we talked about other stuff. Her pony, and her plans to study management at Exeter or Bristol, and her campaign to go out with some boy called Leon.

Grace tried – she really tried – to be the caring friend, asking me how the weekend had been. 'What's she like, Annabel?'

I wrinkled my nose. 'She's super-fake. Trying to be all friendly. I don't want anything to do with her.'

'But you'll have to, eventually, won't you? I mean, when she has the baby ... and aren't they going to get married? When the divorce comes through?'

'Where'd you read that? The *Daily Mail*?'

Grace flinched. 'I'm sorry, Cass. It's just – I worry about you. I'm your friend and I care, and you don't tell me anything.'

'There's nothing to tell,' I said. 'I just want to get on with my life. I'm not someone who likes agonising over all the gory details.'

'Sorry,' she said, but I could tell by her pursed lips and tiny frown that she thought I was odd. I suppose she expected floods of tears.

'It's just the way I deal with things,' I told her, and we quickly moved on to the politics of the netball team, and which university open days we were going to.

And, all the time, Will was sitting with his friends, talking and laughing and entertaining everyone. My skin prickled with irritation. I hated myself for trying to hear what they were saying.

Ten minutes later, Grace went to the loo, and I found my phone. Without thinking too much about it, I texted Will: 'I'm going home in ten. If you get the same bus we could get off a stop early.'

I watched him pause midstream, reach for his phone. Take it out and read the text. Nod his head, give a little smile, glance in my direction.

I ignored him. I didn't even look at the blonde girl at his side. I was in control. I was the Ice Queen.

24: CASS

You'd have thought that people would be used to my dad. After all, most of the people I was at school with had started alongside me in the junior section when we were only five.

But he still turned heads. Every school concert; every parent-teacher evening. Of course, people were even more interested now he'd left us. This was the first time my parents had been seen out together since Dad had left home. Thankfully everyone was too civilized to film us on their phones.

Mum and I had tried to work out a system so they didn't have to spend much time together: maybe if Dad came to history and philosophy, and Mum to economics and English. But he said that'd be more noticeable than if we just pretended everything was OK. 'We'll manage fine,' he told us, all breezy and confident and virtually smelling of Annabel. 'Big smiles all round. Won't be the end of the world.'

'I wish you hadn't come,' I hissed at him as we stood in the queue for my English teacher.

'Of course I came. I care about your education. We may not be together, but we're still your parents, Cass. We'll do everything we can for you.'

'What you can do is stay away!' I whispered. 'Can't you see how difficult it is for Mum?'

Mum wasn't standing with us. She'd spotted her friend, Anthea, and was chatting to her. Mum was wearing lipstick and she'd had her hair done. The heartbreak diet and the Pilates meant that on the surface she looked great. But I knew, and Dad should know, that it was all a façade.

'Cass, this is not the time or the place,' said Dad, keeping his politician's smile in place. 'Now, what was all that from the economics teacher about a failed test?'

'I didn't have time to study, OK? I've been a bit busy recently, propping up the people you left behind.'

'Cass—'

Mum was back with us. 'Everything all right?' she asked.

'Fine,' I said.

'Fine,' said Dad. 'Look, they're about to leave.'

I thought it would be a relief to sit down and talk to the English teacher, Miss Graham. I was totally mistaken.

'Cass, what's happened?' she asked. 'I kept back your latest essay because I couldn't believe you'd written it. I'm used to you getting As and A*s, but this barely scrapes a D.'

I'd totally forgotten about that essay. I'd cobbled it together at the last minute in a study period – the study period I'd set aside for studying for my economics test.

'I don't want to intrude,' said Miss Graham, 'but I know it's been a difficult term for you all. Do you think it's affecting your studies, Cass? Perhaps you should see the head of pastoral care.'

I sat and glowered while Dad went on about how our welfare was their first priority, how both of them were utterly supportive of my education, how keen they were that I should go to Oxford.

'I'm sure this is just a blip,' he finished. 'She's such a hard worker, aren't you, Cass?'

Just for a moment, I contemplated saying 'No'. I considered stopping trying to please everyone, working hard, trying to be a real Montgomery. Forgetting Oxford, and accepting that it wouldn't be a disaster to end up somewhere like Bristol or Exeter. Being *normal*. Getting As and Bs instead of A*s. Leaving school and taking a gap year and just messing around. Just hanging out and having fun. Like Aidan. Like Holly. Like the Cass I would have been if I hadn't got adopted by a couple of high-achieving, pressurising, middle-class ...

'I'm doing too much!'

Mum was looking into the distance, not really listening to the conversation. She came to with a little jump.

'What do you mean, darling?'

'I mean it's too much. A levels are a lot of work. I want to quit the extra stuff. Music and Duke of Edinburgh and being a prefect and all that.'

'But Cass, your personal statement—'

'I don't care about my personal statement!'

Dad looked around. 'Lots of people waiting to see Miss Graham,' he observed. 'Perhaps we need to discuss this with Cass at home. I'm sure she'll be back on track soon. Thank you so much, Miss Graham.' He dazzled her with his smile, teeth gleaming like a cartoon crocodile, and strode out of the room, Mum and I rushing to keep up with him.

People were looking. I knew it even though I couldn't see them. I couldn't see them because of the tears in my eyes.

'Out,' ordered Dad, even though we hadn't seen my history teacher yet. Mum murmured a faint protest, but he was holding the door open for us. We were ejected into the freezing night just as the headmistress arrived.

'Hello, Mr Montgomery! How very nice to see you!'

Mum and I shivered in the cold air, and she tried to put her arm around me. I shrugged her off. 'I'm fine! Leave me alone!'

Dad emerged at last. 'Come on,' he said. 'We need to talk. Let's go home.'

I really lost it then. 'You can't call it home!' I told him. 'You don't live there any more! Because of you there's a "For Sale" sign outside!'

'Well, we can go and sit in a restaurant and you can cry and shout at us in public if you prefer,' he said, smoothly. 'I honestly think it's better to keep things private.'

'Cass, be sensible,' said Mum. 'Of course Dad can come to the house. We need to have a talk, don't we?'

'No!' I said, but Dad just looked at Mum and said, 'I'll see you there.' Then he jumped into his BMW and we got into Mum's Mondeo, and I cried all the way home.

Mum pulled into the driveway, past the massive 'For Sale' sign, and handed me a tissue. I mopped my eyes and blew my nose and tried to think about tsunamis and earthquakes and World War Two and other things that would make my stupid emotions dwindle by their enormity.

'Oh, Cass,' Mum said, 'I'm so sorry, my darling. I wish we could have sheltered you from this.'

'I don't need sheltering!'

'You're bound to feel bad. It's so tough, what you're going through. I've been so bound up in myself, I've neglected you, I'm so sorry.' She had no idea. She had forgotten those weeks when she was a wreck, when she wasn't functioning, when I felt I was going to be left with no family at all. Just because she'd pulled herself together, that didn't mean everything was OK now. She'd shown herself as weak, and that was something I never wanted to see again.

'You haven't neglected me. It's just that I've got a lot on at the moment.'

Mum handed me another tissue. Then she said, 'Cass, do you think, perhaps, you should cut back on your social life? It's just – I can't help noticing – you've been spending a lot of time with that boy recently.'

'Which boy?' I tried to sound unconcerned.

'That boy. The one Ben knows. I can't remember his name. Kate said you went to Oxford with him.'

Mum was embarrassed, trying not to say the word 'black'.

'Oh. Will. What's your problem with him?'

'Well, people do talk, you know. Mrs Featherstone saw }you with him on the train. She couldn't wait to ask me about you. And I have noticed you've been out a lot. And then Mrs Thwaites said she'd seen you with him on the bus.'

The whole town is full of small-minded, bigoted old gossips. I could never take public transport again, but at least no one had seen us in our private experimentation places: the back of the shops in Walker Lane; the bomb shelter in the park. The car park at the library.

None of them were great, really. We'd been talking about finding somewhere warmer. Somewhere indoors.

'They should all mind their own business.'

Dad rapped on the windscreen. 'Are you two going to be in there all night?'

'No, we're coming,' said Mum. 'I know this feels strange, Cass, but we're determined to make it work in a civilised way.'

Dad being in our house was nothing compared to seeing Annabel going into Mum's bedroom, cooking in Mum's kitchen, coming out of the bathroom in Dad's bathrobe. I didn't say so, though.

We all sat round the kitchen table. Mum poured herself a glass of wine. Dad said, 'Better not, I'm driving.' No wine for me, I got offered juice or tea. I didn't want either.

'So,' said Dad. 'This is really quite serious, Cass. You've let things slide. I knew I should have looked at that essay the other day.'

'Look, I don't need you to police my homework, OK? I'm fine. I can manage.'

Thank goodness he didn't speak to the history teacher. I had only handed it in that morning, a full fortnight late.

'What's all this about you giving up your extra-curricular activities? You've always managed fine before.'

'I've got a lot more work now.'

Mum joined in the attack. 'Is the difference this boy? Because if he's disrupting your work, Cass, it's maybe not such a good idea.'

'Oh God! Why can't you keep your nose out! Nothing is happening! He's just a friend!'

'What boy?' Dad again. 'Why are you over-reacting, if he's just a friend? Who is he, anyway?'

It was one thing for Mum to ask questions; Dad butting in was completely, outrageously unacceptable.

'None of your business! I'm not having some secret affair, like you did!'

'Look, Cass, you can't speak to me like that. I'm still your father. I've got the right to interfere if you're messing up your school work, mixing with the wrong sort.'

'You don't know anything about him!' Mum tried to take my hand, but I snatched it away.

'Darling, I'm sure he's a very nice boy. Maybe you could bring him over for dinner one day?'

'He is not my boyfriend! You don't need to meet him!'

Ben came downstairs. He refused to have a babysitter – said he was too old – so when we were all out he was under strict instructions to stay in his room.

He looked alarmed when he saw us all sitting there. 'What's the shouting about? Why's Dad here? Is Cass in trouble?'

Dad ruffled his hair. 'How's it going, kid?'

'Bad, but not as bad as it was before half-term.' Typical Ben. 'How's Annabel?'

Mum's turn to wince. I frowned at Ben. 'Go away.'

'Why?'

'Because Mum and Dad are telling me off.'

'Why?'

'Because I'm not living up to their high standards.'

'Did they find out?' he asked, all wide eyes and innocence.

I froze. Had Will said something to Ben? Had he seen us?

It just wasn't possible to keep a look out all the time; we tended to get a bit distracted.

'Find out what, darling?' asked Mum.

'Shut up, Ben!'

But Ben didn't hear me.

'About Aidan. Cass found her other brother.'

25: AIDAN

Ellie-Mae runs her hands through my hair. Her fingernails scratch my skin as she massages my scalp. It's giving me a headache.

It's model night at the salon, and I came to see Mum, but she's not here. According to Ellie-Mae, Mum hasn't quit three weeks ago.

'I'd have thought you'd know,' she says, letting her finger tickle my neck. 'She wanted more time with her kids.'

'She didn't tell me.'

Ellie-Mae starts washing my hair, rubbing the shampoo into my scalp incredibly slowly. 'Dad wasn't too happy. She left us short-staffed. That's why he's moved me off reception. I don't like washing hair, it plays havoc with my acrylics.'

'Your what?'

'My nails.'

She rinses the shampoo off, wraps a towel round my head. She lets her hand brush my ear.

'Ellie-Mae,' I say, 'I don't suppose you could do me a big favour?'

'Of course!' She bats her eyelashes at me.

'I don't suppose you've got my mum's address?'

'Don't you have it? Can't you ring her?'

'I lost my phone. All my contacts wiped.'

She buys it.

'I might have it in the system . . .'

'I'd be very grateful.' I smile my best smile.

When I'm finished with the haircut and the photo, she slides a piece of paper into my pocket. Without looking, I know it'll have her number on it, as well as Mum's address.

'Ellie-Mae, you're a star,' I tell her.

'What's my reward?'

'I owe you a drink,' I tell her. She beams and I breathe, and I'm out of there.

I'll never go back.

I divert via Rich's bedsit and have a quick drink while he looks up Mum's address on his phone. 'It's not far,' he says, showing me the map. 'Just here, then here. Look. Number twenty-four.'

I drain my beer bottle. 'I'm going round there now. She's got to be in, with the kids and all.'

Rich looks dubious. 'Is that a good idea? Why not ring her or write her a note, or something? I'll help.'

'I've got to see her. She would've told them at the salon not to give me her address if she really didn't want me.'

'Yes, but, Aidan, she threw you out. And what about her old man?'

My throat tightens. My fists clench. 'That was years ago. I was eleven, for Christ's sake. It'll be OK now.'

'Why do you have to go round there? Look, Aidan, let me go for you. I can talk to her, arrange something for you. On neutral ground, like. I mean, she never minded you turning up at the salon, did she?'

'She won't mind me turning up at her house. Not when she hears what I've got to tell her.'

'You're going to tell her about Cass?'

I open another beer. 'Too right I am. She's going to be made up. All those years she didn't know where Cass was, how she was, and now I've found her.'

'Yeah, but what about Cass? Does she want your mum knowing about her?'

'Holly wants to have a tea party. Introduce everyone. It's a crap idea, I know. But it might work. Holly thinks so.'

Rich sighs. 'I don't want you getting hurt, Aidan.'

I throw a pillow at him. 'You big girl. I can take care of myself.'

I finish the beer. It's my third or fourth. I'm perfectly prepared for Mum, head just about clear, nerves nicely numb. Hair newly cut too, clean and shiny. I'll just go round there and knock on the door and say ...

'Rich, why don't you come with me?'

'OK. And if it doesn't go well, we can go for a drink.'

'Yes.'

We get slightly lost on the way so it's forty minutes before we get to Mum's house. It's a semi with a Ford Fiesta parked on the front drive. Just the one car – I hope that means she's not with her bloke, Keith, any more. I kind of forgot about Keith when I decided to come over to Mum's. I've blotted him out of my mind since he threatened to kill me.

Rich rings the doorbell. I bounce on the balls of my feet, getting ready to run if I need to. We notice too late that the curtains are drawn and none of the lights seem to be on.

Someone fumbles with a latch. The door creaks open a crack.

I smile. 'Hey, Mum!'

She's wearing her dressing gown. She's got slippers on. 'What the hell, Aidan?' she hisses, furiously. 'What time do you call this? Are you trying to wake the kids up?'

It's only about 10 p.m. She's making a big fuss about nothing.

'I just thought – I've got some news.'

She opens the door properly. 'What news?'

How many people have mothers who hate them like mine hates me? Luckily, Rich's parents were even more useless than mine and, besides, they're dead. Actually, my dad's dead, too. But I never knew him, and Rich was ten when he lost his mum and dad to a batch of bad heroin. It didn't really bother him much, because he was living with his Nan. When she had a heart attack a year later, that was when he got dumped in care.

'Can't we come in?' says Rich. 'It's kind of cold out here.'

She gives him the glimmer of a smile. 'Hey, Rich,' she says. 'I like your hair.'

Typical that she mentions Rich's stupid red hair, but doesn't notice that I've just had mine cut. She pecks him on the cheek. I get a quick kiss, too. Her lips are dry and cold.

'You're looking well,' he tells her. I keep quiet. She's looking rubbish. She's got no make-up on and her roots need retouching.

She leads us into their front room. Massive flat-screen telly, two black leather sofas, shiny laminate floor, a smoky glass coffee table. She's done well for herself.

She folds her arms. 'Lucky for you, Keith's out tonight,' she says.

So she is still with him.

I look around the room. Everything is sparkling clean. There's a huge picture on the wall, one of those photographs printed on canvas. Their kids. Hers and Keith's. Louis, the boy, must be eight or nine now. Scarlett, the girl, is maybe five. I can't bear to look at it. There's a black leather sofa right underneath and I sit down, so there's no danger that I'll see it. Rich sits down next to me, and Mum sits opposite us. She doesn't offer us a drink or nothing.

'How's Keith?' I ask. 'Still in the same business?' When I knew him, Keith was running a nightclub in Dagenham. I'd liked it, because he wasn't around that much.

'Yes,' she says. 'He's got a few clubs now. He's doing OK. Is that why you're here, Aidan? Looking for money?'

'I'm not looking for money,' I say, 'because I've got a job. And if I was looking for money, I wouldn't be asking Keith for it.'

'Don't get aggressive with me,' she says, though I didn't even raise my voice. 'What's the problem, then? In trouble with the police?'

'I've come to tell you something,' I say. 'Something that I think will make you happy. If you're capable of being happy, that is, you sour-faced bitch.'

'Aidan—' says Rich, but I wave my hand to shut him up.

'You always have to dig at me, don't you? You always have

to make out it's my fault. You never think that you abandoned me when I was six and then, when you thought you were finally ready to be a mum, you just couldn't hack it. If I'm screwed up, it's all down to you. It's all your fault. What happened to Louis, it's down to you.'

'Get out!' she says, her voice shrill with fury. 'Get out!'

'Mummy!'

The little girl's voice from the door shocks us. We all sit there, staring as she runs across the room and dives into Mum's lap.

Mum hugs her. 'It's OK, Scar. Did you have a bad dream?'

She's got Cass's red hair this Scarlett – she's well-named, actually – and Keith's brown eyes. She's peeking a look at me, and I have a brainwave. I dig in my backpack.

'Hey, Scarlett,' I say. 'I've got something for you.' I hold out a lollipop. I keep them on me for Finn's worst meltdowns.

Scarlett reaches over and grabs it. 'Thank you,' she says.

'She's brushed her teeth,' says Mum. 'Give it to me, Scar, and you can have it tomorrow.'

'It's sugar-free,' I tell her. Holly's very fussy about Finn's teeth. Scarlett shakes her red curls and says, 'I keep it safe' and I'm hit by a memory of Cass at around the same age, Cass hugging her teddy and singing and asking me for a story. I'd just started learning to read at school, and I used to remember books off by heart to tell them to her. There was a nice lady at school and she helped me with my reading. I could read then. I wasn't stupid, or confused, or dyslexic, or whatever happened to me later.

I blink. I'm going to have to be careful or the memories will set me off.

'What's your name?' asks Scarlett, and Mum says, 'Never you mind. He's going soon. Get back to bed now, darling.'

'She's got a right to know,' I say. 'I'm called Aidan.'

'Not now,' says Mum.

'I want a story,' says Scarlett, and again I remember Cass.

Mum kisses Scarlett's head. 'It's late now. Off you go.'

'Mum,' I say, as soon as Scarlett's out of the room. 'Mum, I'm sorry. I didn't mean to lose my temper. I just came to tell you something.'

But Mum had spotted Finn's baby wipes and stuff in my bag when I opened it.

'Aidan, what was that stuff in your bag? You've got a kid?'

'No,' I say. 'Leave it, Mum.'

'What do you want then?'

'Well, it's just … it's Cass. I've found her.'

Mum's face is a picture. Her cheeks are puffed out and if her eyebrows went any higher, they'd be hovering in mid-air above her head.

'What do you mean, you've found her?'

'Aidan found her,' says Rich. 'On Facebook. It's definitely her. She looks just like your Scarlett.'

'I've been in touch with her and met her and she wants to meet you,' I say, all in a rush, before Rich can tell any more of my story.

Mum's eyes are full of tears. 'Aidan? I can see Cass again? Oh, Aidan, you darling, darling boy.'

My mum hasn't said anything like that to me since I was eleven. Her birthday, actually. Keith had bought her some bubble bath and told her it was her present from me, and she'd made a big fuss of me when she'd opened it. I'd loved it. I

194

remember thinking I needed to save and plan for her next birthday.

Three days later, she called social services and asked them to take me back.

And now, it's like I'm naked and shivering, and she's wrapped me in a blanket, a soft, warm blanket, the sort you could sleep in for ever.

'She's great, Cass,' I say. 'She's clever and she's pretty and she's doing really well at school.'

'Oh, Aidan,' says Mum. 'I'm so sorry – I didn't realise that was why you came. Thank you, Aidan. Thank you.'

None of us hear the key in the lock. None of us notice the door opening.

So it's a total shock when a bald, stocky guy – Mum's Keith, same as ever – grabs my shirt, sticks his face in mine and yells, 'What the hell are you doing here? I told you to keep away from my family!'

26: CASS

Mum said she was going to ban me from using Facebook.

'How are you going to police that?' I asked. 'Going to take my laptop, too? Because if so, I might as well give up on A levels right now.'

'I'll change the internet code.'

I snorted. She had to ask me for help whenever there was any problem with the router.

'I'm going to report him – Aidan – to Facebook. Surely there's a law against contacting minors on the internet.'

'I bet there isn't.'

'There should be.'

'Give me your phone, Cass,' said Dad. 'I don't think we can trust you with it.'

I kept a firm grip on it.

'I want to see any messages he's sent you,' said Dad. 'I want you to delete his contact details from your phone and your computer.'

'Who are you – Stalin? I'm nearly seventeen years old. This is completely ridiculous!'

'Give it to me!'

'No!'

Mum wiped away a tear. 'When you're eighteen, we can't stop you finding out about your birth parents, even being in touch. But then it'd be done properly. With support. You'd have counselling, and a social worker to help you through the process.'

'I don't need counselling! I don't need a social worker! I like Aidan, and it's good to be in touch with him, and I don't understand why you kept us apart all these years.'

'At least you haven't met him,' said Dad, grimly.

'I have met him!'

Collective gasp.

'You've met him? How irresponsible! You met someone who contacted you over the internet! Cass! Don't you realise how dangerous that could be? What were you thinking?'

'I took a friend with me so I was perfectly safe. I met him in a public place. We had a coffee together. He's my brother! I can't believe you're both being so horrible!'

Dad thundered as though he were back in the House of Commons. 'You took a friend with you? A friend knows about this? Which friend? The boyfriend?'

'He's not a boyfriend!'

'Grace? Megan? How many people know about this?'

'Oh, that's so typical of you! You only care about what other people think!'

'We care about your welfare!'

'If you really cared about that then you wouldn't be selling my home! You wouldn't have split up!'

'Cass! That's completely unfair!'

'Oh, no, it isn't!' *Bang, slam,* I ran up to my bedroom.

Ben was curled up on my bed, hands over his ears, tears pouring down his face.

'Oh, Ben! Don't cry!'

'It was all my fault! I thought you'd told them!'

'I shouldn't have made you keep a secret. I'm sorry, Ben.' I flung my arms around him, gave him a big hug.

'You're so angry all the time, Cass.'

I was angry. I had every right to be angry. I loved Ben, but right then I wanted to let myself be furious, not worrying about him.

'Ben, go to bed. It's late.'

'Will you stop shouting?'

'In a bit. But it's not your fault.'

Mum and Dad hadn't changed their stance when I went back into the kitchen. Dad was on the phone explaining to Annabel that he wouldn't make it back to London that night; Mum was on the internet, trying to look up how to ban me from Facebook.

Dad had cracked open a bottle of red wine. I hoped he'd get done for drink driving.

'Has Aidan asked you for money, Cass?'

Outrageous.

'No, he hasn't. What do you think of him? He's got a job, actually. He works in a shop.'

'Maybe we should call the adoption society ... the social workers ...' suggested Mum.

'No way on earth am I allowing social workers anywhere near my daughter,' said Dad. Just in case anyone thought that I was his priority, he added, 'Can you imagine if the press got hold of it?'

'Cass ... Aidan wouldn't go to the papers, would he?' Mum's face was pale, her voice shaky.

'To say what?'

'Of course he won't.' Dad's frown was ferocious.

'What's he like now?' asked Mum. 'He was such a beautiful child.'

'Why are you scared of him going to the press? What would he talk to them about?'

'He might exploit you ... his knowledge of you ... your family background. It wasn't very happy,' said Dad.

I gazed at him. He was hiding something.

'What are you scared of?'

'I met your mother,' said Mum. 'Just the two of us. She was very clear that it was the best thing for you to be adopted. It was very selfless of her. She accepted that she couldn't look after you, that she needed to get her life together. She trusted us to give you a good home. And we did ... we did ...'

Why was she crying? Dad put his arm around her. 'Don't blame yourself, Susie.'

'Don't blame yourself for what?' My voice was shrill and scared.

Dad cleared his throat. 'Cass, obviously we've kept from you some of the details of your life before we adopted you. It wouldn't have been appropriate to share them with you; we didn't want to burden you with everything. But perhaps now it's time for you to know the full facts. We were going to wait until you were eighteen, ideally until after your A levels, but it's clearly now more urgent.'

'Why is Mum crying? Why shouldn't she blame herself?'

'Well, of course the adoption process was quite drawn out, and we had to give you and Aidan time to get used to the idea that you wouldn't be living together any more. Time to say goodbye.'

'That's not true, Oliver,' said Mum. 'We owe it to Cass to tell her what happened.'

'What? What happened?'

Mum wiped her eyes, blew her nose.

'We were going to adopt both of you. You both came to live with us. But Aidan's behaviour was so challenging that we couldn't cope. We had to make the decision. It was so difficult—'

People talk about out-of-body experiences: when they feel their soul leave their body and hover over it, watching themselves from above. I swear I had that then. I was watching from a long way away, three people sitting at a kitchen table, talking about something so terrible, so dreadful, so unforgivable ...

'You gave him back?'

27: CASS

It reminded me of hearing the news that Grandad had died (a heart attack, very sudden, Christmas Eve when I was eight). All sad faces and hushed voices. And me, unable to believe what I'd just been told, numb and cold and angry.

'How can you even do that? Hand back a child when you've said you'd adopt him?'

Dad's voice was measured, professional, unemotional. It made me want to scream.'It was quite early on in the process. We realised that Aidan had complex needs that couldn't be met in our family. The social workers agreed that they should find a family of his own.'

'That's ... that's disgusting. You separated us. I was all he had.'

'We tried so hard, darling. But he hated us. He didn't want to be part of our family.'

'He was only six!'

'Let me explain, Susie,' said Dad in his best persuasive

politician's voice. 'Cass, I believe in adoption. I've been pushing for it to be available to more children, at a younger age. To speed up the system, find families for children who desperately need them. You know that. It's one of the ironies of this whole situation that I've had to move off my family brief.'

'This whole situation!' I exclaimed, but Mum said, 'Not now, Cass. We need you to understand what happened.'

'It's probably the most important thing that we as a government can do. Saving children from neglect and abuse. Changing the balance in favour of the children's right to a decent home, not a parent's right to mistreat their children.'

'Dad, I don't need a speech. I want to know about Aidan.'

'He didn't want to be adopted by us, that's what it came down to. He constantly tried to disrupt our attempts to bond with either of you. He'd pinch you if he thought you were having a happy time with us. Kick you under the table. We'd try and give you equal attention, one-to-one time, but he'd make sure he got in the way if Mum tried to read you a story, or have a cuddle. I've never seen a kid so cold, so calculating, so determined to be completely unlovable.'

'Oh, Oliver, that's so harsh,' said Mum. 'We weren't enough for him, Cass. He needed so much love, so much reassurance ... We let him down. We couldn't reach him.'

'You say that, Susie, but some children have just been damaged too much. I honestly believe that there is nothing more we could have done for Aidan.'

I think about Aidan, how he kisses Holly when he thinks no one's looking, how he cuddles Finn.

'He's not like that! He's so happy to see me again!'

'Can't you see, Cass? He's doing it again. He's trying to create a rift between you and us.'

'He must have missed you,' said Mum, but even she didn't sound very sure.

'Do you remember that day out, Susie? That day we realised?'

'We asked the social workers, and they suggested giving him more attention,' said Mum. 'So, one day, we asked Granny and Kate to look after you. You were so open to new people that you'd go to anyone. Actually, we worried about that. We wondered if you could form close relationships, or if everything was superficial.'

'Of course I can!'

'Of course you can. It was just a worry then. Look at you with all your friends. Anyway, we spent the day with Aidan. It was wonderful. He seemed much happier, better behaved. We took him to the playground, read stories to him, just had a normal, nice day. We felt like his real parents, began to see how it could be. And then when I put him to bed he said, "Can I go home now?"'

'He must have missed his mum.'

'He hadn't lived with her for years. She abandoned both of you. First with her mother, and then you went into care. She stayed with a man who beat her up rather than look after her children. It's a hard thing to believe, I know, but she made that choice. She gave you up because she couldn't leave him.'

I felt sick. 'My father?'

Dad shook his head. 'I'm sorry, Cass. We kept the details from you. Should we have told you? He was a very troubled man. He was violent, alcoholic ... she left him eventually, but too late to keep her children. As far as I know, your father's been in and out of prison. We wanted to shelter you … we thought we'd wait until you were an adult …'

'I was four! I must have had memories!'

'You had bad dreams sometimes, but you settled so well with us. Especially when Aidan had gone – you hardly seemed to miss him at all. We thought it was better to dwell on the present and the future as you'd gone through so much pain in the past.'

How could I have forgotten Aidan so easily? There must be something wrong with me. I really was the Ice Queen. Right down to my frozen heart.

'Thank you for telling me,' I said. 'And don't worry about me and my work. It's just been a bit of a blip. I'm right back on track.'

Dad sighs.'Cass, I can't let you go on seeing Aidan like this. We have no idea about his life … his state of mind … anything …'

'I'm nearly seventeen. In just over a year I'll legally be an adult. You won't be able to stop me.'

'Cass—'

'He's my brother. Just as much as Ben is. More, actually.'

Dad's face changed. He wasn't Oliver Montgomery, suave star of the government front bench. He was just a tired, middle-aged man, with lines on his forehead and bags under his eyes.

'Cass, my darling, your birth family, they have a claim to you, but they can't possibly love you more than we do. I sometimes think that being your father is my greatest achievement. We took on a little girl – a bright, beautiful little girl – who was being neglected and exposed to violence and all sorts of other horrors – real horrors, Cass – and we gave her the chance to turn into someone so wonderful, so impressive … Please never, ever feel that Ben isn't your real brother, that we're not your real parents. It's just not true.'

Unworthy, that's how I felt. Fraudulent. And resentful that he presented his love in terms of success and achievement. What if I'd failed at being his daughter? Would I have been sent away like Aidan?

'I actually hate your birth parents,' said Dad. 'It's politically incorrect, I know –nowadays we have to understand and not condemn – but when I think about them, how they failed to care for you properly, put their own selfish needs first, I want to string them up.'

'Oliver, they're people with difficult lives, they have their own stories.' Mum put her hand over mine. I wanted to shrug her off, but I didn't have the energy. 'I met your mother, Cass. She genuinely wanted to do the best for her children.'

I pulled my hand away. 'She trusted you, and you let her down!'

Dad sighed, ran his fingers through his hair.

'It was you or him, Cass. Or let the whole thing break down. What would you have done?'

28: AIDAN

'Aidan? Aidan! Wake up!'

My head hurts. In fact, every bit of my body hurts. My eyes are so crusty that for a moment I think they're full of superglue and they're never going to open. My mouth's dry, my tongue is as furry as Finn's favourite toy rabbit, and my stomach is churning like a washing machine.

I made it home, but wasn't sharp enough to unlock the door: I spent the night propped up against the sign Clive has on the pavement. 'SALVAGE' it says. I'm beyond saving, though. I'm damp, I smell, I disgust myself, so God knows what Holly and Finn think.

'What's happened to Aidy?' says Finn, and I start crying again.

'Let's go inside,' Holly says. 'Come on, Aidan. Let's go inside. Before Uncle Clive gets here.'

We go up to the flat. Finn's still staring at me, like I'm an alien or something. Like he doesn't know me. I have to turn my back on him because it hurts to look at his face.

I start babbling. 'I'm sorry ... really sorry ... I just had a few ... it was late ...' Holly's face is stony, though. There's no point. 'I know, I've totally screwed up. I'll go. I'm sorry.'

'Get yourself showered,' she says, pushing me towards the bathroom, 'and then we'll talk.'

The shower is good. Shaving is better. I wish I could shave away all the layers of things that I want to forget. The haircut is sharp. But the face that looks back at me from the mirror is nothing like the one in the photograph yesterday. There are bags under my eyes, a grey-green tinge to my skin, a split lip, a few bruises. No permanent damage.

When I come out of the bathroom, Finn and Holly aren't there. She must have taken him to the child minder. I'd better start packing. She won't want me to stay. You don't get too many chances once you screw up, and I'm always screwing up with Holly, I'm always getting it wrong, and now ... I'm trying not to panic. I'm trying not to remember all the other places I've moved on from. It's just that this was the best, this was home, this was ... I probably won't have a job any more ...

I sit down on the floor and take deep breaths and try not to panic too much. I'm eighteen. I ought to be able to take knockbacks, I'm certainly used to it. I grind my fists into my mouth and concentrate on a happy memory. The day after Holly and I got together and I went and got the word 'Hope' tattooed on my neck, because it was half of Holly's name all wrapped up in something bigger, something I hadn't had much of. She'd loved that.

But that meant without Holly there was no hope.

I'm so wrapped up in my own misery that I don't hear her come back.

'Hey.' Her hand brushes my cheek. 'Hey, Aidan.'

'Holly, I'm so sorry, Holly, I'm going, I won't stay, I'm sorry ... truly I am ... Can I just stay to say goodbye to Finn? Because people say it's better for the kids if they don't say goodbye, but it isn't, it's worse. Believe me, I know.'

'I'll make you a coffee,' she says. 'Come and talk to me.'

It takes some persuading to get me to sit down and drink her coffee, because I don't even want to look her in the eye. I know I've screwed up once too often and I wish that Finn hadn't seen me like that. Lying outside the shop like an old tramp. But she says I owe her an explanation, and I suppose she's right. So I plunge right in.

'I had to have a drink,' I say. 'Something happened, and I was upset and I needed a drink.'

'Did you get beaten up before or after the drink?'

I shrug. 'I don't know. I'd had a few.'

'Aidan, tell me the whole story. Please. I love you, Aidan. If you keep things from me then I can't help you. I don't want you to leave. I love you.'

Holly's too loving for her own good. If I were her, I'd throw me out. But I'm so lucky to have her. I can't tell her the whole story. I can't lose her.

'I went to see my mum. I wanted to tell her about Cass. It went OK until her boyfriend came in. He doesn't like me, and he went mental. It was kind of upsetting so I needed a drink. I had a few drinks. I tried to get home – I did get home – but I must've been more drunk than I realised. I don't remember much what happened after that. I'm sorry.'

If I had £1 for every time I'd apologised to Holly then I could buy us a weekend at a hotel by the seaside. It's a shame it doesn't work like that.

'He hit you? Your mum's boyfriend?'

'He shouted at me. Threw me out. He hates me. I didn't know he'd be there.'

'You should phone the police!' She lowers her voice, takes my hand. 'Aidan, love, did he beat you up when you were a kid? Was that why you went back into care?'

I hate it when she goes into social-worker mode. Like I can explain everything, just like that. Like it's all down to one bad person or one bad thing, when there's so much to explain that it'd take a book. A Life Book. I should have asked Mum if she still had mine.

'He just doesn't like me. He never beat me up. I just drank too much and ... I dunno. Stuff happened.'

Keith was all right before it happened. Really tried to welcome me into the family. It wasn't his fault that I missed Jim and Betty so much. Not his fault what happened to Louis neither. And I can't really blame him for hating me. If someone did to Finn what I did to Louis ...

'Why doesn't he like you?' she says, and just for a moment I think about spitting it out. Telling her what happened. This could be the moment.

But then I'd lose everything. Every last bit of hope.

'He's just a nutter.'

'Aidan, you need to stop drinking. You don't want to let it become a problem.'

'You drink, too,' I point out.

'I have the occasional glass of wine! I don't go out on benders and stay out all night, or pass out on the pavement!

I don't get into fights that I can't even remember and scare Finn by looking like a corpse on the doorstep.'

'Your life wasn't as shit as mine,' I say.

It sounds pathetic even to me. Holly ignores it.

'They have an alcohol counsellor at the practice. He's meant to be very good. There's a long waiting list but I could push you to the top. There have to be some perks to working there.'

I shrugged. 'OK, if you want.'

I've seen counsellors before, at school mostly. I liked it. I got to miss lessons and I had someone listening to me. They'd ask me about how things were and I'd tell them all the gossip – Denise's pregnancy, Neil up in court, Rich trying to top himself – and they'd tut and sigh and say, 'But what about *you*, Aidan, what's happening with *you*?'

I'd been drinking forever: I learned really young that if you're going to have to do stuff you don't want to do, it's easier to be drunk. It relaxes you. You might even start feeling that you do want to do it.

Of course, sometimes you don't know what's going to happen beforehand and you have to get drunk afterwards to block it out of your memory. That's not so good. And you've got to be old enough to take the alcohol. You've got to be twelve at least.

It's not a good way to grow up. You need someone bigger and stronger than you to keep you safe. I never had that, and I couldn't do it for Cass or Rich, but I want to be there for Finn. So if Holly thinks some counsellor can help, well, I'll try it. I'm doubtful, but I'll do anything she wants.

'I wish you'd trust me, Aidan. I wish you'd tell me all the bad stuff that happened to you in the past. I think it would help.'

'There's nothing to tell,' I say. 'I'm just a screw-up. I don't know why you put up with me.'

'Because I know that's not true,' she says. 'So, you told your mum about Cass?'

'Yes. She was happy. She wants to meet her.'

'Did you suggest meeting here? So I could meet her, too?'

I'm shaking with nerves just thinking about it. I'm going to have to trust Mum, that she'll give me a chance, that she won't tell Holly or Cass what happened, that she'll let it lie, that she'll be so grateful to me that finally, finally, she'll forgive me.

Forgive and forget.

'She's up for it. We just have to set a date.'

29: CASS

On the bus the next morning Will texted me:

- I've got an indoor location for this afternoon. Are you interested and available after school?

- Where?

- I'll meet you round the back of Tesco

- OK

- You're not very enthusiastic. Are you OK?

- I'm fine

- Will you be able to stay for long?

- Yes

- Good

I hardly glanced at him when he got off, him and his noisy group of friends. As the bus drove off, though, I did notice that he was walking with the pretty, blonde girl, stooping over to talk to her. Was his arm around her? The bus was too fast to be sure.

Well, so what if he was seeing someone else? We weren't

going out. We were just experimenting. She'd be good cover. I was sure Grace would tell me all about her sooner or later.

Will was waiting for me after school, leaning up against the back wall of the supermarket, reading a paperback. He was wearing an oversized red jumper and skinny jeans, with red Converses. I stood for a moment, just enjoying the view. I didn't know what was going on between Will and me, and I didn't really care. I just liked looking at him, anticipating the moment when we'd carry on with our experiments. And it gave me a buzz, thinking of all the people on the other side of the wall, doing their shopping, while we were meeting in secret just yards away. People we knew. People who'd be interested.

He looked up from his book. 'Hey,' he said. 'How long have you been standing there?'

'I just got here. What are you reading?'

He showed me. *Anna Karenina.*

'You're just a walking cliché,' I told him.

'You should read it.' He flipped to the first page. '"Happy families are all alike; every unhappy family is unhappy in its own way."'

'Ha ha, very funny.'

'I didn't mean it like that. You know I wouldn't make fun of you.'

'Good,' I said.

'You look quite different,' he said. 'I don't know what to make of it.'

Mum hadn't seen me that morning, which was why I'd got away with wearing a super-short skirt that had been lurking at the back of my wardrobe since I was twelve. Then, it had come down past my knee and was actually too big. Now, I could still wriggle it over my hips, and it was just about long enough to be decent. I'd worn it all prepared for another screaming row – they seemed to think I was a tart, I might as well dress like one – but Dad's car had gone in the morning and Mum had left a note on the table saying she'd gone to early morning Pilates.

I was beginning to realise that they were both quite selfish.

I'd got quite a few comments at school, as well. Miss Graham had told me that my skirt was too short and too tight and she didn't want to see me wearing it to school again. 'I'm disappointed in you, Cass.'

Grace told me she loved the tight black cardigan, and it'd be even better with a few buttons undone. 'You're sure to get his attention like that,' she said. 'Your make-up's gorgeous, too. Are you coming to Starbucks after school?'

'No,' I told her. 'I've got some important research to do.'

'I thought I'd change my style a bit,' I told Will, now. 'You told me I was boring.'

'I was talking about your colour palette,' he said, 'but hey, I'm not complaining.'

'Where are we going then?'

'My sister's flat. She's away on holiday and I'm feeding the cats and watering the plants.'

A whole flat to ourselves, with no chance of anyone seeing us. This was taking experimentation to new and dangerous levels.

Excellent.

My phone pinged with a new text. Dad had agreed in the

end that he couldn't confiscate it, that it would cause Mum too much hassle, worrying about where I was, plus I could easily buy myself a cheap pay-as-you go. A small victory. But they were still insisting that they had to meet Aidan themselves, and know about any plans to meet my birth parents.

Why couldn't they mind their own business?

The text was from Mum, wondering why I wasn't home yet.

After-school meeting for prefects. I texted.

When will you be back? I need to talk to you.

Not until late. Duke of Edinburgh volunteering.

Can't you drop it for once?

Just thinking about my personal statement. Don't you want me to go to Oxford?

'Finished?'

'Oh, sorry, Will. Just getting rid of my mum. I bet your parents don't bug you like mine do.'

'Oldest child syndrome. They used to bug my eldest sister all the time. Layla fought all the battles. Then Cleo and Hattie fought them all over again. By the time I got to puberty they were exhausted.'

'I wish someone had done that for me,' I said. Of course, that's what Aidan should have done. He could have been my older brother all my life. Everything would have been different.

'Cass! What did I say?' Will's arms were around me. I gulped, turned a sob into a laugh.

'Nothing – what are you doing? I'm fine.'

'You looked like someone had died – I'm sorry. I got it wrong. Are you coming to visit the cats, then?'

'Of course. It's a difficult job, feeding cats. You need some help.'

Will's sister's flat was in a small modern block, just off the high street. We walked quickly, avoiding the main road, hoping no one would spot us. The entrance hall smelled of furniture polish and bleach.

We climbed the stairs to the second floor. 'Here we go,' said Will. He opened the door. 'Alone at last. Apart from Lily and Barney, that is. Cleo got them from the cat shelter. Do you like cats?'

'I don't mind cats.'

'You're like a cat. Everything's on your terms. I'm much more of a dog.'

'You said it.'

Barney was a sleepy grey tabby, curled up on a cushion; Lily was a black and white bundle of nerves, who took a swipe at Will when he tried to stroke her. He fed them and refilled their water bowl while I prowled around the flat.

It was the opposite of Dad's bland pied-à-terre in London. It was full of colour: a bright pink shawl slung over a purple sofa; a mirror framed with tiny turquoise and gold mosaic tiles; a picture of skeletons dancing in a graveyard, grinning happily under Mexican sombreros. Photos everywhere, a whole wall of them, pinned onto cork tiles, some old family pictures in frames on the wall. One man – serious in a single-breasted suit – looked a lot like a retro-version of Will, except he was scowling into the camera. And I found lots of pictures of Will himself, always smiling, always happy, always surrounded by admiring women, it seemed. I assumed they were sisters, aunties, his mum, but one was definitely Ruby and that blonde girl was there too.

Will's sister Cleo liked candles – they were everywhere, in

various stages of meltdown. She had a jelly bean dispenser on her coffee table and a packet of Chi tea (whatever that was) on the kitchen top. She'd stuck postcards onto her fridge from Mumbai, Scunthorpe, Kingston, Jamaica.

She'd left Will a note, scrawled in red crayon on the bottom of a credit card bill:

Hey bro!

Thanks for feeding my babies! Lily will be feeling a bit needy so make sure you give her a big cuddle and make a fuss of her.

Help yourself to anything in the fridge and there's ice cream in the freezer if you want it.

No parties! Good luck with you-know-what!

Love you Cleo xxxxxxx

Will was looking over my shoulder. 'Ah, ice cream. Cleo's good with the essentials. Do you want some?'

I only wanted him. 'No thanks,' I said.

He leant in towards me. 'It's different being here on our own, isn't it? Do you like it?'

'I don't know yet.'

But I did know. I just didn't want to tell him how much I liked it there, how different it felt to home, which wasn't really home any more; how the bright colours made me think of Aidan and Holly, young and happy and free and in love.

Just for a minute, I imagined being Will's real girlfriend. Meeting his sisters. Hanging out with his friends. Laughing and kissing and not worrying about gossips and standards and grades.

I wanted it so much that I couldn't risk messing it up.

He kissed me, light, teasing, promising. 'What do you think now?'

217

'Not bad. But I don't think a kitchen is really ideal.'

An hour later, we'd experimented in the kitchen, hallway and the bathroom, which was painted bright pink. I found I liked watching myself kissing Will in the mirrors that covered most of the wall space in it – a patchwork of mirrors – some big, some small, all of them showing us ourselves, entwined, my hair bright against his dark skin.

There was a very comfortable sofa in the living room, and we'd tried sitting on it and now we were lying down. I'd peeled off my cardigan, and he'd kissed all along the top of my camisole. He'd discarded the big jumper too, and our bare arms were skin to skin, sending little shivers of excitement all through me.

I pulled Will down on top of me. Curled my legs around him. Pushed my hand under his T-shirt, so I could feel more of his skin.

'Mmm,' he said. 'Wow. Mmmm.'

My hand went lower.

'Oh my God,' he groaned. We kissed again, harder this time, exploring each other's bodies.

I wondered how many other girls he'd brought back here.

'What's the matter?'

'Nothing.' Suddenly I felt foolish and awkward. Had he realised how much I was hiding from him? How hopeless I'd be at a real, close relationship? Maybe we were just doing this so he could laugh about me with his friends. I bet Ruby hadn't been uptight or weird.

'Something's the matter. You went all tense.' He kissed me again, just where the camisole ended and my bra began. I felt my shoulders hunch. 'What is it? You liked that before.'

'I know. Sorry. I just thought of something. It doesn't matter.'

'Yes, it does. Tell me.'

I wriggled around, so I was lying on top of him. On top of all of him. I kissed him, full on the lips, nudging his hand with the top of my thigh.

He got the message, all right. His hands were inside my camisole. His flies were undone. I could do this. This was fine.

We were ... we could ... we weren't.

'I'm sorry.'

I was mortified. I didn't know why he'd suddenly pulled away from me (was I meant to know? Had I done something wrong?) but I felt stupid and ashamed and angry. I pulled down my camisole, re-adjusted my skirt, moved to the end of the sofa where we weren't touching any more.

'That's OK,' I said, buttoning my cardigan with furious fingers. Barney and Lily were staring at me. I narrowed my eyes at them and they ran away.

'I'm ... it's just ...' For once I couldn't read Will's face. Was he embarrassed, or repulsed, or had he suddenly realised that I liked him much more than he liked me? 'I wasn't really expecting things to go so fast, and I thought we should, you know, at least talk first.'

'Oh, is that what you usually do?'

'You what?'

'When you bring girls here. I'm sorry. I didn't mean to rush you or anything.' I was back behind the comfortable veil of sarcasm.

'No, no, you don't understand—'

'Oh, well, I'm sure Ruby and Daisy understood perfectly.'

He looked at me, head on one side, an annoying smirk on his lips. 'Cass, are you jealous of my ex-girlfriends?'

'Of course not. Why would I be jealous of them? Daisy, she's doing a BTEC in beauty therapy, isn't she? And Ruby's a star at media studies, I hear.'

Will's mouth was a tight line. I'd defeated the smirk. 'Well, of course, we can't all be straight A students, can we?'

'Look, if you're intimidated, then back off.' I remembered that's just what he'd done. 'I mean, good, let's not do this any more. It was a stupid idea.'

'I thought you liked it.'

'You thought wrong.'

'I just thought we should talk about it. Before we ... you know ... stop pretending it was just an experiment.'

'It was just an experiment.'

'Well, I don't want to experiment any more. It's doing my head in. Look, I really like you, and I really fancy you. Can't we go out, you know, properly? I can be your boyfriend and buy you little presents and meet your friends and stuff like that.'

'And then dump me, like Daisy and Ruby and all the rest? I don't think so.'

'Cass ... you're getting it all mixed up ...'

'I'm not mixed up. It's you who's changing the rules.'

'Don't be angry! Why are you angry? I just want to stop sneaking around! I want everyone to know about us! We can drop all this secrecy, this experiment stuff, and be a couple. We'll be great together.'

'I'm not angry,' I lied.

'Why are you being like this, then?'

'I thought you liked what we were doing.'

'I do like it! It's because I like it that I want more – a proper relationship. I hate all the sneaking around, not telling people what's going on.'

'Who defines a proper relationship? The kids in Starbucks? It's none of their business. I don't want people knowing everything about me.'

He was silent. 'You don't want to be seen with me?' He'd lost his usual bounce. His voice was softer, more hesitant. 'You don't think I'm good enough for you?'

'No, don't be so stupid,' I said, but he'd got up, gone over to the window.

'What is it, Cass? I'm not rich enough for you? Or maybe it's because I'm not at the grammar school? It couldn't be the colour of my skin, could it? I'm sure you're much too politically correct for that.'

'It's nothing like that,' I said.

'I saw your face when we bumped into your aunt. You were horrified.'

I didn't have the time or energy to work out what to say. All I knew was that I'd been right. I couldn't get close to anyone. I was abnormal. No wonder my birth parents had given me away.

'I'm going,' I said. 'I can't deal with this now.'

'Cass, stay, please. We can't leave it like this.'

'Oh, can't we?' I said. 'Watch me.'

30: AIDAN

I've got two customers in the shop and I think they're thieves. They're doing the things that shoplifters do: picking up stuff and putting it down; talking in whispers and then breaking apart; one wandering around at the front of the shop, the other at the back.

One's a pretty girl, yellow curls tied back off her face with a scarf. The other is a black guy, dressed in grey trackies and a black hoody. I keep my eye on him when she approaches me to ask about some electronic keyboards we've got in. I know the way it works. One distracts, the other nicks.

Rich and I used to do it all the time.

I answer her questions as quickly as possible. Yes, we can deliver. Yes, there's a guarantee. Yes, even though it's second-hand.

'I'll think about it,' she says. Surprise, surprise.

The bell jangles. Perfect. Fantastic. Just what I need: it's Neil, glaring around him as though it's my fault that I've actually got customers.

He leans into my shoulder. 'Get rid,' he mutters into my ear.

I stand up fast. Just his hot breath on my cheek makes me feel dirty.

'We're closing up now!' I announce, but I can't see them. I just get a glance of her spotty scarf – a red flash – and the door slams. They've abandoned me, without even knowing it.

'Lock the door,' says Neil.

'My boss, Clive, might come back,' I tell him, but he shakes his head.

'Went off an hour ago, didn't he? With a nice empty van. Is he teaching you to drive that van, Aidan? Or is the written test too much for you?'

Bastard. I learned to drive ages ago, Clive paid for lessons, let me practise on the van. I could pass my test just like that if it wasn't for the written bit. I've tried and tried to learn it, but whenever I sit the test, the words start whirling around and I have to stop. It makes me feel sick.

'You got anyone watching you?'

I shake my head. There is CCTV, but it's in the yard. Metal gets a good price, especially copper, so we have to guard it pretty carefully. The walls have spikes and barbed wire. It's not helped keep Neil out though.

I'm breathing hard and my heart is thumping. I'm kind of used to it. I've been scared for so much of my life that the moments of calm are the unusual ones.

'I'm trusting you. You wouldn't lie to me, would you, Aidan?'

I shake my head.

'Remember Bobby? You wouldn't want your little boy looking like him.'

Bobby was a kid at the home, a bit younger than me. Neil and his mates decided to carve the letter 'B' into his forehead. I don't know why, and I don't know what happened to Bobby, and I don't know how they got away with it.

I just know that no one's ever going to do that to Finn.

'You're not going to hurt him,' I tell Neil. I look him in the eye. 'You listen to me – nothing happens to him, OK?'

'Not if you do what you're told.'

Neil locates a dark corner at the back, jerks his head. 'Floorboards,' he says.

I fetch a hammer, prise up the nails, lift the boards up. Neil opens his coat, pulls out a package. I know what it is. A firearm. A weapon. A gun.

'You do it,' says Neil. 'Put it in.'

I don't want to touch it. But I do what I'm told. Now my prints are all over the packaging.

At least he hasn't picked up on Cass. At least he's forgotten about her.

It's as though he can read my mind.

'Striking, that girl you had here the other day. Very striking. All that red hair. Natural, is it?'

'Yes,' I say, and he says, 'I suppose you've checked,' and gives me a nudge and a dirty laugh.

I want to kill him.

'Your girlfriend know you're running around with another girl? Looked like jailbait, too. You're a naughty boy, Aidan.'

'Can we just get on with it?'

'She's a bit posh for you.'

'Can we hurry up? Before Clive gets back?'

'Oh, don't worry about Clive. He's stuck in Colchester.'

'How do you know?'

'He thinks he's gone to pick up a load of copper piping, but when he gets there, he'll find nothing. Serve him right.'

'For what?'

'Oh, nothing. Don't bother your pretty head about it.'

My hand curls into a fist, and then uncurls again. We replace the floorboards, bang in the nails, cover them up with a rug that Clive swears is Indian and worth a fortune if he could only find a buyer.

'That's it,' I say, sucking my finger. I've got a splinter. 'They won't be here long, will they, Neil? I'm not sure how long I'll be in this job.'

Clive's a great boss. Really understanding. But, just recently, he's been talking about needing someone who can take on more of the driving. 'I'm sorry, lad,' he said the other day. 'If you can't get through the test, then I'll have to consider the situation. I can't carry passengers. And, while I'm at it, you can stop helping yourself to the stock.'

'It was only a rabbit!' I said. 'For Finn. Its ears were singed, I swear.'

'Ask first,' he said. 'And get studying for that test. I can get a girl in to mind the shop – it's your muscle I need. I'm not a young man any more.'

Neil's voice brings me back. 'You going anywhere, Aidan? Not planning anything stupid, are we?'

'Nah. Just this job ... it's so boring ...'

'Stick with it,' he says. 'I need you here. For a few months, anyway. Agreed?'

'Yeah, well, yeah.'

There's no point in arguing with Neil.

He leaves and I know that I ought to open up the shop again, but I can't. I sink down on an armchair (salvaged from a flooded warehouse, slight water damage) and try and think my way out of this one. I could call the police. I could take the gun out, try and get rid of it. I could persuade Holly to leave London – take her and Finn off to Spain where her dad lives. Maybe he'd help us find jobs, somewhere to live. Maybe Keith would give me money to piss off out of London, out of Mum's life, out of Louis and Scarlett's life for ever.

No. I'm not asking him.

Maybe Cass could lend me money. She must have loads of money.

What am I thinking? I can't ask Cass, and I can't run away from Cass – I've only just found her.

I'm trapped. I'm caught.

I really, really need a drink. If I had any money, I would go to the pub.

I realise that it's kind of ironic that I'm thinking like this, as this morning was my first counselling session.

The counsellor was useless. He talked about his method, which is based on the twelve-step approach used by Alcoholics Anonymous. He said it was proven to work, that millions of people had been helped by it.

'You have to admit that you have a problem,' he said. 'You have to stop blaming other people for it. You have to accept that there is a higher power.'

'I'm not religious,' I say, and he says, 'You don't have to be. Just stop thinking that you are all alone and no one is looking out for you.'

'Is that it?'

'Eventually, you'll need to help other people tackle their addictions, just as I'm going to help you. But first you need to apologise to those people you've hurt.'

I got angry right away.

'What about all the people who hurt me? What about them? I haven't had any apologies from them!'

'This is about you and your actions, Aidan. Not other people's actions, hard as it is to accept.'

'I never hurt anyone!' I said, and he looked at me, but I couldn't meet his eye, so I sat and stared at my trainers, which I'd taken from Clive's shop because he had a whole container full of them – five-hundred shoes – and my old trainers had holes in them.

'Let's work on that next time,' he'd said in the end, after ten minutes of silence.

I really, really want to go to the pub. Clive wouldn't notice, would he, if I just borrowed some money from the till? Just twenty quid. He won't notice.

It's in my pocket before I have time to think.

31: AIDAN

Holly and Finn have made cornflake crispies, lemon drizzle cake, chocolate brownies. 'I think I'm going to do a batch of fairy cakes,' she says, 'just in case they don't like chocolate.'

'But then they've got the lemon cake.'

'I know, but Finn wants to decorate them.'

'Spinders!' says Finn, which I happen to know means those multi-coloured things you put on cakes. Sprinkles, Holly calls them.

I ruffle his hair and say, 'Spinders and cherries, if we've got any left.'

'Aidan, what if she brings her kids? So they can meet their sister.'

'She won't,' I say, and go off to tidy the lounge.

It's a tip. I put Finn's toys back in the toy box, sweep up all the crumbs and hairs and dust from the floor. I chuck out some flowers that have been mouldering in a vase for about a week. I open a window to make it smell fresher. I spray furniture polish on the table and make it shine with a duster.

'You're acting like my gran!' says Holly, carrying in the lemon cake. 'Cleaning every surface.'

'Hang on a minute!' I borrowed a tablecloth from the shop earlier, and it's in my backpack. It's white and lacy, just like one that Betty had. I spread it on the table. 'There!'

'It looks great, Aidan.'

'Shall we sit at the table to eat, or take plates and sit around the room?'

'Shall we just see what people want to do?'

'I want it to be ready, though.'

Betty used to make afternoon tea when her mum and sister came round. She'd set the table, and make sandwiches with no crusts. She cut them in little triangles. I used to help her. The teapot had flowers on it, and a bright pink, knitted tea cosy. She made scones with jam and cream, and we'd all have our own serviette.

'Have we got serviettes? We need them! Shall I go out and buy some?'

'You'll want doilies for the plates next!' Holly's laughing at me.

'No, but really Holly, I think we need serviettes.'

'I'll go and get some,' she says. 'Are you sure your mum won't bring anyone?'

'She won't. Holly, my mum's a bit odd. Don't pay too much attention to what she says.'

'How do you mean?'

'She might say weird stuff. Don't worry about it.'

'I'll go,' she says. 'They'll have serviettes at the corner shop.

You keep an eye on Finn, eh? I left him putting cherries on the fairy cakes.'

I open the door to the kitchen.

'Aidy! Look! Cakes!'

Finn has studded every cake with cherries, crusted them with sprinkles, piled chocolate chips and sugar roses on top.

'More spinders!'

'You've got loads of cherries and sprinkles. Let's put them on a plate.'

He watches me suspiciously as I pick up each cake and lay them out on a plate. The cherries look like massive tumours or boils, scarring the icing.

'Pretty cakes!' says Finn.

'Let's clean you up,' I tell him, getting a wet cloth to wipe his hands and face. But Finn scrambles down from the chair and runs away, laughing like a maniac, wiping his sticky hands on the sofa, clutching at the tablecloth ...

The lemon cake lands *splat* on the floor. A plate follows it, but I dive to catch it mid-air and end up on the floor, clutching the plate to my chest. Finn cackles with delight and gives the cloth another tug.

'Finn! For Christ's sake!' I bellow, as teacups tumble around my ears.

Finn watches open-mouthed as three cups smash onto the fireplace, and he bursts into tears. Good.

'Finn! Look what you've done! Naughty!'

I give him a quick slap on the bum. He needs to know that he can't act like that.

He howls. 'Aidy! I hate you!'

'I hate you too!'

The room is a complete mess. Any minute now Mum could arrive, Cass ...

I'm sweating, my breath is coming in little pants. I wanted everything to be perfect and now it's all ruined.

Finn runs out of the room, wailing for Holly. I gather up the smashed china, try and save the lemon cake. But the sticky drizzle hit the rug. The cake is covered with turquoise fuzz, and the rug glitters with sugar. The chocolate crispies have disintegrated. There's a huge chocolate smear on the tablecloth.

The doorbell rings.

Oh Jesus. I don't know what to do. I try and sweep everything up into one big pile, but I accidentally kneel in a chocolate crispie and it looks like I've got dog shit all over my clean jeans.

The bell rings again.

I give up. I lean back against the sofa, stare at the ceiling, put my hands over my ears and pray for Finn to shut up and whoever is at the door to just go away.

Finn screams louder. The bell goes again.

Finn's gone quiet. The bell seems to have stopped. Maybe they've gone away. They've realised that this was a terrible, terrible idea, and they've gone away and left me. Holly and Finn and Cass and Mum.

They've left me alone, like everyone does in the end.

'Oh my God! What happened!' It's Holly's voice.

'Aidy hit me!'

'Finn! Did you make this mess? Aidan? What happened?'

I don't move. 'I didn't hit him,' I say.

'You did! On my bum!'

'Just a smack. He made all this mess on purpose, Holly. It's ruined. You'll have to tell them ... you'll have to ...'

Finn's staring at me. He opens his mouth and starts yelling again. 'My cakes! My spinders!'

'I'm going to put you down for a nap,' says Holly. 'I'll ... I'll leave you to it.'

'Oh, Aidan.'

That's not Holly's voice. Slowly I look towards the door. Holly and Finn have gone. But I'm not on my own.

'What a mess,' says my mum.

She doesn't say it in a nasty way. She sounds sad for me. She can see that I wanted everything to be nice for her, for Cass. I tried. I really tried.

'I wanted it all nice,' I told her. 'Finn messed it up.'

She comes and crouches down next to me, picks up bits of tea cup. 'It's only china, darling. Nothing that can't be replaced. No bones broken.'

I scrape up a mass of cherries and icing.

'Aidan, is he yours? The little boy? Why didn't you tell me?'

I shake my head.

'You've taken him on, though?'

'I've been around since he was a baby.'

Silence. She's thinking.

'You've taken on a lot, sweetheart. Are you sure you're ready?'

'I want to.' It comes out more fiercely than I expected.

'And everything's gone well?'

She's trying, I can tell, really trying not to upset me, to remind me of why things might go badly. She's on her best behaviour. I wish she could always be like this.

'Yes. It has gone well. Ask Holly.' I want to get down on my knees and beg her not to spoil it, not to say anything that would take Finn away from me. 'Mum, this really means a lot – if you wouldn't say anything to Holly, about - you know – '

'It's a big thing that you're asking me.' She pauses, looks round the room. 'You have made it nice here. I'm happy for you. I'm not going to spoil it for you, Aidan. Just make sure that little boy is safe and sound, you hear me?'

I'm so grateful. 'Yes, Mum.'

She pats my shoulder. 'Tell you what, I'll go out and have a fag and you can clear this up and forget it happened.'

I close my eyes. 'Thanks, Mum.'

She's gone.

32: CASS

I was a bit late getting to Aidan and Holly's flat. A whole hour late, in fact. It was incredibly difficult finding excuses to go out: Mum quizzed me the whole time about what I was doing, where I was going, and Dad phoned and texted the whole time about my homework.

In the end, I lied. I told her I was going out with Will, it was none of her business and I'd be back when I felt like it. We had a short, sharp little fight, and I succeeded in distracting her from the truth. But all the way to London on the train I felt like a bully and a traitor.

All Mum ever did was love her family, and one by one we had all betrayed her.

Maybe it was what she deserved. After all, look what she did to Aidan. Told him that he had new parents and then handed him back. I felt physically sick whenever I thought about it. I didn't want to hear their explanations. How could anyone reject a six-year-old boy because he didn't come up to scratch?

And that led me on to thinking about the mother I was about to meet. Why did she hand over her children for adoption? Or were we taken away from her?

I was so wound up when I arrived at Aidan's flat that I felt too nervous to go in. Instead, I stared into the window of the salvage shop. The display was children's toys: dolls lined up in a row, some furry rabbits. A dolls' house. I wondered where the stock came from, why some things ended up here. It all looked fine, good as new. Was there damage that couldn't be seen? Or was it all just surplus to requirements?

'Cass?'

I jumped, and looked round to meet Holly's eyes.

'I'm sorry – I just got here. I was about to ring the bell.'

'I saw you out of the window,' she said. 'You must be on edge – Aidan's been a bag of nerves all morning and I'm not much better. But she seems really nice, your mum. She's very good with Finn.'

I followed Holly up the stairs, thinking again how unlikely a partner she was for Aidan. He's so tall and good looking, laidback and ... well ... young. There was something a bit mumsy about Holly. It might have been her slightly dumpy figure or her minimal make-up, or the way her thin, flat hair was scraped up in a wispy bun. But I thought it was more about her air of anxious concern, as though she was constantly waiting for you to crumble onto her shoulder.

'*Judgemental, much, Miss Montgomery?*' said Will's voice in my head, and I felt intensely irritated. When would I stop thinking about him? He was making it easy enough – no calls, no texts – at least, there hadn't been any the last time I checked my phone, which was when I came out of Camden Town underground station.

'Here we are,' said Holly, and opened the door.

Little Finn, building his train track again. Aidan on the floor, helping him. A plate of biscuits on the table – they hadn't gone to much trouble, which was good, because I'd been dreading some sort of formal, sit-down sandwiches and cake affair.

And her. Sitting on the sofa, drinking tea from a mug, which she put down as soon as she saw me, put down on the floor, so she could stand up and walk towards me.

'I'm Janette, darling. Your mum. I couldn't mistake you, you've hardly changed at all. You're the spitting image of my Scarlett, too.'

She was shorter than me, but she wore heels with her black skirt and white blouse. Her hair was a soft honey-blonde, brittle with spray. If she really was my mum, then a lot of work had gone into eliminating the ginger. She had on a lot of eye make-up – smoky, shimmering grey – and candy pink lipstick.

My hand was over my mouth. I made a little sound that wasn't quite formed enough to be a word.

'Here, have a look.'

She passed me her iPhone. Two kids smiled out from the screen. One did look like a younger version of me – not just the marmalade hair, but something about the eyes and nose as well. The other, the boy, had straight brown hair and a round face, and was squinting into the sunshine.

'They're lovely,' I said. My voice was shy, nervous. I didn't sound like myself. My voice was shy – less than, short of, its normal sound.

'They're my second chance,' she said, taking back the

phone. 'What can I say to you? I'm so sorry, darling. I thought I was doing the best thing at the time.'

'I understand,' I said, because I wanted it to be true.

'Come and sit down,' said Holly. 'Cass, would you like tea? A top up, Janette?'

'There are biscuits, too,' said Aidan. 'And there would have been cake ...'

I waited for the '... but the shop didn't have any' or whatever, but he left it at that and went back to Finn's train set. No hug for me. Barely a smile. Had this all been a colossal mistake?

'Here we are then,' said Janette. I couldn't think of her as 'Mum'. 'Here we are. After all these years.' A tear rolled elegantly down her cheek. 'What must you think of me? Abandoning you to strangers.'

'They're not strangers ... I mean, I've had a good life. They've always loved me and everything. It's just now ... things have got difficult. You know. But it was never like that before.'

'What's difficult, darling?'

Aidan must have told her about my dad – or hadn't he?

'My dad ... he's left ... I thought you knew.'

'Aidan didn't say.'

I was absolutely certain she didn't mean to criticise Aidan, but I saw his jaw clench and his shoulders hunch just a little.

'Cass's dad is in the papers,' he told her. 'The politician. You know the one.'

Janette gasped. She was a little odd, I decided. Everything about her felt like she was acting a part.

'Him? The one who's run off with a girl half his age?'

I dug my fingernails into my palm. 'Actually, she is twenty-three and he is fifty so, technically, he is more than twice her age.'

'How could he? The bastard.'

Bizarrely, I felt the need to defend him. 'He fell in love, I suppose. Mum said they hadn't really been happy for ages. And he hasn't run away from us ... from my brother, Ben and me. He's still a totally hands-on parent.'

My phone pinged with a text. My heart leapt for a second – *Will*? – but, no, it was Dad. Right on cue. Asking about my economics worksheet.

Holly handed out mugs of tea. Aidan glowered at his – he must be fussy about how he likes tea. His expression, all pout and crinkled forehead, sparked a distant memory. Aidan with just that frown, kicking a table. Kicking our table. In our kitchen. Scaring me, upsetting me, making me wish he would go away.

Did I ever tell him to go away? Did I reject him along with Mum and Dad?

Janette asked me lots of questions. My school, my exam results, my plans for university. Did I remember before I was adopted? She seemed sad, yet relieved when I said no, I couldn't remember anything at all.

'They were bad times.' She looked around at the flat. 'This is lovely. Aidan, you've fallen on your feet. Holly, I hope he behaves himself, I hope he's a good boy ... a good partner ...'

Holly blinked. 'He's wonderful. He's great.' Her hand rested on Aidan's shoulder. 'Finn and I, we're so lucky to have him.'

Aidan ducked his head down.

'I think he's the lucky one, aren't you, darling?' Janette

seemed like she was about to say something else, but no words came. We sat waiting for a long minute, until Holly passed around the biscuits.

'So your mum, Cass, she's having a hard time then?' Janette's voice was straining to sound sympathetic, but she didn't quite get there.

'She's coping. We're fine.'

'I met her, you know. When they first made the placement.' Janette's voice was husky, as though she was trying so hard to sound posh, say the right thing, that it was giving her a sore throat. 'Right in the middle of my troubles. They said it would help if I met her, it would put my mind at rest. She was lovely, you know. Well, of course, you know. And she did put my mind at rest. Told me how much she loved you already. Told me how clever you were – well, I knew that, didn't I? How you liked living out in the country. Has she been a good mum to you, darling?'

'Yes,' I told her.

'Because she was lying, that was the thing. I believed her and it helped me. But she told me the same about Aidan. That she'd love him too—'

'Shut up.'

Aidan's voice was so angry, so abrupt, that Holly spilt a little bit of tea on the sofa, and Finn asked, 'Why you cross, Aidy?'

'What's the matter, Aidan? Cass must know. She must remember. How did they explain it to you, darling? When they sent your brother back into care?'

I opened my mouth, but Aidan got there first.

'Shut up, Mum. No one needs ... this isn't important ... it was a long time ago ...'

239

'I didn't know,' I told him. 'I truly didn't know. I didn't remember and no one told me, no one explained. Only last week. They told me last week.'

Holly's mouth is round with surprise. 'Why didn't you ever tell me, Aidan?'

He tried to make a joke of it. 'If I told you about everyone who didn't want me, it'd probably have put you off.' No one laughed.

Holly turned to Janette. 'He came back to you?'

'Not then,' she said. 'Later he did. After I fought for him. They gave me a lot of guff about how they'd find him someone else, how he needed to be the only child, centre of attention. "Somewhere more suited to his cultural background," they said. What a load of rubbish. He never even knew his dad.'

Her eyes went from Finn to Holly and back again. 'You got left in the lurch too, darling?'

'Finn's dad ... he wasn't ready for the responsibility.'

Aidan exhaled, somewhere between a hiss and a sigh.

'Aidan, please ...' Holly's face was worried.

'Why don't you tell them the truth? Finn's dad was never around. When I met you, you were with that toe-rag Don. I saved you from him. I told him to go away and he never came back. He hit her, but I saved her.' His voice was proud.

'You and Uncle Clive's solicitor and the restraining order,' said Holly. 'Not that I think it's anyone's business.'

Aidan turned away. I felt terrible for him. All he wanted was for Holly to acknowledge what he'd done for her, say thank you for protecting her. She'd managed to make him sound in the wrong.

Janette's eyes were sad under all that make-up, and her mouth might have been sad too, if it hadn't been drinking her tea.

'Let's face it,' she said, putting her mug back down on the floor. 'A good man is hard to find. I'm lucky, I've found one now. But I chose some monsters in the past.'

'I don't actually know anything about my dad,' I said, cautiously. 'Just that he couldn't cope and left.'

'He couldn't cope. Well. That's one way of putting it. What a bastard he turned out to be.'

I winced, but she didn't notice.

'Took it out on me and Aidan, whenever anything went wrong. Jesus, he hated Aidan. Aidan tried to stand up for me, you see, get in the way when Mac was angry.' A tear escaped her eye, and she brushed it away with a brisk flick of her hand. 'First I sent him to my mum. Then we all went. But Mac followed us, wouldn't leave us in peace. So we went to the refuge.'

'Were you ... we ... were we safe there?'

'Sweetheart, you're never safe anywhere. And I was my own worst enemy.'

'It must have been so hard for you.'

'He loved you.' I couldn't work out if Aidan was angry with me or not. His voice was harsh and his face unreadable. 'He loved you, Cass. He did really bad things, but he never hurt you. That was the one good thing about him.'

Janette ignored him. 'I was in love, understand? I was dazzled. I thought it was my fault when he got into a temper. I thought if only I could get it right, he was the only one who could bring me happiness. I lost sight of what was really happening. I'm so sorry, Cass, so sorry.'

'You don't need to say sorry to me, really.' My cheeks were hot with embarrassment. She was apologising to me because I'd had a great life, full of love and privilege and holidays and presents.

Aidan stood up. 'How about me, eh? Where's my apology? He never touched Cass.'

Finn was watching and listening to all this and although he couldn't possibly understand, his eyes were wide and his lip was trembling.

'Finn, do you want to show me your room?' I asked, and he nodded and took my hand and we went and looked at his bed, his toys and his books. Holly is a great mum, I thought. It couldn't be easy for her, but she'd given her kid everything he needs, and even a really caring substitute dad.

Finn and I could hear raised voices from the living room, but we couldn't hear what they were saying, until the door opened and the voices were right by us.

Aidan was nearly shouting, 'I told you this was a bad idea. I told you it was a mistake. I can't take it any more.'

And Holly, 'Just stop and listen, calm down. Let's talk.'

Then the front door, *bang, slam*. Twice.

33: CASS

Finn wasn't happy, and I didn't know how long Aidan and Holly were going to be, so I took him back into the living room, where Janette was sitting, gazing into space. 'Why don't you pick out a DVD, sweetheart?' she told him. We all three sat and watched penguins shuffling around in the snow.

'The male Emperor penguin incubates the egg in the coldest winter weather, while the mother is searching for food.' David Attenborough's voice boomed out.

'I'm dying for a cigarette,' said Janette. 'Do you think they'd mind? Tell you what, pass me that saucer, I'll use it as an ashtray.'

There's a broken saucer just poking out from under the sofa. I was surprised to see it there – Finn could have hurt himself on the sharp edges.

She offered me her cigarette packet.

'No, thank you,' I said.

'Don't smoke, eh? Clever girl.'

She lit her cigarette, inhaled deeply, let the breath go in a cloud of smoke. Strangely, it reminded me of Mum doing yoga: her shoulders relaxed, her forehead unpinched itself. 'I've thought about you, worried about you, wondered about you so much over the years. It's marvellous to see you so grown up and beautiful. My little girl.'

I cringed inwardly, but hoped it didn't show. I reminded myself that she knew me much better than I knew her; she knew about a part of my life that I couldn't remember at all.

'My dad – Mac? Can you tell me anything about him? Now Aidan's not here.'

She took another breath in, smoke out. I tried not to cough.

'He was a bad boy. It took me a long time to realise. I thought it was me, thought as long as I didn't make him angry ...' She waved her cigarette in my direction. 'Got a boyfriend, darling?'

'No.'

'Pretty girl like you? I'd have thought you'd have loads of admirers.'

'I prefer to concentrate on my school work,' I said, and then spent the silence that followed thinking about how prim and stuck up I sounded. *Urgh.*

'I wasn't sensible like you. I messed around at school. Only had eyes for the boys. I liked being noticed, liked the attention. And then Ray came along. I was swept off my feet. I was only fifteen, darling, younger than you. He dazzled me.'

'He was Aidan's dad?'

'Good looking, like Aidan. That wicked smile, that was Ray. His granddad came from Trinidad, that's where he gets his colouring from. Oh, how I loved him. And he looked after me, too. Helped me find somewhere to stay when I fell out with your nan. Gave me money to help with the baby.'

'Oh. Why ... what happened?'

'Went off to war. He looked gorgeous in his uniform. Went to Kosovo, to fight other people's battles for them. He was a hero. Of course, the British Army didn't want to know about me and my baby that hadn't even been born yet.'

'That's terrible!'

'Yeah. Everything went to his wife.'

'Oh.'

I didn't like to ask if Ray and his wife had children, children that were Aidan's brothers or sisters, just as much as he was mine. I imagined our family like a daisy chain, spreading out on a green lawn. Some of the linked stalks were strong and firm, but others were wilting and withered, falling to pieces or failing to connect in the first place.

'So I was in a bad place when I met your dad, Cass. I needed someone to lean on. And he was so strong, Mac. That's what I thought. A strong, protective man. But he had a terrible temper. I left him three times when I was pregnant. Aidan stayed with my mum for a while. But I kept on going back. I loved him, you see, and he loved you. He loved you so much. I didn't feel I could take you from him.'

'Do you think ... should I find him?'

Finn's DVD came to an end, and he came out of his penguin-induced trance.

'Where's Mummy?' he asked.

'She'll be back soon,' I told him. He gazed at me, trying to work out if I could be trusted. Then his lip trembled. 'I want Mummy! Want Aidy!'

'They'll be back soon,' I promised again. 'Do you want to choose another DVD?'

Janette waited until Peppa Pig had started, then she ground her second cigarette into the saucer and said, 'No point looking for Mac, darling. You're better off without him. He sent me to the hospital so many times. Every time he swore it was a one off. Aidan's arm got broken, I covered up for Mac. Said it was an accident. But then things got really out of hand. So I talked to the social workers, and they said they'd find you a new family. And I honestly thought it would be the best thing. For both of you.'

'I think it was, for me anyway. I mean, I've had a really good life so far. I appreciate that chance.'

'Do you? Did I do the right thing?'

I was saved from answering by Finn.

'Where's Mummy?'

'Let's try ringing Aidan,' I said. 'I bet they'll be back soon.'

Aidan's phone went straight to voicemail, so I texted him.

Will you be back soon? Finn asking.

'Let's find you some supper, Finn,' said Janette. She looked at her watch. 'I hope they're back soon. I need to go. Keith's taken the kids out to his mum's in Harlow for the day, but they'll be back by eight.'

'I need to go, too.' There'd be a big row waiting for me if I was very late.

We looked through the kitchen cupboards and found some pasta. Janette boiled the kettle, found a saucepan. 'Ketchup, sweetie? Cheese? Why don't you show me where the bowls live?'

Finn was grizzling, his little fists rubbing his eyes.

'He's tired,' said Janette. 'Have your tea, darling, and then we'll put you to bed. Oh, he does remind me of Aidan at the same age. I was sure he was the dad. It'd be just like him to have a child and not tell me.'

'Why ... you don't really get on, do you?' I was being as tactful as I could be, but she was offended. I could tell by the way her shoulders hunched as she drained the pasta.

'He's impossible. I'm sorry to say it, but it's true. There's a side to him ... he turns people away. He's had so many chances. I'm not surprised your parents found him too much to handle. Do you know, I fought for him for years, fought to get him out of foster care, get him back home again, and then he let me down. He let me down so badly.'

I gave Finn a spoon to eat with, but he threw it onto the floor and picked up the pasta twirls one by one with his hands, which soon became sticky with ketchup. So did his face. He wasn't crying, though.

'How did he let you down?'

'My Louis – he was only small. Only about the same age as little Finn. Aidan was a big, tall boy of nearly twelve years old. As tall as me, he was. I'd worked so hard to make a home for him, Cass. I met Keith, and he's a good man, Keith. He works so hard and we have a good life, and we were happy with Louis. But I couldn't forget Aidan and you. There was nothing I could do about you, but I fought and fought to get Aidan.

Those foster parents, those social workers, they wanted to keep us apart, but I won. I won!'

She wiped away a tear. 'All those years of being judged, thinking I was a bad mum. I can't tell you how it felt. How happy I was. But Aidan didn't want to know.'

'Why?'

She shrugged. 'He's an angry boy. He fought us, he was rude, he turned away. Wouldn't kiss me, cuddle me. Every single day he told me he hated me, hated Keith, hated Louis. Tried to run away a few times. I was at my wits' end.'

I wondered what Aidan was like when he lived at our house. Was he so angry all the time? Is that why they rejected him? I never used to get angry. Never thought I had anything to get angry about. After all, my life was perfect. But now that it wasn't perfect, I felt angry all the time and that scared me. Could it be that I thought I'd disappear if I got angry? Just like Aidan?

'I thought things were getting better. He stopped running away, seemed to calm down. But then, one day, Louis did something to annoy him. He lost it. Lashed out. At a little kiddie, Cass. Just like Finn here.'

I found a tea cloth, wet it under the tap, cleaned Finn's face and hands. He started to cry again. 'Mummy!' he wailed. 'Want Mummy!'

'Keith wouldn't have him in the house after that. I was pregnant again, and he said we couldn't trust Aidan.'

'Couldn't you have given him another chance?' I lifted Finn out of his chair. 'Come on Finn, time for bed, eh?'

Finn's crying was getting louder and I had to strain to hear Janette's soft voice.

'You don't understand—'

'Mummy!' Finn had heard the key in the door. I lifted him out of his high-chair and he rushed to the front door.

'Cass, listen to me.' Janette put her hand on my arm. 'Aidan didn't just slap Louis. He blinded him.'

34: AIDAN

'She hates me,' I tell Holly. 'There's only so long I can take it. I'm sorry, I thought I could do it, but I can't.'

We're sitting by the canal, legs dangling over the water. I didn't expect Holly to come after me when I walked out on the tea party, and I kept telling her that she should go back to Finn, to them. But she stuck with me. 'They can look after him for a bit. He'll be OK. You're important, too.'

So now we're sitting here, shivering a bit, but we're alone and it's quiet and I feel I can talk. Perhaps I can even explain.

'Why would she hate you? She's your mother. Aidan, you've got it wrong. There's nothing to hate about you.'

'You don't know the whole story.'

'You can tell me.'

How can I tell her? She's going to hear what I did to Louis and she's going to look at Finn, his beautiful face, his perfect eyes, and she's going to tell me to get out and go away and never come near them again.

'The thing is, I was really happy before she took me away. I loved living with Betty and Jim. They were ... they were safe, Holly. They gave me a home, and I thought it might be for ever. They said ... they were talking about ...'

I never talk much about Betty and Jim because I get too choked up. But now I can't stop myself.

'They seemed to really love me, you know? And I don't think they would have got rid of me. They'd done a lot of fostering and they knew ... they seemed to know what I needed ...'

'We could get in touch with them,' says Holly. 'They'd be happy to see you again, wouldn't they?'

I shake my head. 'I screwed it up. I don't know how, but I did.'

She leans against me, plants a kiss on the back of my neck, just where I've got the tattoo that tells the world that I had hope. A tattoo I can't see. And I'm glad of that. Because I'm not sure that I could take being reminded of it, day after day, if this doesn't work.

Finn is my hope. Finn is my chance. Finn is everything to me. And Holly as well, of course. I love her. I really do.

'You're not going to screw it up with us,' says Holly. 'I don't think you know how much we love you.'

Holly and Finn come together, that's the thing. She gave me an instant family, and that's what I wanted, that's what I need, that's what I can't risk losing.

'I'm not ... I try ... I do love you so much, Holly.'

'I wake up in the morning and I see you in my bed and I think, what did I do to deserve you, Aidan? You're the best thing that ever happened to me. You're ... I don't think you realise what a failure I felt up to the point where we got

together. Failed as a daughter, failed at school … university …
got pregnant like a loser, by a man who didn't want to know.
And then Don – that bully – I had to be supported by my
family. Didn't even have a job. And then you came along. You
changed my luck. I pinch myself sometimes, to make sure it's
real, that you really are mine.'

'I'm nothing special, Holly.' I don't like it when she goes on
like this. She sees the outside me, she ought to know that it
doesn't match the mess inside.

She wrinkles her nose. 'You are special. You've got that …
that look. People notice you.'

Whatever that look is, it's brought me more trouble than
good. I used to pray for something – spots, maybe – that
would put people off. Stop them wanting me, because they
wanted me in a bad way. But if Holly likes the way I look, I
suppose it's worth having.

She's not finished. 'My mum was disappointed in me; my
dad doesn't give a toss. I was always near the bottom of the
class at school. I always felt like a failure. But now I've got you.'

'I'm nothing much. Not like Cass.'

'You haven't had her advantages. She's great, but she's a bit
scary isn't she. So cool and confident. I'd have been terrified of
her at school.'

I put my arm around her. 'Remember that day? When Finn
was staying at your mum's?'

Don't ask me when I got the idea of getting with Holly.
We'd been flatmates for a few months, and we were kind of
used to each other. We liked some of the same things – folk
music, quiz shows, hot curries – and we both loved Finn. I was
doing more and more stuff for her, babysitting and shopping

and all that crap, and I loved it. We were playing houses. We were pretending to be a family.

And then I started fancying her. She'd walk out of the bathroom, wrapped in a towel, and I'd feel turned on. Her arm would brush against mine, and I'd want to grab her, kiss her. I started staring at her lips, wondering if I could, if she'd ... But I never did anything. I just worried she'd notice the staring.

Rich and I had had a drink one night, and I'd told him about it, and he'd laughed at me. 'You're crazy! Of course she'd be interested. Who isn't interested in you?'

'That's really not helpful.'

'She'd be grateful! I mean she's not all that ...'

'Don't talk about her like that.'

Rich looked at me. 'You really do like her, don't you?'

'She's my friend. I do really like her.'

'Tell her.'

'I can't. It's too much of a risk. She might think I was being weird and inappropriate. She might throw me out.'

Rich patted my arm. 'You can come and live with me in my beautiful home.'

And we'd laughed a lot, because he'd just moved into the bedsit, which was seven foot by ten, and he went out drinking every night he could afford it so he wouldn't have to sit and look at the same four walls.

Anyway, I'd gone home and I was tiptoeing around, because I hadn't wanted to wake up Finn or Holly, but when I made a coffee and took it to sit in front of the telly, there she was. Sitting in the dark, wrapped in a duvet, surrounded by sad little heaps of scrumpled tissues.

'Holly? What's the matter? Is Finn OK?'

She'd blinked at me. 'Aidan? Oh, I'm sorry. Nothing's the matter. Finn's staying at Mum's. I just got thinking ... about stuff ...'

'You've been crying.' I was only stating the obvious, but it set her off again.

It was the perfect excuse. I wrapped my arms around her, and she cried more and more, but then I very gently kissed the top of her ear, so if she didn't like it she could move away and we could both pretend it hadn't happened. But I worked out she did like it, because she tilted her face towards mine, and I kissed her eyebrows, and she didn't even try to stop me, so I kissed all over her face, licking away the salty tears, and she stopped crying altogether. And I made her happy. And she made me happy, and we'd made each other happy, more or less, ever since, which was a bloody miracle in my life.

She still cried for no special reason sometimes, and it was more awkward if Finn was around, but I could always kiss her, and we both remembered that night, and sometimes it helped her feel better.

'You're so kind,' she says now. 'So loving. You're everything to me. I'm very lucky.'

I take a deep breath.

'Holly, I ought to tell you about when I lived with Mum.'

'Tell me.'

I can't find the words. She kisses my neck, holds my hand. 'Come on, it can't be that bad.'

'I hated being there. I wanted to be with Jim and Betty. I tried to get back to them, but I couldn't ... I was eleven, Holly, and no one listened to me ...'

She waits. 'What happened?'

'I can't explain it. I remember once – round about this time of year, when everyone wears those paper poppies for the soldiers—'

'Yes ...'

'My mum had one of those and I thought, that's what I'm like, all black and evil inside.'

'Oh, Aidan. You were only a kid. You can't have done anything that bad.'

Under my feet the water of the canal is still and dark and welcoming. It's getting late. I can hear traffic and people laughing and water and the beating of my heart.

I'm going to have to tell Holly somehow.

'I was only eleven,' I start, but my phone buzzes, not for the first time, and she says, 'Is that Cass texting you? We've been out ages. Finn doesn't really know them.'

'We'd better get back,' I say. 'Finn needs us.'

We scramble to our feet. Holly's still holding my hand.

'What were you going to tell me?' she asks.

I'm trembling. I want to tell her, to trust her, but I can't. I can't. She'd throw me out and I'd have nothing. I can't let that happen.

'Oh, nothing. My mum, she's just a bit difficult, that's all. Makes up things. Bad memory. Don't believe all she tells you.'

Holly looks confused, but her mind is on Finn, and deep down, I think, she doesn't want me to tell her bad stuff. She doesn't want to spoil her vision of the perfect boy in her bed. I do that enough as it is, every time I get drunk.

So, we go home.

35: CASS

Aidan apologised to me when he got back. To me, and to Janette. He was sorry for walking out. He'd got overcome with the emotion of it all. 'It's not easy,' he said.

Janette kissed him – how could she? – and said, 'It's all right, darling. I know how it goes.' Then she gave me an unexpectedly fierce hug, enveloping me in the scent of smoke and hairspray. Her body felt thin and fragile, easily breakable. So different from Mum's warm strength. But even Mum could crumble. Maybe the whole idea I had of a mother – someone to protect you – was wrong.

She left and I didn't know what to say to Aidan; I could hardly look at him. 'What did you do to our brother?' I wanted to ask him. 'How could you do that? How do you live with yourself?' But I couldn't bring myself to ask him. I didn't know him well enough.

When I saw Holly cuddling Finn, singing him a song to get

him to sleep, I wondered if I should tell her. She couldn't possibly know. Why would she let Aidan anywhere near her son? But surely it was up to Janette to tell her? 'I honestly don't know what to do about that girl,' she'd said to me. 'Should I tell her what Aidan did? No wonder he never came clean with me, never told me about the boy.'

It wasn't my story. It wasn't my business. So I said goodbye, smiled, pretended everything was fine.

When I got home, Mum was on the phone and Ben was already asleep. So I didn't have to talk to anyone.

The next day I got up early for school and caught the bus so I'd be there in good time for the prefects' meeting. My hair was back in its usual neat ponytail. I wore a white shirt and a black skirt. No make-up. I wanted to get back to normal; I had a lot of studying to catch up on.

It took until lunchtime to realise that Grace was upset about something, and it took Megan to tell me.

'What's wrong with her?' I was sure it couldn't be anything much. Just an essay crisis or a boyfriend disaster, something Grace thinks is important, but isn't really. I was bored just thinking about it, but she'd been trying so hard to be there for me, I knew I had to reciprocate.

'I don't know what her problem is,' said Megan. 'She wants to talk to you, but she thinks she can't because you've been so secret squirrel about Will Hughes.'

I'd been quite happy eating my spaghetti bolognaise, but right then it turned into a writhing plate of maggots. I dropped my fork. 'What? What did you say?'

'Oh, come on, Cass, it's the worst-kept secret in town. Abigail Ellis saw you two walking through the park, and

Isabella Thomas said she saw you going into some house in Denton Road. Her grandma lives there—'

'We were just going to feed some cats,' I said. 'He runs a business. Pet care and private investigations. I might be going to work with him.'

'Pet care and private investigation? Really?'

'There's a lot of money in it,' I said, brisk and smiling, so she could think it was a joke if she wanted, but she wouldn't realise that I was dying inside. 'He's quite an entrepreneur. I might invest. Where is Grace, anyway?'

'She's out by the pond.'

The pond is outside the science labs, and it's where the Year Sevens watch frogspawn turn into tadpoles and frogs. In the summer, lots of people take their packed lunch and go and sit on the benches and pretend they're on the beach by sunbathing as many bits of themselves that they can expose.

In the winter, though, there was just Grace, sitting moping on a bench, throwing her sandwich bit by bit at the one lonely duck. I sat down on the bench next to her.

'You're not meant to do that. It'll ruin the pond's eco-system, and I don't think they eat peanut butter.'

'Like I care.'

'Well, Miss Worth will care, and it's not really fair to the biology department. Or the duck.'

'Oh, that is so typical, Cass. You're more bothered about a stupid duck than about me! I'm meant to be your best friend, but you've hardly spoken to me for weeks.'

'I've just had a load of work to do.'

'Yeah, right, whatever, Cass.'

'What's going on with you? Come on, Grace, you can tell me.'

'I'm not telling you a thing. You don't tell me anything.'

'Oh God, it's not a game, Grace. It's not, "I'll tell you and you can tell me." My parents have split up, OK? It was all over the newspapers. Isn't that enough for you?'

'I don't believe you. I know there's something going on with Will. And you've just been so ... so ...'

'So what?' I snapped.

'So closed up. Like, if I ask you how you are, I'm intruding. So I feel I can't tell you anything, or ask you anything, and we're meant to be friends, Cass. I don't know what's going on.'

I was about to feed her the pet care and private investigation story, but then I paused. Maybe she was right. Maybe I did push people away. If families were so unreliable, maybe I needed my friends.

'I'm sorry. There has been loads going on. So much that I don't even know where to start.'

'Try me.'

The bell rang for the end of lunch. Saved by double history.

Grace stayed put.

'Come on, we've got to go.'

'I'm not going.'

'You've got to come. We're going to miss history.'

Grace put her lunchbox back into her bag. The duck nibbled a bit of crust and then swam away fast.

'If you walk away from me now, that's our friendship over, Cass Montgomery.'

'You've gone mad! Can't we just talk after school?'

Grace pouted. 'No.'

'What's up with you, anyway?'

'Liam dumped me.'

Who's Liam?

I improvised. 'Oh, no, Grace, that's awful. The bastard. What happened?'

She snorted. 'That so proves you don't pay a blind bit of notice to anything I tell you. Who's Liam, Cass? I've never heard of him.'

My face was hot. 'Oh, well, if you're going to play tricks on me, I might as well go to history.'

'Go on, then. Run along, Little Miss Perfect.'

To my horror and surprise my eyes filled with tears. 'I don't know what I'm meant to have done.'

'You're meant to be my best friend, but you never talk to me. You don't let me in, Cass. You've gone through so much with your mum and dad splitting up, but you don't talk to me at all, don't even let me come round. I feel like you don't value me or our friendship at all. I can't tell you stuff if you don't tell me anything.'

I was blinking back tears, torn between guilt – she wasn't wrong – and fury. How dare she make me cry at school! I'd be a laughing stock.

'I told you all about Leon – that's his name, by the way – that day in Starbucks, and you just nodded and smiled and went on total auto-pilot, and I thought never mind, actually it's quite sweet because she's looking over at Will and he's looking at her, and that's just what she needs when everything's been so rubbish for her at home ...'

'Oh! No ... Grace—'

'And you never asked once about how it went—'

How what went?

She read it on my face. 'The audition! Duh!'

'Oh right, the *audition*.'

'For the drama production. Joint with the boys' school. Oh, Cass.'

'I'm sorry, I'm sorry, I'm sorry,' I burbled. 'You are right. I am wrong. I've been a terrible friend. How did it go?'

'Hopeless. I got Lady Macbeth, but Leon didn't get Macbeth. He's Banquo. We hardly have any scenes together.'

'Oh, well, but Grace, there's not a lot of scope for romance in the Macbeths anyway.'

'Shared rehearsals. Practising lines.'

'I bet you can do that anyway.'

'If he's interested. It's all right for you, Cass. You play it so cool all the time that all the boys see you as a challenge. I'm sure I just come over as desperate.'

'No, you don't, because you're not desperate. Leon's probably just shy because so many boys like you.'

'I am desperate. I'm desperate to stop feeling like a reject just because I haven't got a boyfriend. I wish I could be as self-sufficient as you, Cass. You're so confident. You don't need a boy's approval to feel good about yourself.'

'I'm not so confident. I don't even know who I am any more.'

The school was quiet, everyone was in class. I supposed if anyone saw us they'd assume we had a free study period.

'My life is one big mess,' I told her. 'I still can't believe Mum and Dad have split up. I hate him so much, but I still really love him, too. And I'm dreading having to move house. I feel like I've got so much security bound up there. It's stupid, I know, but I sometimes think I care more about the house than any-thing else.'

'You don't really, though,' said Grace. 'I mean, it's probably easier to feel upset about the house than about your parents.'

Which parents, I want to ask. My parents now, or my parents then? The mother I hardly know any more. The father who kept secrets from us. And Janette - what about her? And Mac, violent, scary Mac?

'And then my brother ... he contacted me. Not Ben, I mean my birth brother. I hadn't seen him since I was really little.'

'Oh my God! Cass! How did he contact you?'

'Facebook.'

'How do you know it's really him? It could be some pae-dophile!'

'No, it's really him. I've met him. And it's the strangest thing ... it's great actually, but it's turning everything around. It's making me question everything about my life. I can't explain it ... it's making me think about who I am and what I want, and I just don't have any answers, Grace.'

'What's he like, your brother? Have you met him?'

'He's ... I don't know. He's different from anyone I ever met. He seems nice.'

He seems nice, but he attacked a little child. And I'm not sure if his girlfriend knows. And she has a child, exactly the same age ...

'I can't believe you didn't tell me!'

'I couldn't. It was all a secret. I didn't even tell my parents or Ben, until, well, last week and I didn't want to then. It all just came out.'

Grace seized my hand, gave it a squeeze. It was the sort of

gesture that I'd have pulled away from normally, but I didn't mind this time. It was cold, sitting on the stone bench, and I appreciated the warmth.

'Oh, Cass, I am so here for you.'

'Thanks, Grace.'

'And what about Will? I know there's something going on.'

I wasn't going to say anything about Will, because, frankly, what was there to say? Nothing was going on. We'd messed around a bit, but I hadn't heard from him for ages. I'd avoided the bus, and the one time I'd spotted him in town, he'd been with the blonde girl. I was almost sure he'd moved on.

'There's nothing going on,' I said, completely and totally accurately.

'Are you sure? Because everyone says . . . '

That's when I surprised myself. 'There's nothing going on,' I insisted, 'because I've totally messed up.' Suddenly, stupid, unnecessary tears were pouring down my face. 'And I don't know what to do!'

Grace patted me on the knee. 'Trust me,' she said. 'Tell me the whole story.'

36: AIDAN

She knows. Cass knows. I could tell by her face when I walked in the door.

That's typical Mum. She makes me think that everything's OK, that I can trust her to keep my secret. And then she spills it to the one person whose opinion I care about as much as Holly's. To the one person who might tell Holly, as well.

I can't decide what to do. Tell Holly? I was going to, out by the canal, but now I'm here, back in the flat, holding Finn in my arms, I can't bear the idea of losing everything.

Should I ring Cass? Go and see Mum? It's tearing me apart. My mind is all over the place. Which is why I overslept. Which is why Clive's pissed off with me.

He starts right away. Why am I late? What's going on?

'I'm sorry! I'm sorry!'

'You only live upstairs! What's going to happen when you're further away?'

I freeze inside.

'What do you mean?'

'I'm thinking of putting the flat on the market. I mentioned it to Holly the other day. I can't afford to keep it on, not at the rent you two pay. I'm sorry and all that, but I've given you a pretty good start. Now you'll have to find somewhere new. I'm sure it'll be fine. You can always stay with Juliet in the interim. She's got plenty of room, she's rattling around in that house.'

Holly's mum lives in Primrose Hill and true, her house has room for all of us, and more. But every time I've been there, she's been as cold as the house – which is permanently freezing, because she doesn't ever put the heating on.

The minute I walk in, she's fussing about dirt on the floor: 'Can you just take your big boots off, Aidan?' and criticising my manners. 'Would you like a spoon with your pasta, Aidan?'

And then she starts with the awkward questions and the snarky comments.

'So, Aidan, did you actually get any qualifications at school? None at all? Not even the easy kind?'

'So, Aidan, when did you enter the care system? You were five? How tragic.'

'So, Aidan, do you think you'll want to spread your wings one day? You're very young to be tied down to a partner and someone else's baby.'

'So, Aidan, it must be hard to adjust to family life when you've never really experienced it.'

The worse thing is that Holly can't even hear it. If I say one thing afterwards, then Holly tells me I've got it wrong and her mum is just trying to get to know me and understand, and actually she's dead impressed at how hard I work to overcome my difficult past.

That'll be why Holly hasn't told me about the flat. She'll be working out how to persuade me to stay at Juliet's. Or she's planning to move in there, just her and Finn. She just hasn't got around to telling me yet. She's just going to move out one day and I won't even get to say goodbye ...

'Wakey, wakey!' Clive's clicking his fingers in front of my face. 'What's the matter? You knew the flat wasn't for ever. Don't worry, we'll sort you out with something.'

He's off on another job. 'You can study for your written driving test while I'm out,' he says, shoving the book in my hand. 'Time's running out, Aidan. You've got a new test coming up next week. You'd better pass it.'

I'm never going to pass it. The job's going to end. Holly's going to leave me. I'll never see Finn again. And the flat ... the flat ...

Clive goes out into the yard. I'm not meant to use my phone when I'm in the shop, but it's ringing and it's Rich and I haven't spoken to him for ages.

'Hey, Rich?'

'Hey! Long time, no speak!'

A little bit of my brain registers his bright, happy tone. I'd better call in about a week to check he's OK. That's how the cycle usually goes.

'You sound good, how's it going?'

'Oh my God, Aid, I'm in heaven!'

'Well, you've tried to go there enough times.'

He ignores the dig. 'This is The One! I can't tell you!'

He's going to anyway.

'What's his name?'

'Brendan. I told you about him before.'

I'd probably zoned out. There have been so many Ones over the years.

'Rich, don't get too excited. Remember Tom ... and that little shit, Joel ... and Hans didn't work out too well, did he?'

'Aid, you're bringing me down! Shut up! Brendan's different.'

'How?'

'He's like, he's really caring. We talk. He's introduced me to his parents.'

'What? When? Why?'

'Last weekend. They were so lovely, Aid. His mum gave me a hug and said she hoped I'd feel part of the family. Said she'd never seen Brendan like this about any other guy.'

I swallowed. 'That's great, Rich. What does he do, this Brendan?'

'He's so cool. He's a student.'

'A student of what?'

'Dance. He's all muscle and stubble and curls.'

'Oh. Well. I just hope he's not like the others.'

'He's totally not. He really loves me. We're talking about moving in together. He's got a flat in Walthamstow.'

'Oh, right. Well, I'd better go. I'm at work. Can't sit around talking all day.'

'I've got a job, too. That's how we met. I'm working at an organic café in a dance studio in Crouch End. It's such a cool place, and I get free food, and that's how I met Brendan, because he teaches zumba there, three times a week, and he came into the café for a smoothie and we just looked ... we saw each other ... and I've never felt like this before, Aidan. This is it, he is The One.'

'Got to go!' I say. 'Customer! Bye.'

I don't know why I feel so sad and empty. I ought to be happy that Rich is in love, right? He deserves to find someone: we all do. It's just that I'm about to be homeless and loveless, and he's going to be living in bliss in Walthamstow. He won't need me any more. I'll be all on my own.

The bell jangles. I look up, startled.

'Well!' says Neil. 'Good morning to you.'

37: AIDAN

'Go away, Neil.' I do my best to sound strong and tough. 'It's not a good time.'

'Expecting someone, are we? That girl, maybe, with the hair. Ginger. What was her name? Cass?'

'I don't know. She's not really a friend.'

'Cass Montgomery. Daughter of Oliver Montgomery. You've heard of him, haven't you, Aidan? Oh, no, hang on, I forgot. You don't read much, do you? Probably missed the news about how he's been caught with his pants down.'

I hold my breath. Had Neil worked out that Cass was adopted, that there could be a connection? No. He didn't say anything. I shrug. 'She could be anyone. You know what it's like.'

Neil smirked. 'I'm sure you have a lot of success with the ladies, Aidan. Maybe your girlfriend would like to hear about this Cass, how you got off with old Monty's daughter. How you're mixing with the rich and famous.'

He's so off-track that I actually laugh. 'You've got to be joking.'

'Maybe the papers would like to know about it?'

'You're just scum,' I tell him. 'No one's going to listen to you.'

'You're very cocky,' he says. 'Think your girlfriend's so into you that you can get away with anything? Well, I suppose a single mother is grateful for anything she can get. She's looking a bit old for you, though, Aid. Old and knackered. No wonder you take the chance to get some posh totty when you can.'

'What the hell do you think you're doing in my shop?' Clive's bark shocks us both. I nearly fall off my chair. 'I know you, don't I?' Clive's not scared of Neil. He's right in his face.

Neil's always had a bit of a swagger about him. But now he's chewing his thumbnail, staring at his boots.

'You only lasted here two days. Had to let you go. Fingers in the till, wasn't it? What's your name, again?'

Neil mutters something under his breath.

'Neil, weren't it? I remember you, Neil Johnson. Seem to remember telling you not to come back here again.'

'Aidan asked me. Said he needed help with something.'

'This true, Aidan?'

I don't know what to say. I want to tell him no, no it's not true, but Neil's there and he's looking at me, and kind of squinting his eyes and I know the kind of thing he'll do if I don't lie for him. Broken fingers is just the start of it.

'Yeah. Sorry, Clive. I didn't know.'

Clive's eyeballing Neil. I never knew he could look so scary. 'You listen to me, lad. My niece is Aidan's girlfriend and if I ever hear you speak about her with less than total respect, you'll feel the consequences. I let you off with a warning, last time, remember?'

Neil mutters something so low that I can't make out if it's actual words or just the kind of growl that a cornered dog might make.

'What? Speak up!'

'Yeah,' said Neil. 'But it was Aidan who—'

'What's that?'

'Yes, sir.'

'That's better. Now, you get out. And I never want to see or hear you near this place again, got it?'

'Got it.'

I am amazed. My mouth is hanging open. One minute Neil was big and scary and had so much power over me that I was like an ant under his Timberlands, the next he's shuffling out like a little boy. All from a telling-off from Clive, ordinary Clive in his sweater from Debenhams, with his bald head and his little paunch and his glasses. Clive who looks like every dad in every crap Christmas advert you've ever seen. Boring old Clive who's straight as a die – whatever that means – and doesn't even swear.

Clive, who's looking at me like I'm a stranger.

'Did I hear right? That toe-rag Neil is your mate? You were slagging off our Holly, laughing about her?'

'No, Clive, wait, that's not right.'

'You've been having it off with a bit of posh totty, in secret?'

'No, no, Clive, that's all wrong.'

'You bet your life it's wrong. You bet your life it's very wrong. What's Holly going to say when I tell her?'

I'm scared witless. I just stare at the floor.

'How d'you know Neil Johnson?'

'From the children's home. Clive, it's not like he said—'

'My brother, Bernie, and me, we went into those homes together. Barnardo's. None of your private rooms and televisions and en suites in those days. We were all together in dorms. It was a hard life. You kids today, you're soft. You have no idea.'

I open my mouth. I close it again.

'Me and Bernie, we came out, we were both determined to make something of our lives. I did it the legit way; Bernie, not so much. That's why he buggered off to Spain. Costa del Crime, you know. Holly, bless her, she's got no idea. Don't want you telling her, neither.'

'I won't . . . I didn't do anything, Clive.'

'I thought you really cared for her. God knows, she's not had it easy.'

'I do – I do really care.'

'Well, watch it. Don't invite your friend here again.'

I want to explain that Neil's the last person I ever want to see again, but the words are flying around in my head, the way they usually do on the page.

Clive's getting nearer. His head is getting bigger. I can smell aftershave and alcohol – although it's coming from my breath, not his, I had an emergency can of lager for breakfast – and all I can think of is Mac, his big, red face, and the way he shouted, and the sudden smack of his hand on my ear or my head, and the way that anything I did, any sound or movement or just looking at him could set him off . . .

I flinch away from Clive, so suddenly that I fall off my chair for real. I can't breathe. I'm gasping for air.

'Whoa!' he says. 'What's going on?'

I can't speak.

'What's the matter? Are you ill?'

'I'm ... I'm OK. I'm all right. I'm sorry.'

Clive's shaking his head. 'Someone's spooked you bad in the past, eh? Take a moment. Calm down. Want to take a break? Go up to the flat?'

'I'm OK.'

He leaves me alone for a bit, makes coffee, comes back and hands me my mug. He's remembered I take two sugars. He's put some ginger nuts on a plate.

I hate how badly I've let Clive down. I've let Neil hide a gun in his shop. I owe him so much, and I've done this terrible thing. The coffee nearly chokes me. I can't take his biscuits.

'What's the problem, Aidan? What's going on?'

'Nothing. He's ... Neil ... he's not a friend of mine. He just came in the shop. Honest.'

'He's scum. You don't need to associate with trash like that. Call the police if he comes back, tell them he's shoplifting.'

'OK.'

'You thought I was going to hit you?'

'No ... not really ... you just reminded me of someone.'

'Listen, son, you got the chance of a new life. Bad stuff has happened to you in the past, but that's gone now. You pick yourself up and you start again. You've got a chance here, you make the most of it.'

'I will ... I am ...'

'You got nothing to be scared of, Aidan, as long as you treat Holly and Finn right and keep out of trouble. Work hard and stop stealing from me. If you need new trainers, just ask me. Think you can do that?'

I wish it was as easy as he seems to think it is. 'Yeah, Clive.'

'Good lad. Right, time for you to go and stack the new boxes in the yard.'

'Yeah, OK.'

I feel better with the coffee inside me, but as I stack the boxes all my troubles come back, rolling around my head.

Neil's not allowed to come into the shop, but I still have his gun.

Clive's going to sell the flat.

Holly might already know what I did to Louis. Cass almost definitely does know.

I'm losing everything.

I might lose Finn.

38: CASS

Ben's birthday. Always a problem, every year, but this year it promised to be a disaster.

My birthday parties were easy. All my friends at our house. We used to play party games and eat jelly and ice-cream. More recently we'd eat pizzas and watch DVDs. For my sixteenth I'd had a proper party – a disco in the church hall. For my seventeenth – well, it was months away. I had time to decide, and I felt I was getting a bit old for birthday parties. But Ben never had enough friends for a party. One year Mum invited a load of kids anyway, and watched as they played football in the garden while Ben sat in his room, headphones on, reading *The Hobbit*. Dad shouted at him to join in and Ben cried. After that, we'd made a new birthday tradition. Just family. A film, or bowling, and a meal out.

But this year, Dad wasn't really part of our family any more. And Ben had ideas of his own.

'I want to invite people and I want a barbeque.'

'It's November,' I pointed out. 'Not exactly barbeque weather. And who's going to barbeque?'

'Dad.'

Dad always took charge of our barbeque. I'd forgotten that was another part of family life that had ended, *bang, smash,* just like that.

'Dad doesn't live here any more.' Time to be brutal.

'He'll come for my birthday!' Ben looked amazed that I even thought that perhaps this year Dad could be missing from the birthday celebrations.

'Look, Ben, things are different now.'

'He'll come. It's my birthday!'

I talked to Mum and she agreed that it was hard for Ben to be expected to celebrate without Dad. 'Besides, it'll be his last birthday in this house. It's nice that he wants to have a party. Do you know who he's inviting?'

I did not. Ben was very secretive about it. It was only when Mum started asking how much food she should buy that I got the list from him.

Dad

Annabel

Annabel's sisters, Imogen and Rosie

Aunt Kate and Granny

Emily and Phoenix and Sebastian

Will Hughes

'Ben!'

'What?'

Where to start?

'Ben, you can't invite Annabel, let alone her sisters. How would Mum feel?'

Ben considered this for a few minutes, frowning as though he was working out a tricky maths problem.

'Mum might like them. They're all nice. Rosie has got a toy poodle called Apricot. She could bring her.'

'Ben, Annabel has ruined Mum's life! She stole her husband! You can't just start expecting everyone to get on!'

Ben's mouth drooped. 'But it would be better if everyone did get on, Cass. The vicar said ...'

'No one listens to him! Who are Emily, Sebastian and Phoenix? And what's Will Hughes doing on this list?'

'I love Will. He found my friends, Emily and Phoenix and Sebastian. We all like *Lord of the Rings* and *Doctor Who*, and Emily plays chess and Phoenix is teaching me to program, and Sebastian like bowling, and we're all friends now, and it's all thanks to Will. I love him and I thought you were his friend.'

'He found you all those friends?'

'I actually like going to school now, because I've got friends. And none of them play football. And we all like barbeque.'

'But maybe Will won't want to come to a twelve-year-old's party.' I was panicking now. He wouldn't turn up, would he? If he did, I'd have to thank him, because finding friends for Ben ... I mean, no one had ever been able to do that. It was the best thing that anyone could possibly have done for our whole family and when I thought about Will doing that, really thinking about Ben and caring enough to bother, I felt warm and tingly. I nearly phoned up Grace to ask her advice, should Will turn up. But he wouldn't, I was sure. He'd forgotten all about me. I just hoped he wouldn't see Ben's invitation as some cringingly clumsy attempt engineered by me to get back together with him.

Anyway, Mum looked at the list and said, firmly and bravely, that Ben must be allowed to pick his own guests. In the end, though, Annabel, Imogen and Rosie sent him cards and letters saying how much they would have loved to have come, but their step-brother was getting married and they all had to go to his fiancée's hen weekend at a spa hotel in Cornwall. Imogen and Rosie sent him gift tokens, and Annabel said, 'I'll give you your present when I see you!!!!'

So I only had Dad and Will to worry about, and Will probably wouldn't turn up. Surely he'd realise it wasn't appropriate?

He didn't. A good hour after everyone else, there he was on our doorstep, holding a lumpy parcel.

'Oh. Hi. Is that for Ben? I'll give it to him, I said, which wasn't at all the polite, dignified, serene welcome that I'd planned that morning.

'Hi, Cass.' Will wasn't smiling, which was so unusual for him that he felt like a completely different person: someone who'd had a personality transplant, or a terrible tragedy in his life. Maybe something had happened. Maybe someone had died, or had an accident or something? He wasn't wearing his usual colourful clothes either. He was in jeans and a black hoodie, no Dora bag, no fake glasses. On Will, this looked liek a funeral suit.

'Are you all right?' I asked him.

'I'm fine.'

'Oh. Good.'

'And you? How are you, Cass?'

'I'm absolutely fine.'

I quickly turned my back on him because I had a massive lump in my throat, and led the way through the house.

'Ben's in the living room. Here.' I showed Will the door. 'The barbeque's in the garden, but it's really cold, so we're bringing the food into the kitchen. That door there. But you might not want to stay that long. I mean, it's mostly just Ben's Year Seven friends, and family.'

Will paused at the door to the living room. 'I'll stay as long as you want me to. Not long, I suspect.'

I opened my mouth to reply, but he'd gone. I heard squeals of delight from Ben and Emily and Phoenix and Sebastian over the rumblings of the DVD of *The Fellowship of the Ring*, and then the door closed in my face.

Mum and Granny and Aunt Kate were making salads in the kitchen, and I really didn't want to be quizzed about my A levels, so I went out into the garden to see how Dad was getting on. At least it wasn't raining, although it was so cold that I could see my own breath.

'Trust Ben!' said Dad. 'A barbeque on the coldest day of the year so far. Oh well. I suppose I'm lucky to be on the guest list.'

He did that thing, the waggling eyebrows thing, and instead of feeling irritated and angry and full of hate, I remembered when it was the funniest joke in the world that my dad could waggle his eyebrows. When I was about six and I'd stopped having nightmares and I'd almost for-gotten – had forgotten, really – that there was ever another Mum and Dad, that there ever was an Aidan, that once upon a time I lived in a very different sort of home. I had a new Mummy and Daddy and I loved them and they loved me and everything was all right and we were going to live happily ever after.

Dad looked at me, and he must have seen something in my face because he said, 'I'm sorry. I'm so sorry, Cass. I never meant to hurt you. I think this divorce is harder for you than anyone.'

I actually gasped out loud. 'What the hell? What about Mum?'

'The way it happened was terrible, I admit that. But we haven't really been happy for years. Not as a couple. We both knew it, but we never addressed it. We were too busy putting on a brave face, thinking about you children.'

'Mum's lost everything! Her home ... her job ... '

'She didn't have a job.'

'Being your wife was a full-time job!'

'Yes, and she never let me forget it. Cass, we were a good team. But the magic went out of our relationship a long time ago. Maybe it's inevitable when you've been married a long time. But I know I'm happier now, and I'm convinced your mum can be happier, too. But can you be? What about you, Cass?'

'What about Ben?' My voice was a growl.

'Ben's blossoming! He's made friends, he's really taken to Annabel. Your Mum and I agree that he's coped surprisingly well. Ben isn't too involved in other people's emotions. And that's actually helped him cope with all of this.'

'Oh, so you did it for Ben's sake?'

'Try not to be childish, Cass.'

I felt childish, that was the problem. I wanted to shout and scream and stamp my foot. But that was unthinkable, with Mum and Granny and Aunt Kate just metres away. With Will Hughes in my house.

'Is he here, the boyfriend?' asked Dad, expertly rolling the sausages over. 'I'd like to meet him again.'

'He's not my boyfriend. He's Ben's peer mentor or buddy or something.'

'He found all these friends for Ben, I gather. Took them all bowling. Seems like a nice lad.'

How did Dad know all this stuff when I didn't? Obviously I'd been too busy moping to notice anything that was going on around me.

Mum came out into the garden. 'Any food ready? If you put the sausages into this dish, Cass, I can keep them warm in the Aga.'

'Just a few minutes,' said Dad. 'Susie, thank you for including me in this. I appreciate it isn't easy.'

'It wouldn't be the same without you,' she said.

I exploded.

'How can you say that! He's smashed our family to pieces! We're having to sell the house! It's never going to be the same again!'

'Calm down, darling, remember we've got guests,' said Mum.

'I don't care!'

A cough. Will's cough.

'Umm, I just came to say goodbye. Thanks for inviting me.'

Mum and Dad looked at me – red-faced, fighting back tears, furious – and Will, an unbouncy, frowning, almost unrecognisable version of Will.

SALVAGE

'Won't you stay and have some sausages?' said Mum.

Will looked at me. I couldn't speak.

'I'd better go,' he said.

I heard the faint sound of the doorbell. I ran to answer it.

There standing on the doorstep was the last person on earth I expected to see.

Aidan.

39: AIDAN

Holly knows. Holly can't look at me. She tells Finn to hush and watch Peppa Pig when he shrieks 'Aidy!' and rushes to hug me after work.

I pick him up and breathe in his smell – ketchup and baby shampoo – and rub my face on his fuzzy hair and think about carrying him down the stairs, jumping on to a bus, running away and taking him with me.

But Finn can't be without Holly. He needs to be with her. Just because I didn't get much of my mum when I was growing up doesn't mean I can take Finn away from his.

'Aidan, we need to talk,' she says, over Finn's head, once he's settled in front of the telly again.

'Is it about . . . is it bad?'

'Of course it's bad. You must have been so scared that I'd find out.'

Just for a moment I think that means that she understands. That she's kind enough to give me the benefit of the doubt.

Then I catch her eye and there's no love, no warmth, nothing left for me.

'I wanted to tell you – I tried.'

'But you didn't try hard enough.'

'Who told you? Mum?'

She shook her head. 'Not your mum, no. You'd have thought she would. She should understand how a mother ... how I'd feel ... why I needed to know.'

So it must have been Cass.

Betrayed by my sister.

'Holly, I'd never hurt him, you know that. You must know that.'

Finn stirs, turns, looks at us. 'Aidy?'

'It's OK, Finn. Nothing's happening. Aidy and I are just going to the kitchen to make a cup of tea, OK?'

His thumb is in his mouth. 'OK,' he says.

In the kitchen she lets the tears go. I want to hug her, kiss them away. I can't. She wouldn't let me.

'Holly, it was an accident, I swear. I was only eleven. I never meant to hurt him.'

'You blinded a little boy!' She's shaking. 'You did that and you never told me! You came into my life, and you – I let you! – be a dad to Finn, and all the time you had this awful secret ... it was only seven years ago ...'

'I was a kid.'

'I'm trying to understand. I'm trying Aidan, but it's a shock ... such a shock ... I wish you'd told me right at the start. You kept it secret!'

'Because I knew you'd be like this! And I didn't *blind* him, it was only one eye ...'

'I can't believe you think that makes it better!'

284

'I'm just trying to explain ... how it was—'

'What else haven't you told me, Aidan?'

What else haven't I told her? I take a deep breath.

There's the life I led before I was taken into care. Not just Mac, but the filthy places we lived, and the parties, the way people came in and out of our house, strange people, sick people. Sleeping in my clothes because I was cold, and I had no pyjamas. All that stuff. Cass is lucky she doesn't remember.

There's the day Mum gave us up for adoption.

There's all the stuff that happened to me at foster homes and in children's homes, the people who touched and hit and made me feel like I was nothing.

How it felt to be rejected again and again and again.

There's the fear I carry, day after day, that she's going to stop loving me.

And today's the day.

'I'll go,' I say. 'Just let me say goodbye to Finn.'

'Aidan, I just need time ... to take this in. To understand. I need to think ...'

But I don't stop to listen to her. I go into the living room, lift Finn up into my arms, hug him as tight as I can without hurting him.

'I've got to go, Finn, but I will see you again. I promise I will. I love you and I will always love you and you never did anything wrong, OK? You never did anything to stop me loving you. I'm going to go now, but it's not down to you. Take care. Look after Mummy for me.'

He's screaming and crying and I'm crying too, and I can't stand to hear it so I don't even stop to pack anything. I just put him down again, and brush past Holly. I grab my backpack and I go.

I tumble down the stairs and then I think of the gun. How can I leave it in Clive's shop? What if the police find it? Then he'd know and Holly would know that I was weak and bad and dishonest. And they'd tell Finn that, and he'd believe them.

So I open the door to the shop and I get Clive's tool kit and I prise up the floorboards. I pick out the gun and I push it right down to the bottom of my bag.

Then I help myself to some things that might be useful. A hoody. Some jeans. A toothbrush – Clive just had a load of stuff in from a burned down pharmacy in Wembley. A rabbit, to remind me of Finn's.

I get all the way to Rich's bedsit and I'm thumping on the door before I remember Brendan. Bloody Brendan and his flat in Walthamstow. Rich'll be there, all loved up, thinking he's going to live happy ever after.

There's no such thing as happy ever after.

I pull out my spare key and in I go. It's empty. No more posters on the wall, no duvet, no cans of cider.

It's like Rich is dead. This is what I feared I'd find, all those times. But he's not dead. He's allowed himself to hope. He should know better. I should've known better.

I slide my hand round to the back of my neck and I remember how it hurt, the needles going in and out, marking my skin, writing my happiness onto me for ever. What a stupid thing to do. What a joke.

I sit down on the bed, and I stare at the walls and I think about the gun in my bag.

And I think about Cass. She has everything and I have nothing. And she took everything away from me.

40: CASS

'Aidan? What are you doing here? How did you know where I lived?'

'I got a train. Then, I dunno, I just walked around. I asked people.'

There was something wrong with him. He wasn't smiling, or looking me in the eye. He was mumbling, twitchy. His eyes were bloodshot, his face pale and unshaven. Drugs, said a voice in my head. Drink. Danger.

'Look, it's not such a great time—'

He laughed. Aidan laughed. But it wasn't the sort of laugh that meant he was happy, or he thought something was funny.

'I remember this house, you know. I remember that clock. I never saw anything like it again.'

The clock is eighteenth century and it belonged to Mum's great-grandfather. It's a quite rare sort of pendulum clock, with a pattern of stars on the clockface and a mahogany case. It's beautiful. I'm not surprised he remembers it.

'Look, Aidan, maybe we'd better go and get a coffee in town?'

'I just need to ask you something. It won't take long. I just need to understand.'

'We could go to Starbucks or somewhere,' I said. Then I thought of all the people I might know in there, seeing me with Aidan, and I cringed a bit. And then hated myself for it.

'I won't be long.' He looked around. 'It's so weird to be here.'

'Cass?' It was Mum's voice. 'Who was it, darling? The food's ready.'

I was frozen. Should I invite him in?

Ben and his friends burst into the hallway. Phoenix, with his camouflage trousers and his ponytail; Emily, little and plump, wearing glasses so thick they look like swimming goggles; Sebastian, who had a woolly hat and stripy shirt just like *Where's Wally*. They gathered behind Ben like a freaky little welcoming party.

Aidan blinked. 'I didn't know you ... your parents ... they foster kids?'

'No ... it's just ...'

'Hey!' said Ben. 'Who are you?'

'I'm Aidan.'

Ben totally took this in his stride.

'I'm Ben. Are you Cass's brother, Aidan? I'm Cass's brother, too. Have you come for my birthday?'

Aidan blinked. I realised I hadn't warned him about what Ben was like.

'Not really,' he said. 'I'm sorry. Cass. I just—'

'Come and have barbeque! Mum said I couldn't have any because it was November and too cold, but I said we could get

Dad to come home and make it, and he has. And we're going to eat it inside and there is cake, as well.'

'What's the hold up?' Mum appeared at the kitchen door. 'Come on, get your food while it's hot.'

Ben and Phoenix and Emily and Sebastian ran past her.

'Come on, Cass,' she said.

Then she saw Aidan.

'Oh! It's . . . Aidan, is it you? Is it really you?'

Aidan swayed a little, and I thought he was going to fall. He put his hand on the doorframe to steady himself.

'Susie?'

She laughed, a funny, choking little laugh. 'You always called me Susie. Oh, Aidan, it is good to see you.'

'It is?'

'Cass, can you go and see that everything is all right? And ask Dad to come out here, too? I think we need to talk to Aidan, don't we, Aidan?'

'Susie,' he says again. 'I remember this house. I lived here.'

'You did live here, Aidan, you did. For three months. We need to talk to you, Aidan.'

'I need to talk to Cass.'

'We can go to Starbucks,' I said, and then I realised that Will was standing in the hallway. His mouth was actually wide open as he looked from Aidan to me and back again.

'Bye, Will,' I said quickly.

Will scrunched up his eyes. 'Cass – I just thought . . . if I could have a word?'

'Look, another time, OK?'

'But it's really urgent now.'

'It's just not a good time, as you can see,' I said.

Aidan was staring at Will. 'I've seen you before,' he said. 'I know I have.'

'No, Aidan, you can't have,' I said, but then remembered how Will had been there the first time I met Aidan. 'Look, I can explain. Will was just there to keep an eye on me. Just in case, you know, you were some dodgy internet person. But you weren't.'

'Cass – I – look, maybe I will stay a bit longer,' said Will. 'Those sausages looked awesome.' And he went back into the kitchen.

'Aidan, dear, won't you join us? We're just about to eat – Ben's party, you know . . . '

'I don't want . . . I just wanted to tell Cass something—'

'We can go in the study,' I said. 'We can talk there. Just you and me. We could take some food in there, if it's OK with you, Mum?'

Mum's usually a bit strict about not eating anywhere except the kitchen table or the dining room, but she said, 'That sounds like a good idea. You go along and I'll bring you some food. And then we can talk later, Aidan? I know Oliver will want to see you again. How tall you are! Oh, I'm sorry, what a stupid thing to say . . . '

'This way.' I showed Aidan into the study, and shut the door behind us. 'Look, you sit there and I'll grab another chair.'

But Aidan just sank to the floor, sitting with his back against the wall, and after a moment or two I did the same. Side by side. Brother and sister. Maybe we sat here together like this twelve years ago.

Aidan was thinking the same thing. He stared at the wall

where I'd put my revision whiteboard. 'There used to be a picture there. A man and a ship.'

'Dad's great-great-great uncle. He served with Nelson.' Dad took the picture with him when he left. A little bit of family history that the new baby's going to have. It doesn't matter to me, I suppose, but I feel upset on Ben's behalf. Nelson should belong to him.

'Oh,' he said. 'I think I remember.'

'Aidan, are you all right?' Stupid question. He obviously wasn't.

'Cass, I needed to say ... to say that you mustn't feel bad.'

'Sorry?'

'I understand. I think I understand. I've been staying at Rich's, two days now, and thinking, trying to work out ... I'm not as clever as you, Cass, and maybe it takes me longer.'

'Don't say that, Aidan, please—'

'I just couldn't understand why you would do that. Why you would tell Holly. But you did and I see it was probably for the best because she had to know sometime, and for all you knew I'd told her anyway, and that's what I should have done.'

I tried to untangle this statement. 'I'm not sure I under-stand ...'

'You told Holly about Louis. About what I did. Mum told you and you told Holly and now she's angry with me, because I never told her. And I just came to tell you not to feel bad, and not to feel guilty and I'm still so happy that I met you again.'

He took a deep breath, looked up at the ceiling.

'Aidan—'

The door opened. Will, carrying two plates full of burgers, sausages, salads, bread. It seemed excessive, somehow, all this food. I remembered Aidan's idea of a tea party – a few biscuits on a plate.

Here we were, showing off how much we had.

'Thanks,' I said, scrambling up to take the plates from him. 'Look, Will, it was kind of you to come for Ben's party, but I need to talk to Aidan now—'

'We need to talk.'

Was he winking at me? What on earth was going through his mind? I was seriously rethinking any ideas I still had about trying to make things up with him. The boy was socially inept.

'Look, it's not the right time.' Of course, he wanted to tell me he was seeing that blonde girl. That was why he'd come to the party. Well, he'd have to wait.

Will leaned towards me and I got a whiff of that sweet Will smell, and I was slightly overwhelmed with wanting to throw my arms around him. Luckily, I was carrying two plates of food and couldn't.

'I think he's got a gun.'

Will's whisper was so soft, and directed straight at my ear, so his breath tickled it, and for a second I was so distracted that I thought I'd heard wrong. I must have heard wrong.

But Aidan heard too. And Aidan reacted. He grabbed his bag and he pulled out something that was definitely a gun.

A gun.

A real gun. Here in my home, with a houseful of people. My brother with a gun.

And the gun was pointing at Will.

41: AIDAN

I don't know what's going on any more.

I've spent so long thinking about Cass and why she might have told Holly, trying to make it OK, so I don't have to hate her; trying to make it so I can just do what I need to do and not hurt anyone, not make her feel bad.

That's the problem when you care about people – you care about what they feel, and I don't want to care any more, I just want it to finish, because I can't take the pain any more, of being told goodbye, of never feeling at home, never knowing when the end's going to come.

This is how Rich must feel. I've never felt it before.

I thought and thought about Cass and why she told Holly, and I thought she must know the right thing to do, being so clever and all, going to Oxford University, and I can't even read, but now I don't know what to think.

Because she's with some guy who's got to be working for

Neil. He was there in the shop, I remember now. He was dressed just the same, in a black hoody. I thought he was on the steal, but he wasn't, he can't have been, he just sank into the shadows when Neil brought the guns. They must've been together, they must've been working together, and now he's here, he's with my sister – you can tell they're together by the way they look at each other.

Neil's following me, he's on to me, the sooner I do it the better, and then he can't hurt me no more, no one can hurt me, and I can't hurt no one else.

So all I have to do is turn the gun around, turn it around, so it's pointing at me, and not him, and fire.

But what if this guy is Neil's way of getting at me through Cass?

And what about Finn? What if he finds out? And he thinks I didn't care about him; that I didn't care enough to stay around? He won't remember me saying goodbye, he'll just think I didn't care enough.

After all, my dad didn't care enough to stay around for me. Every story Mum tells about him is different. Every story she tells is a lie. I asked Jon, my social worker, outright one day, and he found out for me. My dad killed himself. He was in the army – my mum hadn't lied about that – so it was kind of an easy thing to do. He shot himself. Maybe he'd just heard that Mum was expecting.

I hate him for it, but I hate Mac more. Better to have the kind of dad who kills himself than the kind who kills other people. Mac killed some random guy and they put him in prison for a long, long time. And that told me that some random guy mattered more than Mum and me.

'Aidan,' she says. His daughter. Mac's daughter. 'Aidan, put the gun down.'

Mac loved her. I was safe if I was holding her, holding the baby. I used to carry her around like a lucky charm. I thought she'd bring me luck again. But instead she's destroyed me, and I might have destroyed her. Neil might destroy her.

'Aidan, I didn't tell Holly. I didn't tell her anything. Why don't we ring her, Aidan? Why don't I talk to her?'

I shake my head. 'My phone's dead.'

'I can charge it for you! Or give me her number.'

'I can't talk to her.'

'Aidan, put the gun down, please, and then we can talk properly. We can sort this out. Holly loves you so much.'

'Not any more.'

'I don't believe it!'

'I'm not telling lies.'

She's lost her cool confidence, my sister. She's pink and scared and there are tears in her eyes. I don't want her to be scared. I hate her being scared.

But Neil's man is here. He'll want the gun. He's the threat.

'Cass, this guy, he's a gangster, he's a criminal, he's going to hurt us.'

The guy squawks, 'No – I –' but I lift up my hand with the gun, to show I mean business.

'Aidan, he's not, really not – you're mixing him up with someone else. He's called Will and he's in the sixth form, and he's really a very nice person.'

The guy nods, 'It's true! Honest!' and I don't know what to think.

'You were in my shop. You were with Neil.'

'Was he the guy who made you hide the gun? I was there. I'm sorry. I was spying on you. But I wasn't there to help him. Don't you remember? I had a girl with me.'

'What girl?' Cass sounds like she's going to cry.

'Just a friend - honest, Cass, she's just a friend. Katie. She's at Bonny's. She helps me out with jobs sometimes. We've been friends – just friends - for ever. You've got to believe me.'

I don't believe him.

'You're working for Neil. He's got you spying on me, spying on my sister.'

'No, really. I'm nosy, that's my problem, and I was worried about Cass, and I thought I should check you out. She said she'd met someone dodgy there, and I thought I could find out who he was.'

'Aidan, truly, he's not working for whoever you think – he is nosy, Aidan, please ...'

The guy is sweating. 'I'm sorry! Please will you put the gun down?'

I'm weak and weary and I can hardly hold the gun any more, it's so heavy. Maybe he isn't Neil's man. Maybe Cass is safe.

But if she really didn't tell Holly, who did?

Cass is still holding two plates of food, and I'm starving hungry and all I want is to eat and sleep, but obviously I can't, not now. Not now I've scared them. There's no place for me here. There's no placde for me anywhere!

'I'll go,' I say. 'I'm sorry.'

But Cass shakes her head. 'Aidan, give me your phone. We need to call Holly.'

42: CASS

Aidan was slumped against the wall, breathing heavily, as though he was in pain. His eyes were closed. The gun dangled from his hand. I slid down the wall next to him again. I thought about grabbing it, but decided against it. What if it went off as I took it?

'Why don't you go and find my dad?' I said, as softly as possible to Will. He nodded, and left the room.

I leant my head against the wall. Now that he'd had a gun pointing at his head, what was the likelihood that Will would ever want to see me again? Zero.

Aidan was crying now, big, gulping sobs, but making no noise. It was the saddest thing I'd ever seen. It was as though he'd taught himself to cry in silence, so no one would hear him.

I put my arms round my brother. He moaned, but didn't move. 'It'll be all right, Aidan, I know it will. It'll be all right. I never said anything to Holly. I trust you, Aidan. And I've seen you with Finn. You'd never hurt him.'

He put his head on my shoulder and we just sat there, side by side. And I remembered. I remembered crying, and Aidan rocking me in his arms. I remembered Aidan carrying me. I remembered Aidan telling me not to worry, he would get me back to Mum and everything would be safe. He would take care of both of us. He'd protect me. He'd protect Mum.

I'd found my big brother again. I'd found myself.

The door opened. I thought it would be Will, but it was my dad. He looked down at us like a bemused giant.

Then he sat down on the floor right in front of us. Next to the gun. If Aidan realised, if he panicked – anything could happen.

'Well. This takes me back.'

'How?' Aidan wasn't crying any more, but he was shivering, like a scared greyhound.

'The two of you. You used to huddle up like that when you were little. On the floor. Under the table, sometimes.'

Now I was the one crying. Tears, streaming down my nose.

'Aidan, I've been talking to Will. He's told me that you need my help.'

Aidan made a noise that wasn't quite a word.

'He told me that you've found a firearm and very responsibly you've come to me, as a government minister and friend of the family, to put it in the right hands. Is that correct?'

Aidan blinked and nodded.

'Very good. That is the gun in question?'

Aidan let out a deep breath and nodded.

'Perhaps you'd like to give it to me? I will pass it to one of the ministerial protection officers and I can assure you that they won't take any action against you. '

No response. Aidan seemed to be processing everything incredibly slowly.

Dad waited patiently, and then changed tack. 'Will also seems to think that you're in some sort of trouble? With your partner?'

'She won't ... I need to explain ... I can't ...'

'I thought we could ring her,' I said.

'In my experience, it's never a good idea to rush into these things when you're in an emotional state,' said Dad. 'Better to talk it out first here, and calm down. What's your girlfriend's name, Aidan?'

'Holly.'

'What's happened with Holly?'

'Someone's told her something ... something that I did. That I never told her. And now she thinks bad things about me, that I'm going to hurt her son, maybe her, and I wouldn't, I really wouldn't ... but she won't listen.'

'Have you ever hurt her? Or her son?'

The look on Aidan's face is so painful that I can't bear it.

'Of course he didn't!' I tell Dad. 'Just look at him! He loves them both so much!'

'I want Aidan to tell me, Cass.'

'I would never hurt them. They're everything to me.'

'So why would Holly think you could hurt them?'

'I did something bad to my brother. When I was eleven.'

Dad's voice is coaxing, persuasive. The gun still dangled from Aidan's hand. 'Tell us, Aidan. If you can tell us, it'll be easier to tell Holly.'

'I was – it was when I went back to live with my mum. I didn't want to, I was happy where I was before. I had foster parents who wanted to adopt me. But she went to court and got me back.'

'She must have loved you very much to do that.'

Aidan thought about this.'I wish she didn't love me then.'

'We can't change the past,' says Dad.'And we can't always control our emotions.'

'She'd had another baby by then. Louis. He was about two.' His voice wobbled.'About the same age Finn is now.'

'What happened to Louis?'

'Louis was bothering me. Wanted me to read to him. I couldn't read. Ever since I'd left Jim and Betty's, I'd been rubbish at my new school. I couldn't get the words together. I told him no, and he shouted at me.

'I lost it. It hadn't happened for ages – Jim and Betty really calmed me down, but I was so on edge, missing them and scared of Keith. I'd never known my mum with a man who didn't beat her or me, and I was waiting all the time for the moment when he went on the attack.

'But it was me who lost it first, not Keith. I swiped Louis across the face. I never meant to hurt him, honest to God, but I had a biro in my hand, and it caught him. In his eye. He screamed, this scream that went on and on and there was so much blood and I just wanted him to be quiet, but he wouldn't. He couldn't. And then it came true what I'd feared because Keith did shout and shout and when the ambulance had come and taken Mum and Louis away, he smacked me, hard. The strange thing was, I felt better then. It was better being hit than being scared, waiting for it. And it was like ... Keith had been nice to me, but I was waiting, always waiting for him to show what he was really like. And I'd made him do it.'

Dad shook his head, just a little bit. I tried not to make any noise at all.

'The social worker came the next day. My stuff was already packed. Keith did it for me and loads of things got left behind. Things from Betty and Jim. Their address and phone number. Mum was still at the hospital with Louis. She never even said goodbye.

'Keith talked to the social worker and told her I was not welcome there again. But I didn't care because I thought ... I thought ...' Aidan's breath was shallow and his face was pale, and for a moment I thought he might be going to break down altogether. 'I thought I would be going back to Jim and Betty, you see? To my foster parents? But we didn't go there. We went to the children's home and she took my bag out of the car, and I asked her, and she said ...'

He trailed to a halt.

'What did she say, Aidan?'

'That Betty and Jim were looking after a baby, so it weren't appropriate, it weren't safe ... I weren't safe.'

'Oh, Aidan, that's so unfair.' I squeezed his hand.

'I didn't believe it. I still don't. They wanted to punish me. Or Betty and Jim, they heard what happened and ... I'd let them down, you see. They didn't want me any more.'

I can count on the fingers of one hand the times I've seen my Dad cry. He's old-fashioned and he's very English and he just doesn't do it. He didn't cry when he told us about Annabel and he didn't cry when he packed up his stuff and left. But when Aidan said that, a big tear trickled down Dad's face.

'It's tough, isn't it, when all you've known is rejection,' Dad said. 'That's why we need to make sure that Holly hears this story from you, just as you've told it, so she understands that it was an accident. Nothing but an accident. A child loses his temper in a stressful situation. Understandable that Keith and your mother were very upset … although in my opinion, nothing justifies violence towards a child … But Holly has nothing to fear from you. And nor have any of us. That's right, isn't it, Aidan?'

Aidan looked at him, his eyes huge and almost childlike, and then he handed Dad the gun. 'Can you really just get rid of it for me?'

'I know someone who can. Now, Susie is very keen to hear all about your life, and Ben is desperate to meet you, and you look as though you need some food. Could you bear some family time before I drive you back to London?'

Aidan nodded, slowly. He and Dad stood up. I stayed where I was. I'd hardly breathed for about half an hour, and I felt too weak to move.

'If you want,' Dad continued, 'I'll come with you and stay with you while you talk to Holly … if you think that would help? To make sure she understands the whole story. And if things don't go so well, you can come and stay at my flat, and we'll figure things out. You're not alone, Aidan. We let you down once, and I'm truly sorry for that, and we're here for you now.'

I didn't know whether to be grateful to Dad for being so kind and helpful to Aidan, or angry with him. Because he was doing it out of guilt, I know. And he had a lot to be guilty about.

Or maybe he was doing it for me?

Dad looked down at me. 'Come and join us, Cass, when you're ready.'

They left me, and I cried for a bit longer, and then I blew my nose and scrubbed at my eyes and felt thankful that I wasn't silly enough to wear mascara because it would have been all over my face.

The door opened again. I thought it might be Mum, but it was Will. Of course. Will and a piece of birthday cake, and me with piggy pink eyes and a scarlet nose.

'Hey. I thought you might not be up to crowds of people.'

'I missed Ben blow out his candles?' Oh, no. Tears again.

'He'll live. He's really happy with all his mates here. It's nice to see.'

'It's thanks to you.'

'Not me. He did all the hard work of actually making friends. I just introduced them.'

'You did more than that. I don't know how to thank you.'

Will offered me a tissue. 'It's totally clean, honest.'

I accepted it gratefully. 'I'm sorry. I don't know what's the matter with me.'

'It's OK. That was pretty scary. I've never had a gun pointed at me before.'

'I'm sorry! I'm so sorry.'

'That's OK. It was nice of you to speak up for me. I'm glad you think I'm ... what was it? "A very nice person." That was special.'

It was my moment.

Naturally I screwed it up.

'I don't know if you're making fun of me or not.'

'I wouldn't dream of making fun of you,' he said. Then he took my hand. 'Cass, can we try again?'

Obviously, most girls would have just smiled and laughed

and fallen into his arms, and that's what I wanted to do. But I had to make sure what I was doing.

'When you say try again, what do you mean, exactly?'

Will grimaced and said, 'Did you ever think of studying law? Because I think it might suit you much better than history. What I *mean* is anything you want. If you want to go on sneaking around in secret, that's fine. If you want a full-on, official, devoted boyfriend, then let it be me. I can't stop thinking about you, and I miss you so much and I really want to get it right. Just let me know what you want, Cass Montgomery, and I'll do my best.'

I didn't know what to say, so I pulled out my phone. Clicked on the Facebook app, found my profile.

'Look,' I told him, as I clicked the keys.

'"In a relationship with Will Hughes",' he read. 'Blimey. When you make your mind up, you really go for it.'

'Is that OK?' I wasn't sure if I'd just made a monumental fool of myself. Looking again at the phone in my hand, I saw that five people had already liked my changed status.

'It's a lot more than OK,' he said, as his nose bumped against mine and finally, finally . . .

The door burst open one more time and I heard Ben's amazed voice. 'Cass! Why are you kissing Will?'

43: AIDAN

It was Neil who told Holly. He made up all sorts of lies about it, all sorts of stuff that made me sound like a monster, like a mutilator. Once Cass's dad took me back to the flat, and we sat down and talked, it got sorted out. It was OK. He was actually really helpful. A bit like a counsellor or a social worker. Better.

Holly understood. She believed me. And we started talking in a different way. Because once she knew the very worst thing about me, I believed she'd stay with me, I believed she loved me, and that made me feel completely different about myself. I began to see how I could stop drinking. And then I did it. I stopped, just like that. Stood in the pub with Rich and ordered a Coke. It felt strange and stupid, but I did it once and then I did it again.

Even though it's Christmas Eve. I've set myself the target of an alcohol-free festive season. I actually think I can do it. I just say no, again and again and again, and every time I feel stronger and proud of myself. And if I don't feel strong enough, then I call Cass's dad and just talk it through a bit with him. He promised he'd be there for me, and so far he has been. I like him a lot. I haven't mentioned my benefits, though. Just in case.

Kicking the drinking really helps with the reading and writing. My brain seems clearer, the words stay still. I think I'd underestimated how much alcohol I'd been having and what it was doing to my brain.

I'm still seeing the counsellor. Still picking through the past. It's like when I'd help Betty unravel her knitting wool. Sometimes it'd go fast and easy and sometimes there'd be knots to untangle. I was good at that then, and I'm discovering I can do it now.

I asked my social worker about Betty and Jim and he contacted them for me. I got a letter from Betty, all wobbly handwriting and lavender scent. They missed me, she said. They always wondered how I'd got on. They'd love to see me again. I didn't have any trouble reading Betty's letter, not the first time, nor every time since then. They want me to go and visit. They want to meet Holly and Finn.

I can't believe how much difference a letter can make.

So I'm writing one of my own. Rich is here to help me with it. Rich's new boyfriend is great, never gets jealous, never minds if I come over to see Rich, never hits on me, really seems to care about him. It's a weight off my mind.

Brendan's gone out tonight, and Holly's at zumba class, so it's just me and Rich babysitting Finn. I'm going to drive Rich home when she comes back. No more late buses for us, now I've finally passed my test.

'So,' he says, 'you talk, I'll write. Who is it to?'

'Louis,' I say. 'My brother. You know.'

'"Dear Louis,"' he says, writing it down. 'Go on.'

'"I am your half-brother. I am called Aidan and I came to live with you when I was eleven and you were two. I didn't want to be there. I was angry all the time because I wanted to be with my foster parents, Jim and Betty. They were like my real mum and dad. Not your parents."'

'Slow down,' says Rich. 'I can't write that fast. I'm not sure how great my spelling is, either.'

'Spelling doesn't matter so much,' I told him. 'Susie said that. Cass's mum, you know. She said that what matters is communicating. Relationships break down when you stop. It doesn't matter how you do it.'

'Now you're sounding like a social worker,' says Rich. 'Carry on with your letter.'

'"I am sorry. I am sorry that I hurt you. I didn't mean to."'

'You did mean to,' says Rich.

I think for a moment. '"I didn't mean to hurt you so much. I didn't mean to hurt your eye like that. I just lashed out because I was angry and there was a pen in my hand and it caught you at just the wrong place. I never planned it, I never did anything like that again, and I have felt bad about it for years and years and years."'

I'm all hot and sweaty. Rich looks at me and says, 'Are you really up for this?'

'I am. Keep going. "I know your dad says he can never trust me, but I am different now. I've gone through a lot. And I've got a family of my own now, and I can understand how your dad feels because if anyone hurt my little boy, then I'd hate them too."'

'That's good,' says Rich. 'What about your mum?'

'"I never got on great with my" – no, make that "our" – "with our mum, but I think it's better now. A lot of time has passed. I'm sorry it's taken so long to say sorry."'

'That doesn't sound right.'

I'm exhausted. 'Just put, "Have a happy Christmas and new year. And maybe I can see you one day. I'd like to know you. But only if your mum and dad say it's OK."'

Later on, when Holly is home and I've delivered Rich to Walthamstow and I've parked the van outside our new flat – Clive decided to invest in a conversion in Kentish Town – I pull out my phone and I ring Cass. She answers right away.

'Hey,' I say. 'Happy Christmas.'

'Hello.'

We're getting past that day. We haven't seen each other since. We're just talking on the phone, gradually putting it back together again. It might take months, or years. I want to make it happen.

'What are you doing for Christmas?' she asks.

'We're going to Holly's Uncle Clive,' I tell her. 'He invited us. Her mum's going to her sister in Birmingham and she wanted us to go too, but then Clive said we should come to them. I haven't done it for years, that big, family Christmas thing.'

One day, I think, I'll make a massive turkey dinner for everyone in my life. For Holly and Finn and Clive and his family. For Cass and Will and her brother, Ben. Rich and Brendan. Even Mum and Keith and Scarlett and Louis. Maybe even Jim and Betty. And everyone can have serviettes and everything will be perfect.

Cass seems really happy with this guy Will. She says they

might take a gap year between school and university, travelling, doing some volunteer work. 'And I might not go to Oxford after all,' she told me last time we talked, 'I might look at other places. And I don't think history's for me. I'm looking at law courses.'

'I'm going to start college,' I told her. 'Clive said he'd sponsor me to do a course in accounting. He wants to train me up, give me more responsibility. I'm not sure if I can do it, but I'll give it a try.'

'Of course you can do it,' she told me. 'You'll be fine.'

Now she's going on about Christmas, about how she's going to miss her dad.

'It's going to be different this year,' she says. 'I can't believe my dad's not going to be there. It's going to be strange. But Will and I are going away for New Year. His sister's in pantomime in Newcastle, so we're staying up there.'

'Every year it's different for me,' I tell her. 'You get used to it. As long as the turkey's there, it works.'

Then she laughs and I laugh and it's good. It's OK.

Sometimes I think we all get too worked up about families. I mean, we're all related somehow, aren't we? What is a family, anyway? Is it Cass and Ben and her split-up parents? Is it Rich and Brendan, happy-ever-after in Walthamstow? Is it Cass and me? Holly and Finn?

I always wanted a perfect family, but I'll settle for a normal one.

I'm full of hope.

ACKNOWLEDGMENTS

Jacqui and Karen. Thank you for sharing your experience and wisdom, without your help this book would never have been written.

Anna found my Kindle in a busy shopping centre and her husband and son, Russell and Michael, turned detective to return it to me. Then Russell told me about his shop – The Salvage Store in Burnt Oak, north London – and I had a title and a setting for the idea that I was hoping might turn into a book. Thank you!

Jenny Savill, so much to thank you for, I'm very lucky to have you as my agent.

Sam Smith, your devotion to *Salvage* was above and beyond anything I could have expected. And thanks to everyone else at Atom, for making me feel so welcome, especially Kate Agar.

Anne Joseph, Amanda Swift, Anna Longman, Lydia Syson, Jennifer Gray, Becky Jones, Fenella Fairburn, Keris Stainton, Tamsyn Murray, Luisa Plaja, Susie Day, Cat Clarke and Sophia Bennett, thank you for your support, advice, and the laughs. Special thanks to Fiona Dunbar and Valerie Kampmeier who read multiple versions of *Salvage* and gave me excellent critiques.

Thank you to my family: Mum and Dad, Deborah, Alun, Jeremy, Josh, Avital and Eliana. Phoebe and Judah, you will see more of me now, I promise. Most of all, thanks to Laurence, love of my life. Twenty years, who'd have thought it?